NanoScience and Technology

NANOSCIENCE AND TECHNOLOGY

Series Editors:
P. Avouris B. Bhushan D. Bimberg K. von Klitzing H. Sakaki R. Wiesendanger

The series NanoScience and Technology is focused on the fascinating nano-world, meso-scopic physics, analysis with atomic resolution, nano and quantum-effect devices, nano-mechanics and atomic-scale processes. All the basic aspects and technology-oriented de-velopments in this emerging discipline are covered by comprehensive and timely books. The series constitutes a survey of the relevant special topics, which are presented by lead-ing experts in the field. These books will appeal to researchers, engineers, and advanced students.

Atomic Force Microscopy, Scanning Nearfield Optical Microscopy and Nanoscratching
Application
to Rough and Natural Surfaces
By G. Kaupp

Applied Scanning Probe Methods VI
Characterization
Editors: B. Bhushan and S. Kawata

Applied Scanning Probe Methods VII
Biomimetics
and Industrial Applications
Editors: B. Bhushan and H. Fuchs

Roadmap of Scanning Probe Microscopy
Editors: S. Morita

Nanocatalysis
Editors: U. Heiz and U. Landman

Nanostructures
Fabrication and Analysis
Editor: H. Nejo

Fundamentals of Friction and Wear on the Nanoscale
Editors: E. Gnecco and E. Meyer

Nanostructured Soft Matter
Experiment, Theory, Simulation
and Perspectives
Editor: A.V. Zvelindovsky

Charge Migration in DNA
Perspectives from Physics, Chemistry, and Biology
Editor: T. Chakraborty

Lateral Alignment of Epitaxial Quantum Dots
Editor: O. Schmidt

Applied Scanning Probe Methods VIII
Scanning Probe Microscopy
Techniques
Editors: B. Bhushan, H. Fuchs, and M. Tomitori

Applied Scanning Probe Methods IX
Characterization
Editors: B. Bhushan, H. Fuchs, and M. Tomitori

Applied Scanning Probe Methods X
Biomimetics
and Industrial Applications
Editors: B. Bhushan, H. Fuchs, and M. Tomitori

Semiconductor Nanostructures
Editor: D. Bimberg

Multiscale Dissipative Mechanisms and Hierarchical Surfaces
Friction, Superhydrophobicity,
and Biomimetics
By M. Nosonovsky and B. Bhushan

Michael Nosonovsky
Bharat Bhushan

Multiscale Dissipative Mechanisms and Hierarchical Surfaces

Friction, Superhydrophobicity, and Biomimetics

With 112 Figures

 Springer

Dr. Michael Nosonovsky
Stevens Institute of Technology, Department of Mechanical Engineering
Castle Point of Hudson, Hoboken, NJ 07030, USA
E-mail: michael.nosonovsky@stevens.edu

Professor Dr. Bharat Bhushan
Ohio State University
Nanoprobe Laboratory for Bio- & Nanotechnology and Biomimetics (NLB^2)
201 W. 19th Avenue, Columbus, OH 43210, USA
E-mail: bhushan.2@osu.edu

Series Editors:

Professor Dr. Phaedon Avouris
IBM Research Division
Nanometer Scale Science & Technology
Thomas J. Watson Research Center
P.O. Box 218
Yorktown Heights, NY 10598, USA

Professor Dr., Dres. h.c. Klaus von Klitzing
Max-Planck-Institut
für Festkörperforschung
Heisenbergstr. 1
70569 Stuttgart, Germany

Professor Dr. Bharat Bhushan
Ohio State University
Nanoprobe Laboratory for Bio- &
Nanotechnology and Biomimetics (NLB^2)
201 W. 19th Avenue
Columbus, OH 43210, USA

Professor Hiroyuki Sakaki
University of Tokyo
Institute of Industrial Science
4-6-1 Komaba, Meguro-ku
Tokyo 153-8505, Japan

Professor Dr. Dieter Bimberg
TU Berlin, Fakultät Mathematik/
Naturwissenschaften
Institut für Festkörperphysik
Hardenbergstr. 36
10623 Berlin, Germany

Professor Dr. Roland Wiesendanger
Institut für Angewandte Physik
Universität Hamburg
Jungiusstr. 11
20355 Hamburg, Germany

ISSN 1434-4904
ISBN 978-3-540-78424-1 Springer Berlin Heidelberg New York

Library of Congress Control Number: 2008923742

Springer is a part of Springer Science+Business Media.

springeronline.com

© Springer-Verlag Berlin Heidelberg 2008

Typesetting: Data prepared by VTEX using a Springer LATEX macro
Cover: eStudio Calamar Steinen

Printed on acid-free paper   SPIN: 12041466   57/3180/vtex - 5 4 3 2 1 0

Preface

This book is intended to serve as an introduction to a developing field of engineering biologically inspired surfaces with hierarchical structures. Recent advances in micro- and nanoscience reveal a growing number of surfaces with hierarchical structures, that is, with nanoscale details superimposed on the microscale details, sometimes superimposed on larger macroscale details. Such hierarchical structures are required for certain functions, such as achieving extremely low or extremely high friction and adhesion, and water-repellency. Friction, adhesion, and wetting are complicated processes, which involve effects at different scale levels with different characteristic scale lengths. Engineers are trying to mimic nature in order to design artificial surfaces with desirable properties, referred to as bioinspired or biomimetic surfaces. The field is referred to as biomimetics.

Our purpose is, first of all, to present the qualitative picture of physical phenomena, rather than to provide rigorous mathematical derivations or many technical details, which may be found in the references. We concentrate upon such issues as scale and dimension, linearity and nonlinearity, and the fundamental physical mechanisms and effects involved in the phenomena under consideration. This allows a reader who is not familiar with the field or not a specialist in surface science to grasp quickly the essence of the processes and the issues discussed. On the other hand, we felt it necessary to present a brief discussion of modern analytical and experimental methods and approaches used in mesoscale and multiscale science and recent trends in the development of the surface science and multiscale modeling.

The book is divided into three parts. The first part is devoted to the solid–solid dry friction, which is a traditional subject of study of tribology. In this part, we cover topics such as the statistical and fractal characterization of rough random surfaces and solid–solid contact, which have been developed over the past 30 years and are used widely in engineering. We discuss the measurement techniques and equipment that allows scientists to study surfaces at nanoscale resolution—including scanning probe microscopy, which emerged in the early 1980s. Our emphasis is on the multi-scale, hierarchical nature of the dissipation mechanisms, which are becoming evident as more and more data about the nanoscale friction are obtained.

In the second part of the book, we study the solid–liquid friction and wetting of rough surfaces, as well as related capillary phenomena. Rough water-repellent or superhydrophobic surfaces, which are often found in biological systems, in many cases have a complicated hierarchical structure that is required for certain functionality, such as nonwetting, low solid–liquid friction, high friction and adhesion. Leaves of water-repellent plants, such as the lotus, constitute an example of these surfaces. Their surfaces are extremely hydrophobic, and a droplet can flow over them with low energy dissipation. However, the mechanisms involved in the process are complicated and have different characteristic length scales, so the surfaces should also be hierarchical. Roughness-induced superhydrophobicity and the "lotus-effect" have been studied extensively during the past decade with the number of articles in peer-reviewed journals growing exponentially since the early 2000s. This is because the technology that allows us to produce an artificial lotus leaf surface became available. However, there has been no single book that covers the theory of superhydrophobicity, the observation and characterization of natural superhydrophobic surfaces, and the methods of production and characterization of artificial superhydrophobic surfaces. This book's purpose is to cover this gap in the literature.

Another example of natural hierarchical surfaces is the gecko foot, which has an ability to achieve very high adhesion (so that it can climb upon a vertical wall) and detach from the surface at will. These abilities are known as smart adhesion. Smart adhesion, along with other functional hierarchical biological surfaces, such as the shark skin and the moth eye, are studied in the third part of the book. These functional biological surfaces inspired engineers to design artificial surfaces with similar properties. Biomimetic hierarchical surfaces are discussed in that part of the book along with other practical issues, such as techniques to experimentally study the wetting of rough surfaces.

The book is written with a broad multidisciplinary readership in mind. It can serve as a supplementary textbook for a graduate course in surface science, tribology, or nanotechnology. It can be used by engineers and scientists who want to familiarize themselves with the basic concepts of nanotribology and biologically inspired surfaces. The authors hope that the book will be useful to a broad audience of readers from various backgrounds.

We thank our colleagues, Dr. Stephen M. Hsu, Dr. Seung-Ho Yang and Dr. Huan Zhang from the National Institute of Standards and Technology (NIST) in Gaithersburg, MD; Mr. Yong-Chae Jung and Dr. Tae-Wan Kim at the Ohio State University (OSU) in Columbus, OH; and Ms. Caterina Runyon-Spears from the OSU and others who helped in preparation of this book. The book was written partially while one of the authors, Dr. Michael Nosonovsky, was a National Research Council postdoctoral research fellow at NIST. However, none of the equipment, results, or commercial products mentioned or presented in this book should be treated as endorsed or approved by NIST.

November 2007

Michael Nosonovsky
Bharat Bhushan

Contents

Preface v

Nomenclature xiii

Glossary xv

Abbreviations xvii

Part I Surface Roughness and Hierarchical Friction Mechanisms

1   Introduction 3
    1.1   Surfaces and Surface Free Energy 3
    1.2   Mesoscale 5
    1.3   Hierarchy 7
    1.4   Dissipation 7
    1.5   Tribology 9
    1.6   Biomimetics: From Engineering to Biology and Back 11

2   Rough Surface Topography 13
    2.1   Rough Surface Characterization 13
    2.2   Statistical Analysis of Random Surface Roughness 17
    2.3   Fractal Surface Roughness 20
    2.4   Contact of Rough Solid Surfaces 23
    2.5   Surface Modification 25
        2.5.1   Surface Texturing 25
        2.5.2   Layer Deposition 25
    2.6   Summary 26

3   Mechanisms of Dry Friction, Their Scaling and Linear Properties . . . 27
    3.1   Approaches to the Multiscale Nature of Friction 28
    3.2   Mechanisms of Dry Friction 31
        3.2.1   Adhesive Friction 31

	3.2.2	Deformation of Asperities	38
	3.2.3	Plastic Yield	39
	3.2.4	Fracture	39
	3.2.5	Ratchet and Cobblestone Mechanisms	39
	3.2.6	"Third Body" Mechanism	40
	3.2.7	Discussion	40
3.3	Friction as a Linear Phenomenon		40
	3.3.1	Friction, Controlled by Real Area of Contact	41
	3.3.2	Friction Controlled by Average Surface Slope	43
	3.3.3	Other Explanations of the Linearity of Friction	44
	3.3.4	Linearity and the "Small Parameter"	45
3.4	Summary		45

4 **Friction as a Nonlinear Hierarchical Phenomenon** 47
	4.1	Nonlinear Effects in Dry Friction		47
		4.1.1	Nonlinearity of the Amontons–Coulomb Rule	47
		4.1.2	Dynamic Instabilities Associated with the Nonlinearity	48
		4.1.3	Velocity-Dependence and Dynamic Friction	49
		4.1.4	Interdependence of the Load-, Size-, and Velocity-Dependence of the Coefficient of Friction	50
		4.1.5	Stick–Slip Motion	51
		4.1.6	Self-Organized Criticality	52
	4.2	Nonlinearity and Hierarchy		53
	4.3	Heterogeneity, Hierarchy and Energy Dissipation		55
		4.3.1	Ideal vs. Real Contact Situations	55
		4.3.2	Measure of Inhomogeneity and Dissipation at Various Hierarchy Levels	55
		4.3.3	Order-Parameter and Mesoscopic Functional	59
		4.3.4	Kinetics of the Atomic-Scale Friction	59
	4.4	Mapping of Friction at Various Hierarchy Levels		61
	4.5	Summary		62

Part II Solid–Liquid Friction and Superhydrophobicity

5 **Solid–Liquid Interaction and Capillary Effects** 65
	5.1	Three Phase States of Matter	65
	5.2	Phase Equilibrium and Stability	67
	5.3	Water Phase Diagram at the Nanoscale	69
	5.4	Surface Free Energy and the Laplace Equation	72
	5.5	Contact Angle and the Young Equation	73
	5.6	Kelvin's Equation	76
	5.7	Capillary Effects and Stability Issues	77
	5.8	Summary	79

6   Roughness-Induced Superhydrophobicity   81
6.1   The Phenomenon of Superhydrophobicity   81
6.2   Contact Angle Analysis   85
6.3   Heterogeneous Surfaces and Wenzel and Cassie Equations   86
    6.3.1   Contact Angle with a Rough and Heterogeneous Surfaces ..   86
    6.3.2   The Cassie–Baxter Equation   87
    6.3.3   Limitations of the Wenzel and Cassie Equations   90
    6.3.4   Range of Applicability of the Wenzel and Cassie Equations .   92
6.4   Calculation of the Contact Angle for Selected Surfaces   96
    6.4.1   Two-Dimensional Periodic Profiles   96
    6.4.2   Three-Dimensional Surfaces   100
    6.4.3   Surface Optimization for Maximum Contact Angle   105
6.5   Contact Angle Hysteresis   107
    6.5.1   Origin of the Contact Angle Hysteresis   107
    6.5.2   Pinning of the Triple Line   109
    6.5.3   Contact Angle Hysteresis and the Adhesion Hysteresis   110
6.6   Summary   112

7   Stability of the Composite Interface, Roughness and Meniscus Force .   115
7.1   Destabilization of the Composite Interface.....................   115
    7.1.1   Destabilization Due to Capillary and Gravitational Waves ..   116
    7.1.2   Probabilistic Model   121
    7.1.3   Analysis of Rough Profiles   122
    7.1.4   Effect of Droplet Weight   123
7.2   Contact Angle with Three-Dimensional Solid Harmonic Surface ...   126
    7.2.1   Three-Dimensional Harmonic Rough Surface   126
    7.2.2   Calculations of the Contact Areas   128
    7.2.3   Metastable States   129
    7.2.4   Overall Contact Angle   130
    7.2.5   Discussion of Results   131
    7.2.6   The Similarity of Bubbles and Droplets.................   133
7.3   Capillary Adhesion Force Due to the Meniscus   134
    7.3.1   Sphere in Contact with a Smooth Surface   134
    7.3.2   Multiple-Asperity Contact............................   136
7.4   Roughness Optimization   137
7.5   Effect of the Hierarchical Roughness   141
    7.5.1   Hierarchical Roughness   141
    7.5.2   Stability of a Composite Interface and Hierarchical
            Roughness..   142
    7.5.3   Hierarchical Roughness   145
    7.5.4   Results and Discussion   148
7.6   Summary   151

8   Cassie–Wenzel Wetting Regime Transition   153
8.1   The Cassie–Wenzel Transition and the Contact Angle Hysteresis . . .   153

8.2   Experimental Study of the Cassie–Wenzel Transition 157
8.3   Wetting as a Multiscale Phenomenon 163
8.4   Investigation of Wetting as a Phase Transition 165
8.5   Reversible Superhydrophobicity 166
8.6   Summary 166

9   **Underwater Superhydrophobicity and Dynamic Effects** 169
9.1   Superhydrophobicity for the Liquid Flow 169
9.2   Nanobubbles and Hydrophobic Interaction 171
9.3   Bouncing Droplets 172
9.4   A Droplet on a Hot Surface: the Leidenfrost Effect 175
9.5   A Droplet on an Inclined Surface 176
9.6   Summary 177

Part III   Biological and Biomimetic Surfaces

10   **Lotus-Effect and Water-Repellent Surfaces in Nature** 181
10.1   Water-Repellent Plants 181
10.2   Characterization of Hydrophobic and Hydrophilic Leaf Surfaces . . . 184
       10.2.1   Experimental Techniques 184
       10.2.2   Hydrophobic and Hydrophilic Leaves 185
       10.2.3   Contact Angle Measurements 186
       10.2.4   Surface Characterization Using an Optical Profiler 187
       10.2.5   Leaf Characterization with an AFM 190
       10.2.6   Adhesion Force and Friction 192
       10.2.7   Role of the Hierarchy 196
10.3   Other Biological Superhydrophobic Surfaces 197
10.4   Summary 197

11   **Artificial (Biomimetic) Superhydrophobic Surfaces** 199
11.1   How to Make a Superhydrophobic Surface 201
       11.1.1   Roughening to Create One-Level Structure 202
       11.1.2   Coating to Create One-Level Hydrophobic Structures 204
       11.1.3   Methods to Create Two-Level (Hierarchical)
                Superhydrophobic Structures 205
11.2   Experimental Techniques 206
       11.2.1   Contact Angle, Surface Roughness, and Adhesion 206
       11.2.2   Measurement of Droplet Evaporation 207
       11.2.3   Measurement of Contact Angle Using ESEM 207
11.3   Wetting of Micro- and Nanopatterned Surfaces 208
       11.3.1   Micro- and Nanopatterned Polymers 208
       11.3.2   Micropatterned Si Surfaces 211
11.4   Self-cleaning 227
11.5   Commercially Available Lotus-Effect Products 228
11.6   Summary 229

12   Gecko-Effect and Smart Adhesion 231
   12.1   Gecko 231
   12.2   Hierarchical Structure of the Attachment Pads 233
   12.3   Model of Hierarchical Attachment Pads 236
   12.4   Biomimetic Fibrillar Structures 237
   12.5   Self-cleaning 239
   12.6   Biomimetic Tape Made of Artificial Gecko Skin................. 240
   12.7   Summary 241

13   Other Biomimetic Surfaces 243
   13.1   Hierarchical Organization in Biomaterials 243
   13.2   Moth-Eye-Effect 244
   13.3   Shark Skin 246
   13.4   Darkling Beetle 246
   13.5   Water Strider 247
   13.6   Spider Web.. 247
   13.7   Other Biomimetic Examples 248
   13.8   Summary 249

14   Outlook 251

References 255

Index 271

Nomenclature

a—contact radius; width

A_a, A_r—apparent and real areas of contact, respectively

A_{SL}, A_{SA}, A_{LA}, A_F—solid–liquid, solid–air, liquid–air, flat contact areas, respectively

b—distance

c, C—constants

Ca—capillary number

d—distance

D—diameter, fractal dimension

E—elastic modulus

E_b—energy barrier

E_{tot}—total energy

f_{SL}, f_{LA}—fractions of the solid–liquid and liquid–air interfaces under the droplet

f_0—adhesion stress

F—friction force

F_{cap}—capillary adhesive force

g—gravitational constant

h—height; position of the interface

H—height; hardness of a softer material; film thickness

H_p—component of friction due to surface roughness and plowing

H_r—component of contact angle hysteresis due to surface roughness

k—the Boltzmann constant

k_{nj}—stiffness

K—kurtosis

l—length

l_c—capillary length

l_N—scale lengths

L—sampling length

m—mass; mean

N—number of contacts

N_I—number of springs

p—probability

P—pitch; pressure

P_0 is the atmospheric pressure

P_{sat}—saturated vapor pressure and P is the actual liquid pressure

Q—heat

r—radius

R—radius

R_p—mean asperity peak radius

R_1, R_2—principal radii of curvature

R_f—roughness factors

R_k—Kelvin radius

R_p—peak radius

Re—Reynolds number

S—entropy; space between neighboring fibers

S_f—spacing factor

t—length of the triple line

T—temperature; total energy

T_c—critical temperature

V, $\vec{V}$—velocity

V—volume

W—normal load; work of cohesion; work of cohesion

W_E is the elastic energy

ΔW—energy barriers between the two states

We—Weber number

z—separation distance

z_0—equilibrium distance

α—slope

β—kinetic coefficient

β, β^*—correlation length

γ—surface free energy, surface tension

γ_{SL}, γ_{SA}, γ_{LA}—solid–liquid, solid–air, and liquid–air interface energies, respectively

δ—droop of the droplet; Tolman's length

$\nabla\varepsilon$—strain gradient

η—density of asperities per unit area; packing density; order-parameter

θ, θ_{adv}, θ_{rec}, θ_0, θ_{adv0}, θ_{rec0}—contact angle, advancing and receding contact angles for rough and flat surfaces, respectively

θ—state parameter in dynamic friction models

θ_0—normalization parameter

K—curvature

λ—gradient coefficient; periodicity of a surface profile

μ—coefficient of friction

μ_L, μ_G—liquid and gas viscosities

ρ—density of liquid

σ—surface tension; standard deviation

σ_Y—yield stress

τ—contact line tension; normalized temperature

τ_f—shear strength at the interface

ψ—plasticity index

ω—frequency

Glossary

Asperity is a roughness detail of a surface. Even nominally flat surfaces have some roughness, so asperities are present at virtually every surface. An asperity is characterized by height, width, tip radius of curvature, etc. For fractal surfaces, the concept of asperity is controversial, since the fractal topography implies that the surface consists of the same asperity repeatedly superimposed on itself at different magnification, so there is no way to determine where one asperity ends and another one begins. Asperity may be defined as a roughness detail that participates in the contact and forms a contact spot.

Barbs are a series of branches fused to the rachis of a feather. The barbs themselves are also branched and form the barbules.

Biomimetics (bionics, biognosis, etc.) is the application of methods and systems found in living nature to the study and design of engineering systems and modern technology.

Carbone nanotube (CNT), fullerene, and graphene are allotropes of carbon (other carbon allotropes are diamond and graphite) with unusual properties. They were discovered since 1980s and they are promising for nanotechnology applications. CNTs are cylindrical molecules with very high length to diameter ratios. Fullerenes (the most common example is the C_{60} molecule) are spherical carbon molecules. Graphene is single-sheet monolayer of carbon.

Cornea is the transparent front part of the eye that provides most of an eye's optical power.

Critical point specifies the temperature and pressure at which the liquid state of the matter ceases to exist. As a liquid is heated within a confined space, its density decreases while the pressure and density of the vapor being formed increases, so that their densities become equal at the critical temperature. Near-critical states have unusual properties, in particular, the correlation length in these states can become infinitely large and physical properties are related by power laws (the critical exponents).

Cuticle of a plant is a protective waxy covering produced by the epidermal cells of leaves, young shoots and other aerial plant organs.

Elytron (pl. elytra) is a modified, hardened forewing of certain insects.

Epidermis in plants, the outermost layer of cells covering the leaves and young parts of a plant.

Fractal is a rough or fragmented geometric object that can be subdivided in parts, each of which is at least approximately a reduced-size copy of the whole.

Frustule a unique cell wall made of silica (hydrated silicon dioxide) into which diatom cells are enclaved. Diatoms are one of the most common types of unicellular phytoplankton.

Lamella is a thin plate-like structure, often one amongst many lamellae very close to one another that appears, in particular, in the traction surfaces of geckos.

Lotus-effect is the ability of very rough surfaces for self-cleaning and extreme water-repellency (superhydrophobicity).

Micro/nanoelectromechanical systems (MEMS/NEMS) are small devices that involve mechanical elements, sensors, actuators and electronics on a common silicon substrate through microfabrication technology. Typical MEMS devices include actuators, switches, sensor systems, micromirrors, etc.

Microfluidics is a multidisciplinary discipline that deals with the behavior, precise control and manipulation of microliter and nanoliter volumes of fluids.

Microtrichia are very small protuberances on cornea of moth eye.

Moth-eye-effect is non-reflective ability of a surface with a certain submicron structure, imitating the moth eye.

Papilla (pl. papillae) (papillose epidermal cells) are microscopic bumps on the surface of many water-repellent plant leaves.

Placoid is a special type of scales covering the shark skin that form small V-shaped bumps.

Self-organized criticality (SOC) is a property of certain dynamical systems that have a critical point as an attractor. Their behavior thus displays characteristics of the critical point of a phase transition, but without the need to tune control parameters to precise values. A small perturbation in such system can have a long-lasting effect. It has been speculated that the complexity and hierarchy in nature arises from the SOC.

Seta (pl. setae) is a stiff hair or a hair-like structure. Setae on gecko's footpads are responsible for its ability to cling to vertical surfaces.

Shark-skin-effect is drag reduction can in turbulent flow due to microscopic ridges upon the skin of a shark.

Sol-gel is a wet-chemical technique for the fabrication of materials (typically a metal oxide) starting from a chemical solution containing colloidal precursors.

Spatula (pl. spatulae) are substructures of a seta in a gecko foot.

Spinodal limit is the limit at which the difference between gas and liquid ceases to exist. For water, the pressure corresponding to the spinodal limit at a given temperature can constitute the tensile strength of metastable liquid water.

Stick-slip is a spontaneous jerking motion that can occur while two objects are sliding over each other. The reason for the stick-slip is that the static coefficient of friction is usually greater than the kinetic coefficient of friction.

Stiction is sticking together of two solid bodies, especially components of microdevices, due to adhesion and static friction.

Trichomes are fine outgrowths on plants that have diverse structure and function.

Water-strider-effect is the ability of the water strider to walk upon a water surface without sinking using the surface tension force due to the hierarchical structure of its legs.

Abbreviations

AFM—atomic force microscope
AKD—alkylketene dimmer
BCH—brucite-type cobalt hydroxide
CBD—chemical bath deposition
CNT—carbon nanotube
CVD—chemical vapor deposition
DI—deionized
DMF—dimethylformamide
DMT—Derjagin–Muller–Toporov
DNA—deoxyribonucleic acid
ESEM—environmental scanning electron microscope
GL—Ginzburg–Landau
GSED
HAR—high aspect ratio
ITO—indium tin oxide
JKR—Johnson–Kendall–Roberts
LA—lauric acid
LAR—low aspect ratio
LBL—layer by layer
MD—molecular dynamics
MEMS—microelectromechanical systems
NEMS—nanoelectromechanical systems
NIST—National Institute of Standards and Technology
NLBB—Nanoprobe Laboratory for Bio- & Nanotechnology and Biomimetics
OSU—Ohio State University
OTS—octadecyltricholorosilane
PAA—poly(acrylic acid)
PAA—porous anodic alumina
PAH—poly(allylamine hydrochloride)
PDF—probability distribution function
PDMS—polydimethylsiloxane
PE—polyethylene
PET—poly(ethylene terephthalate)

PF_3—tetrahydroperfluorodecyltrichlorosilane
PFDTES—perfluorodecyltriethoxysilane
PFOS—perfluorooctanesulfonate
PMMA—polymethylmethacrylate
PPy—polypyrrole
PS—polystyrene
P-V—peak-valley
PVD—physical vapor deposition
PVS—poly(vinylsiloxane)
RH—relative humidity
RMS—root-mean square
SAM—self-assembled monolayer
SEM—scanning electron microscope
SOC—self-organized criticality
STM—scanning tunneling microscope
SU8—thick photoresist for MEMS
TMS—tetramethylsilane
UV—ultraviolet

Part I

Surface Roughness and Hierarchical Friction Mechanisms

"God created the solids, the devil created their surfaces."
Wolfgang E. Pauli

1

Introduction

Abstract In the introduction chapter, the subjects and definitions of the surface science and tribology are discussed, as well as their relations to the concepts of hierarchy, mesoscale, energy dissipation and biomimetics.

1.1 Surfaces and Surface Free Energy

Surface science is defined as the study of physical and chemical phenomena that occur at the interface of two phases (solid–liquid, solid–gas, solid–vacuum) or of different substances of the same phase (solid–solid, liquid–liquid) [6]. Various properties of matter (e.g., the density, ρ) can change rapidly at the interface. It is therefore convenient to assume that the interface is a geometrically two-dimensional surface in a sense that every point at the interface can be characterized by only two parameters. In reality, every interface has a nonzero thickness and the bulk properties change gradually at the interface; however, the thickness is so small compared to the two other dimensions that it can often be neglected.

An important characteristic of every surface or interface is the surface free energy, γ. In the bulk of the body, chemical bonds exist between the molecules and certain energy has to be applied in order to break the bonds. The molecules that do not form the bonds have higher potential energy than those that form the bonds. Molecules at the surface do not form bonds at the side of the surface and thus they have higher energy. This additional energy is called surface or interface free energy and is measured in the energy per area units, that is, in the SI system, J/m^2 or N/m. In order to create an interface (e.g., to form a vapor bubble inside boiling water), the energy should be applied which is equal to the area of the interface multiplied by the interface free energy. For the stable existence of the interface it is required that the free energy of formation of the interface be positive, so that accidental fluctuations do not result in the dispersion of one material into the other. The opposite example of an interface, which does not offer opposition to the dispersion, is that between two gases or between miscible liquids [6]. Any system tends to achieve a position that

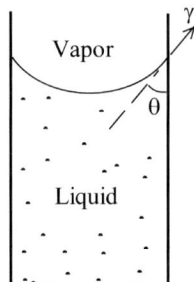

Fig. 1.1. Capillarity effect. Concave meniscus in a tube with water contact angle, θ, of less than 90°. The surface tension force

corresponds to a minimum energy. This is the reason why bubbles and droplets tend to have a spherical shape.

The concept of surface tension was introduced in 1805 by Thomas Young (1773–1829) and almost simultaneously by Pierre-Simon Laplace (1749–1827), while the idea of free surface energy was suggested by Josiah W. Gibbs (1839–1903) in about 1870, and it remains the foundation of the surface science. The accurate thermodynamic definition of the surface free energy involves a distinction between the concepts of the Gibbs free energy (the useful work obtainable from an isothermal isobaric thermodynamic system) and Helmholtz free energy (the useful work obtainable from a closed thermodynamic system). However, for processes that occur under constant pressure and temperature, the difference between these concepts is not significant, so in this book we will speak just about the surface free energy.

The obvious manifestation of the free surface energy is found in the capillarity effect (Fig. 1.1), defined as the ability of a substance to draw another substance into it [6, 283]. When the size of a liquid droplet or a channel is much smaller than the so-called capillary length, given by $l_c = (\gamma/\rho g)^{1/2}$ (where $g = 9.81$ m/s² is the gravitational constant, γ is the free surface energy, and ρ is the density of liquid), the surface energy dominates over the gravity potential energy and the corresponding capillary forces dominate over the weight. The surface tension force is the force which should be applied to the solid–liquid–air contact line (the triple line) to expand the solid–liquid interface. The surface tension is measured in N/m and in many senses it is equivalent to surface free energy. For water at room temperature, $\gamma \approx 72$ mN/m, $\rho \approx 1000$ kg/m³, and $l_c \approx 2.7$ mm.

The interest in surface science is stimulated by the current advances in nanoscience and nanotechnology. Although the volume of a body is proportional to the third power of its linear size, the surface area is proportional to the second power of the linear size. With decreasing size of an object, the surface-to-volume ratio grows and surface effects dominate over the volume effects. This is why for small objects, all surface phenomena, such as capillarity, adhesion, friction, etc., become increasingly important.

1.2 Mesoscale

The length scale less than 1 mm but larger than 100 nm is considered microscale, and the length scale less than 100 nm but larger than the atomic scale is considered nanoscale [8, 36]. In a different manner, the scale length larger than the atomic scale (i.e., 1 nm or less) but smaller than the macroscale (i.e., between 1 nm and 1 mm) may be called mesoscale. There are several types of objects that are usually considered by physicists as mesoscale objects: systems that are of submicron size in at least one dimension, such as nanoparticles; soft condensed matter materials (foams, gels, polymer melts), which are characterized by a mesoscopic length scale; and systems in a near-critical state (near a phase transition point), which posses mesoscopic "correlation length" characterizing spontaneous fluctuations [8].

Nanoscale systems, even with a typical size of several nanometers, involve hundreds and thousands of molecules and can often be considered as a continuum system. However, many of their physical properties are different from macroscale bulk properties. For example, the yield strength and hardness are known to be higher when measured at the nanoscale compared to the macroscale values [42, 51, 159, 230]. The reason for that is believed to be the fact that a solid material consists of a large number of submicron-sized grains, domains, and defects. It is much easier to deform material when the size of the deformation is greater than the size of a grain. At the nanoscale, there are no such defects and material is much stronger. This effect is taken into account by strain-gradient plasticity theories [159, 230].

Another example of how properties at the nanoscale differ from those at the microscale is found in the phase transition, such as boiling/condensation and melting/freezing. At the macroscale, water is known to boil at 100 °C and to freeze at 0 °C; at the nanoscale, however, the situation may be quite different [346]. This is because in order to transform into a different phase, an interface should be created (e.g., a vapor bubble inside the bulk of liquid) which requires additional activation energy. In order to grow, however, the size of a bubble must be greater than a certain critical size, so at the nanoscale the bubble would not be formed. Furthermore, due to the capillary effects and small radii of curvature of nanodroplets and nanoscale water columns, the pressure inside water volumes (the so-called Laplace pressure) may be significantly different from the ambient, e.g., it may be negative (tensile strength) [249, 346].

Adhesion or molecular attraction between bodies in contact is a very important effect at the nanoscale. Adhesive force is a generic name for different forces that can pull together small bodies. Physically, the adhesion force involves the relatively weak and long-range (nanometers) van der Waals electric forces; the relatively strong and short-range chemical bonds between molecules, the electrostatic force; and the meniscus attractive force caused by condensed water bridges near the contacts.

Although many concepts of mesoscale physics and thermodynamics were developed a long time ago, only recently has a high degree of generality—originating from the existence of the mesoscale—been recognized. There are several universal methods that allow physicists to deal with the mesoscale. These methods include the

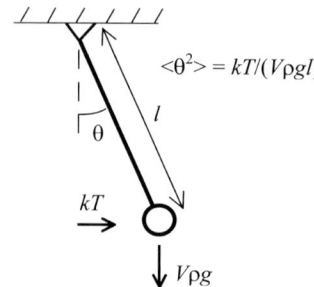

Fig. 1.2. Deflection of a nanoscale pendulum as a result of thermal fluctuations

renormalization group theory, scaling theory, Landau–Ginzburg mesoscopic functional, percolation, etc. [8].

A characteristic feature of mesoscale systems is that fluctuations, or spontaneous deviations from equilibrium, can play a significant role in them [8]. For example, consider a small microscale pendulum of length l with a mass $m = V\rho$ with its position characterized by the angle θ (Fig. 1.2). Collision with molecules results in the mean square position of the micropendulum being equal to

$$\theta^2 = kT/(V\rho gl), \tag{1.1}$$

where k is the Boltzmann constant and T is temperature. Thus, unlike the macroscale pendulum, the typical position of the micropendulum will be different from $\theta = 0$ due to the thermal random fluctuation.

The concept of mesoscale is closely related to the critical phenomena, i.e., phase transitions. Close to a critical point of any kind, the fluctuations become so large that they exhibit macroscale behavior [8]. Critical point is a point at the phase diagram where distinction between two phases vanishes. For example, the critical point of water is at around 374 °C and 218 atm, and the distinction between liquid and gas water at these conditions disappears. Asymptotically close to the critical point, the physical properties obey simple power laws, called the scaling laws. The physical basis for the scaling theories is in the divergence of a mesoscopic characteristic length scale known as the correlation length of the fluctuation. Powerful physical techniques, such as the renormalization-group theory and the Landau–Ginzburg functional, which were originally formulated for phase transition, have been proposed to calculate the critical exponents of the scaling laws and to study the near-critical behavior. The phase transition between a "disordered" and "ordered" state implies that an order-parameter can be identified, which is equal to zero for the disordered phase and different from zero for the ordered phase. Then, power exponents for the scaling laws can be determined. The phase transition approach has also been applied to the molecular-scale friction [104]. In solid mechanics, the common "phase transition" is that between the elastic and plastic phases. Modern theories of plasticity intended for the micron and submicron scale (the strain-gradient plasticity) postulate mesoscale length parameters, which result in the scale dependence of the yield strength and hardness at the mesoscale [122, 159, 230]. The strain-gradient plasticity approach can be applied for the study of scale effect on friction [42].

1.3 Hierarchy

The concept of hierarchy is different from the concept of scale in that hierarchy implies a complicated structure and organization. Studying hierarchical systems often requires a multidisciplinary approach. Investigation of hierarchical surfaces involves mechanics, physics, chemistry, biochemistry, and biology. Hierarchical surfaces are built of elements of different characteristic length, organized in a certain manner (Fig. 1.3). This organization leads to certain functionality. Many examples of these surfaces are found in biology and will be considered in this book.

An important class of hierarchical systems are the fractal objects. The so-called self-similar or fractal structures can be divided by parts, each of which is a reduced-size copy of the whole. Unusual properties of the self-similar curves and surfaces, including their noninteger dimensions, were studied by mathematicians in 1930s. The word "fractal" was coined in 1975 by Benoit Mandelbrot, who popularized the concept of self-similarity and showed that fractal geometry is universal in nature and engineering applications [214]. The fractal concepts were applied to the rough surfaces. In the late 1980s and through the early 1990s, the fractal geometry approach was introduced into the study of engineering rough surfaces by Gagnepain and Roques-Carnes [119], Ling [205], Majumdar and Bhushan [212, 213], and others.

1.4 Dissipation

Many physical processes result in irreversible energy dissipation. Examples include plastic deformation, friction, and viscosity. The energy during dissipative process is converted into heat. The second law of thermodynamics, formulated by Rudolf Clausius (1822–1888), states that the heat, Q, cannot of itself pass from a colder to a hotter body. The mathematical formulation is that the entropy S, defined as $dS = dQ/T$, can only increase (for irreversible process) or remain constant (for reversible processes). When heat dQ is transformed from a system at temperature T_1 to that at T_2, the change of entropy is $dQ(1/T_2 - 1/T_1)$, so the entropy grows when heat is transmitted from a hot to a cold body. The second law has a statistical nature and states that a system tends to transfer from a more ordered state to a less ordered state, which is statistically more probable. The state of thermodynamic equilibrium, at which the temperatures of the contacting bodies are equal ($T_1 = T_2$), corresponds to the less ordered, most probable state and to the highest entropy. At the nanoscale, when the typical energy of the system is comparable with kT, the second law can be violated due to small fluctuations, which can lead to local reductions of the entropy of the system [8].

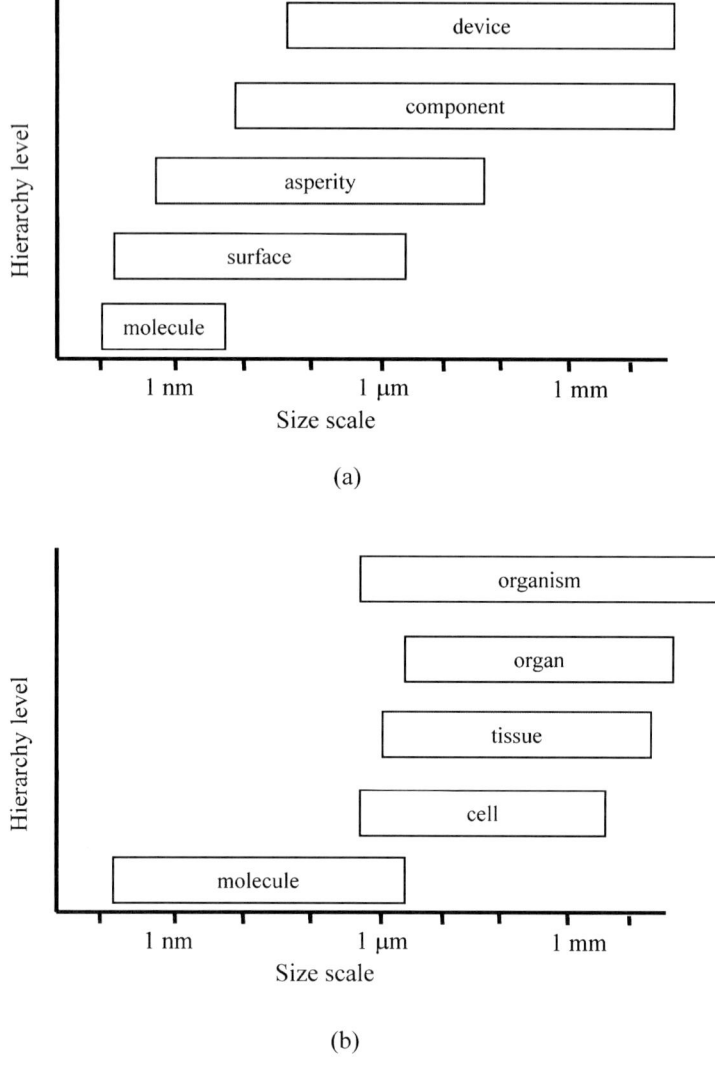

Fig. 1.3. Scale size and hierarchy levels in **a** engineering devices and **b** biological organisms [246]

Throughout this book we will deal with dissipative processes such as friction; however, our focus will be upon mechanical processes, which are usually slower than thermodynamic processes and thus in most cases we will assume that the system is at a thermodynamic equilibrium at a constant temperature T. In this case, the amount of dissipated energy is an appropriate measure of dissipation.

Classical thermodynamic systems, studied by Clausius, are characterized by an increasing disorder [6, 283]. Ilya Prigogine (1917–2003) and his students showed that some thermodynamic systems may lead to an increasing order and self-organization [271]. These so-called dissipative systems are thermodynamically open systems that operate far from thermodynamic equilibrium and can exchange energy, matter, and entropy with the environment. The dissipative systems are characterized by spontaneous symmetry breaking and formation of complex structures, where interacting particles exhibit long-range correlations. Examples of such systems are the Bénard cells in boiling liquid and oscillating chemical reactions. As in the case of the mesoscale thermodynamics, many of these systems were known a long time ago; however, the universality and generality of the processes involved in these systems was understood only through the works of Prigogine. It is believed that this ability for self-organization of physical systems led to the formation of complex hierarchical chemical and biological systems. Nonequilibrium dissipative systems may lead to the hierarchy, and their investigation involves the study of instability and asymmetry. Self-organization is related to an enormous reduction of degrees of freedom and entropy of the macroscopic system, consisting of many nonlinearly interacting subsystems, which macroscopically reveals an increase of order. The flow of heat, entropy, and material away from the interface during dry friction and wear can lead to self-organization so that the so-called "secondary structures" can form [113].

An important example of self-organization that was studied extensively by P. Bak [18] is the so-called "self-organized criticality," which implies that a system can spontaneously achieve a critical-point-like behavior. The related broad field is embraces two disciplines: Complexity Science, where the systems under study are between the perfect order and complete randomness, and Synergetics, an interdisciplinary science that explains the formation and self-organization of patterns and structures in systems far from thermodynamic equilibrium. These disciplines employ theoretical concepts similar to those of the mesoscale physics, e.g. the Landau–Ginzburg functional and the order-parameter [141].

1.5 Tribology

Tribology is defined as the study of contacting surfaces in relative motion [30, 32]. As opposed to surface science, tribology is an application-oriented discipline, which studies adhesion, friction, lubrication, and wear, and involves mechanics, physics, chemistry, materials science, biology, and other related areas. The word "tribophysics" (from the Greek word *tribos* "to rub") was coined in the 1940s by David Tabor (1913–2005), who then worked in Australia [101]. The first official use of the term "tribology" was in 1966 in the British governmental report on research in this area (the so-called "Jost report").

Although the word appeared only about 40 years ago, mankind has paid attention to friction and lubrication since ancient times. There is evidence that water lubrication was used in ancient Egypt as early as 2300 BC and 1800 BC, shown in two examples in Fig. 1.4. Oil lubrication is apparently mentioned in the Bible in the King Saul story (1020 BC), animal fat could be used as a lubricant for ancient Egyptian and Chinese chariots, and there is a list of lubricants in the treatise of Roman au-

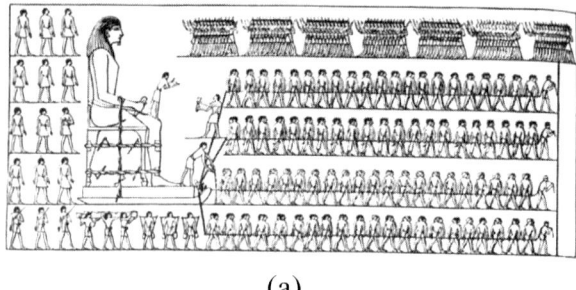

(a)

(b)

Fig. 1.4. Tribological technology in the ancient world. **a** Painting from El-Bersheh, circa 1880 BC, showing transportation of a giant statue. Man in front of the statue pouring liquid from a jar. Some historians suggested that the liquid served ceremonial purposes, whereas some engineers suggested that this is one of the first recorded cases of lubrication [101]. **b** A tomb painting from Theba (Egypt) shows manufacturing of leather-covered shields, with finished shields appearing at the upper left (2nd millennium BC). The Bible mentions using olive oil lubrication for smoothening leather-covered shields in the end of the second millennium BC [236]

thor Pliny the Elder (1st century BC) [101, 236]. The emergence of modern study of friction and lubrication is related to the activity of Leonardo da Vinci (1452–1519), Guillaume Amontons (1663–1705), and Charles August Coulomb (1736–1806), who formulated empirical rules of friction (Fig. 1.5) [224].

Modern tribology concentrates on issues such as rough surface topography, contact mechanics, adhesion, mechanisms of dry and lubricated friction, hydrodynamic (thick film) and boundary (thin film) lubrication, bearings, lubricant chemistry and additives, wear, surface texturing, and medical and biotribology.

Since the 1990s the new field of nanotribology has emerged due to the advances of nanotechnology [31, 34–36, 50]. The idea of nanotechnology was suggested in 1960 by physicist Richard Feynman (1912–1985), who pointed out that there is a physical possibility for manufacturing very small devices, which would be able to perform many tasks considered earlier impossible. The field of nanotechnology emerged and began to grow in the 1990s, stimulated by such discoveries as carbon

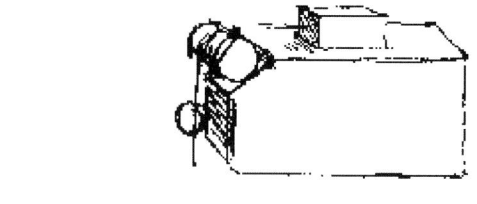

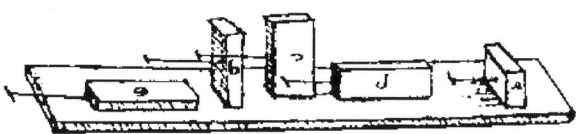

Fig. 1.5. Sketches by Leonardo da Vinci of his devices used to investigate friction laws [224]

nanotubes, C_{60} molecules (fullerenes), graphene (graphite monolayer), and quantum dots. Today it is one of the fastest growing research areas [36, 267]. Although presently the number of practical nanotechnological products is limited, according to some predictions the field may lead, in the next decade, to a technological revolution comparable with the one caused by the emergence of the transistor and microelectronics in the second half of the 20th century. However, right now nanoscience and nanotechnology remain an area of active fundamental and applied research. As it explained earlier, all kinds of surface effects are very important for small devices and therefore a significant part of the nanotechnological research is in nanotribology [38].

Nanotribology concentrates on the study of adhesion, friction, lubrication (in particular, by self-assembled molecular monolayers), and wear at the nanoscale [31, 34, 35, 50]. The main instrument currently employed for nanotribological research is the atomic force microscope (AFM) and its variations (friction force microscope and various other scanning probe microscopes). In the AFM, a small cantilever (typical length is 100 μm or less) with a very sharp tip (typical radius is 10–30 nm) can scan the surface of a sample, the position of which is controlled with great accuracy by a piezoelectric element. In contact with the surface or under the effect of forces (such as the adhesion force), the cantilever can bend and its deflection can be measured with a reflected laser beam. If the stiffness of the cantilever is known, it can be converted into the force acting upon the tip and thus small adhesion and friction force can be measured with high accuracy.

1.6 Biomimetics: From Engineering to Biology and Back

Biomimetics means mimicking biological objects in order to design artificial objects with desirable properties [25]. Another term used sometimes is bionics. The word was coined in the 1960s by biophysicist Otto Schmitt (1913–1998), though biological objects have been actively studied by physicists and chemists before that. However, biomimetics goes further than just biophysics and bioengineering, which only study biological objects, since its objective is to imitate the objects' desirable properties. The idea behind biomimetics is that nature's technical solutions—achieved by millions of years of evolution (or, maybe, given by God, depending on scientist's personal convictions)—are perfect or at least better than those which contemporary engineering technology can suggest. This concept may be applied to various areas of engineering, e.g., artificial intelligence and neural networks in information technology are inspired by the desire to mimic human brain. The existence of biocells and deoxyribonucleic acid (DNA) serves as a source of inspiration for nanotechnologists who hope to one day build self-assembled molecular-scale devices. In the field of biomimetic materials, there is also a whole area of bioinspired ceramics based on sea shells and other biomimetic materials.

In the field of biomimetic surfaces, a number of ideas have been suggested so far [24, 25, 132, 279]. These include the lotus-leaf surface, which has superhydrophobic and self-cleaning properties; the gecko foot, which has very high and adaptive adhesion; the moth eye, which does not reflect light; shark skin, which can suppress turbulence while moving underwater while; the water strider leg, which stays dry atop a pool of water; the darkling beetle, which collects dew using hydrophilic microspots; and the sand skink, which reduces friction using nanothresholds. The common feature found among many of these surfaces is that they have hierarchical roughness with rough details ranging from nanometers to millimeters. This observation inspired us to study multiscale frictional dissipative mechanisms in combination with hierarchical surfaces.

2

Rough Surface Topography

Abstract Approaches to solid surface topography characterization are discussed in this chapter, including experimental methods used in the conventional, nano-, and biotribology. Basic concepts of the statistical and fractal analysis of random rough surfaces and surface contact are reviewed. Common ways of surface modification, such as texturing and layer deposition, are discussed.

In this chapter, rough surface topography will be discussed with emphasis on the traditional engineering surfaces and their multiscale nature. Biological and biomimetic surfaces will be examined in detail in the third part of this book.

2.1 Rough Surface Characterization

A solid surface (or, more exactly, solid–liquid, solid–gas, or solid–vacuum interface) has complex structure and properties depending upon the nature of the material and the method of surface preparation. All solid surfaces, both natural and artificial, irrespective of the method of their formation, contain irregularities. No machining method can produce a molecularly flat surface on conventional materials. Even the smoothest surfaces, obtained by cleavage of some crystals (such as graphite or mica), contain irregularities, heights of which exceed interatomic distances. Engineering surfaces often have different types of random derivation from the prescribed form: the waviness, roughness, lay, and flow (Fig. 2.1). The waviness may result from machine vibration or chatter during machining as well as the heat treatment or warping strains. It includes irregularities with a relatively long (many microns) wavelength. Roughness is formed by fluctuation of the surface of short wavelengths, characterized by asperities (local maxima) and valleys (local minima). Lay is the principal direction of the predominant surface pattern, ordinarily determined by the production method. Flows are unintentional, unexpected, and unwanted interruptions in the texture [30, 32].

    The distinction between various roughness features is somewhat conditional and may depend upon application and upon the resolution of the measuring equipment.

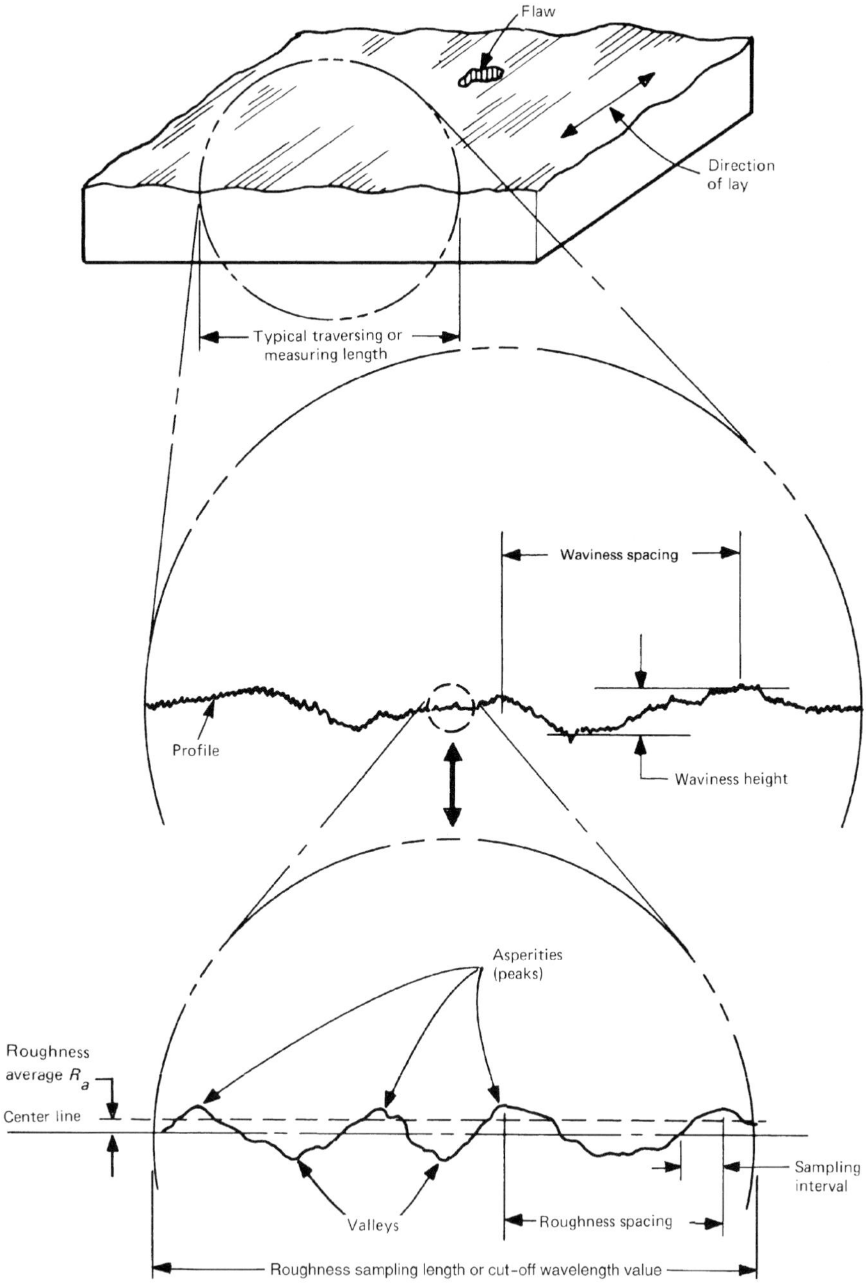

Fig. 2.1. Rough surface texture [32]

It is generally not possible to measure all the features at the same time. As will be discussed in the following, the very definition of the "asperity" involves serious problems. This is because a feature that may be a maximum of the surface profile at a given measurement resolution may involve numerous asperities and valleys when scrutinized at a higher resolution.

Various instruments are available to measure surface roughness. Mechanical (contact) and optical (noncontact) profilers are used to measure macro- and microscale roughness [30, 32]. The mechanical stylus method involves the amplifying and recording of vertical motions of a stylus tip displaced at a constant speed by the surface to be measured. The stylus is mechanically coupled mostly to a linear variable differential transformer or to an optical or capacitance sensor. As the stylus is scanned against the surface or the sample is transported relative to the stylus, an analog signal corresponding to the vertical stylus movement is amplified, conditioned, and digitized. The resolution of the profiler depends upon the dimensions of the tip—the sharper the tip is, the more fine details of the profile can be captured.

Optical methods are based upon measuring the light reflected from the surface. This includes the specular reflection methods that are used in glossmeters, diffuse reflection (scattering) methods, and optical interference methods that are used in various commercially available interferometers. Noncontact methods do not damage the measured surface, which is possible in the case of contact methods.

Several methods have been developed to measure roughness at the micro- and nanoscale. The family of instruments based on scanning tunnel microscopy (STM) and atomic force microscopy (AFM) is called scanning probe microscopy (SPM). In the STM, which was developed in early 1980s, a sharp tiny metal tip is brought very close (0.3–1 nm) to the sample surface. As the voltage between the tip and the sample is applied, the tunneling current is measured, which is proportional to the gap between the tip and the sample. As the tip is scanned against the sample, the sample height profile can be measured with subnanometer resolution [31, 34].

The AFM combines the principle of the STM and the stylus profiler. In the AFM, the tip (with the radius of few nanometers) is placed at the end of a long (dozens of micrometers) stiff cantilever (Fig. 2.2). The cantilever deflection is measured by determining the position of a laser beam reflected from the cantilever surface. In the contact mode, the tip scans the sample and the height map can be obtained with subnanometer resolution. In the noncontact mode, the van der Waals adhesion force acts upon the tip and results in cantilever deflection. As the stiffness of the cantilever is known, the deflection can be converted into the force unit (with subnanonewton resolution). The AFM can operate in ambient air as well as in vacuum [31, 34].

Scanning electron microscopy (SEM) can also be used for studying surface features; however, it has several limitations. First, it is difficult to obtain quantitative data from the SEM, and second, the field of view in SEM is limited. The use of the SEM requires placing the specimen in vacuum. In addition, a conductive coating is required to insulate samples [30, 32]. For biological specimens, there is the technique known as environmental scanning electron microscopy (ESEM), which allows one to conduct measurements in controlled humidity and pressure conditions.

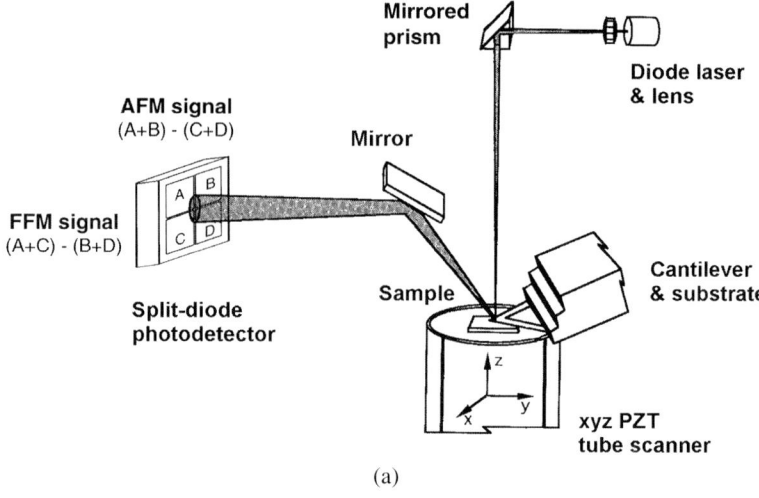

(a)

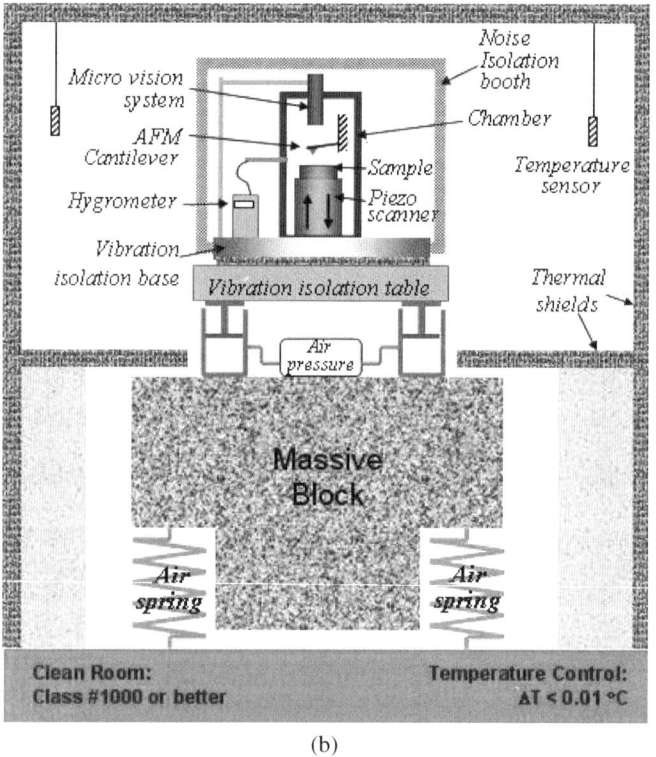

(b)

Fig. 2.2. Atomic force microscope (AFM). **a** Principle of operation. A sample mounted on a piezoelectric tube (PZT) scanner scanned against a sharp tip and the cantilever deflection is measured using a laser beam [32]. **b** Vibration isolated clean-room setup for the AFM used at the NIST (credit to Dr. S.H. Yang, NIST)

Modern methods of surface structure analysis include X-ray spectrometry, Raman spectroscopy, electron diffraction, and others.

In addition to surface irregularities, the technical solid surface itself involves several zones or layers, such as the chemisorbed layer (0.3 nm), physisorbed layer (0.3–3 nm), chemically reacted layer (10–100 nm), etc. [30, 32]. In the chemisorbed layer, the solid surface bonds to the adsorption species through covalent bonds with an actual sharing of electrons. In the physisorbed layer, there are no chemical bonds between the substrate and the adsorbent, and only van der Waals force are involved. The van der Waals force is relatively weak (under 10 kJ/mol) and long range (nanometers) as opposed to strong (40–400 kJ/mol) and short range (comparable with the interatomic distance of about 0.3 nm). Typical adsorbents are oxygen, water vapor, or hydrocarbons from the environment, which can condense at the surface. While the chemisorbed layer is usually a monolayer, the physisorbed layer may include several layers of molecules. The chemically reacted layer is significantly thicker and involves many layers of molecules. The typical example of the chemically reacted layer is the oxide layer at the surface of a metallic substrate.

2.2 Statistical Analysis of Random Surface Roughness

There are several quantitative parameters commonly used to characterize random solid surface roughness, i.e., a random derivation from the nominal (prescribed) shape. These parameters include the amplitude (or height) parameters and the spatial parameters [30, 32, 316] (Fig. 2.3). The most commonly used amplitude parameter is the root mean square (RMS) or the standard deviation from the center-line average. For a 2D roughness profile $z(x)$, the center-line average is defined as the arithmetic mean of the absolute value of the vertical deviation from the mean line of the profile (Fig. 2.4)

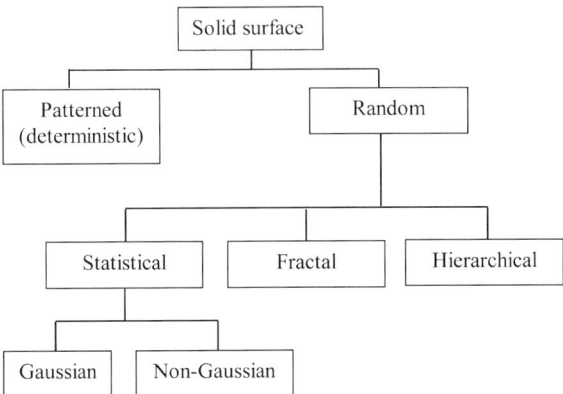

Fig. 2.3. Typology of rough surfaces (based on [32])

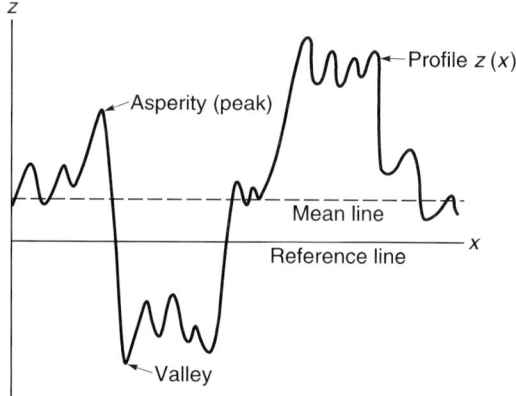

Fig. 2.4. Schematics of a rough surface profile [32]

$$R_{\mathrm{a}} = \frac{1}{L} \int_0^L |z - m| \, dx, \tag{2.1}$$

where L is the sampling length,

$$m = \frac{1}{L} \int_0^L z \, dx. \tag{2.2}$$

The square RMS is given by

$$\sigma^2 = \frac{1}{L} \int_0^L (z - m)^2 \, dx. \tag{2.3}$$

Since different rough surface profiles can have the same RMS, additional parameters are required to characterize details of surface profile. Two additional statistical parameters are the skewness and kurtosis, which are given in the normalized form by

$$Sk = \frac{1}{\sigma^3 L} \int_0^L (z - m)^3 \, dx, \tag{2.4}$$

and

$$K = \frac{1}{\sigma^4 L} \int_0^L (z - m)^4 \, dx. \tag{2.5}$$

A surface with a negative skewness has a larger number of local maxima above the mean, whereas for a positive skewness the opposite is true. Similarly, a surface with a low kurtosis has a larger number of local maxima above the mean as compared to that with a high kurtosis. Note that we defined these parameters for a 2D profile, but they can easily be generalized for a 3D surface [30, 32].

The cumulative probability distribution function, $P(h)$ associated with the random variable $z(h)$, is defined as the probability of the event that $z(x) < h$, and is written as

$$P(h) = \text{Probability } (z < h). \tag{2.6}$$

It is common to describe the probability structure of random data in terms of the slope of the distribution function, known as the probability density function (PDF) and given by the derivative

$$p(z) = \frac{dP(z)}{dz}. \tag{2.7}$$

The integral of the PDF is equal to $P(z)$, and the total area under the PDF must be unity [30, 32].

In many practical cases, the random data tend to have the so-called Gaussian or normal distribution with the PDF given by

$$p(z) = \frac{1}{\sigma\sqrt{2\pi}} \exp\left(-\frac{(z-m)^2}{2\sigma^2}\right), \tag{2.8}$$

where m is the mean and σ is the standard deviation. For convenience, the Gaussian function is often plotted in terms of the normalized variable $z^* = (z - m)/\sigma$ as

$$p(z^*) = \frac{1}{\sqrt{2\pi}} \exp\left(-\frac{z^{*2}}{2}\right). \tag{2.9}$$

The Gaussian distribution has zero skewness $Sk = 0$ and kurtosis $K = 3$ [32] (Fig. 2.5).

The Gaussian distribution is found in nature and in technical applications when the random quantity is a sum of many random factors acting independently of each other. When an engineered surface is formed, there are many random factors that contribute to the roughness, and thus in many cases roughness height is governed by the Gaussian distribution. Such surfaces are called Gaussian surfaces.

In order to represent spatial distribution of random roughness we use the autocorrelation function, defined as

$$C(\tau) = \lim_{L\to\infty} \frac{1}{\sigma^2 L} \int_0^L \left[z(x) - m\right]\left[z(x+\tau) - m\right] dx. \tag{2.10}$$

The autocorrelation function characterizes the correlation between two measurements taken at the distance τ apart, $z(x)$ and $z(x + \tau)$. It is obtained by comparing the function $z(x)$ with a replica of itself shifted for the distance τ. The function $C(\tau)$ approaches zero if there is no statistical correlation between values of z separated by the distance τ; in the opposite case $C(\tau)$ is different from zero. Many engineered surfaces are found to have an exponential autocorrelation function

$$C(\tau) = \exp(-\tau/\beta), \tag{2.11}$$

where β is the parameter called the correlation length or the length over which the autocorrelation function drops to a small fraction of its original value. At the distance β, the autocorrelation function falls to $1/e$. In many cases the value $\beta^* = 2.3\beta$ is used for the correlation length, at which the function falls to 10% of its original value [30, 32].

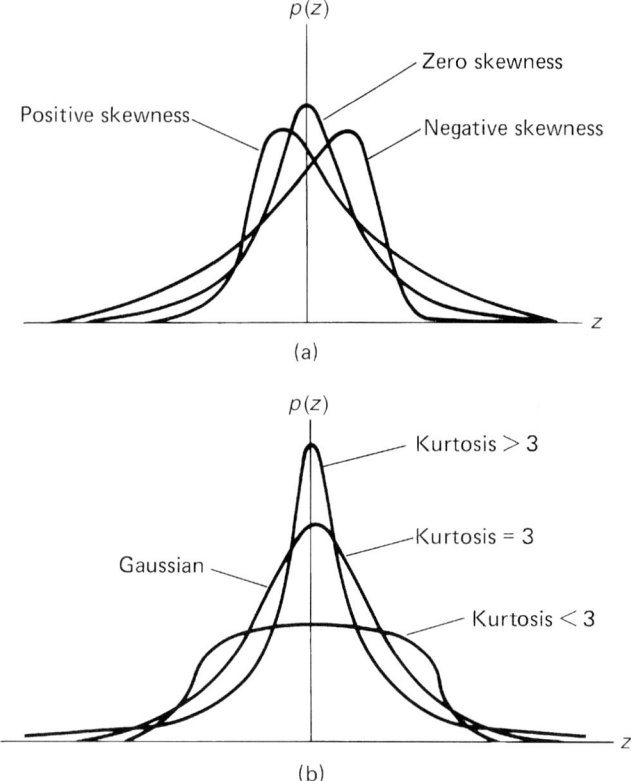

Fig. 2.5. Typical **a** skewness and **b** kurtosis [32]

For a Gaussian surface with the exponential autocorrelation function, σ and β^* are two parameters of the length dimension which conveniently characterize the roughness. While σ is the height parameter that characterizes the height of a typical roughness detail (asperity), β^* is the length parameter that characterizes the length of the detail. The average absolute value of the slope is proportional to the ratio σ/β^*, whereas the average curvature is proportional to β^*/σ^2. For a Gaussian surface, σ is related to the RMS as $\sigma = (\sqrt{\pi/2})\,R_a$ [30, 32]. These two parameters, σ and β^*, are convenient for characterization of many random surfaces. Note that a Gaussian surface has only one inherent length scale parameter, β^*, and one vertical length scale parameter, σ, and thus it cannot describe the multiscale roughness.

2.3 Fractal Surface Roughness

A measurement of the roughness parameters, such as σ and β^*, shows that they are sensitive to the scale, that is to the resolution of a measuring device (the sampling interval or the short wavelength limit) as well as to the scan size (the long wave-

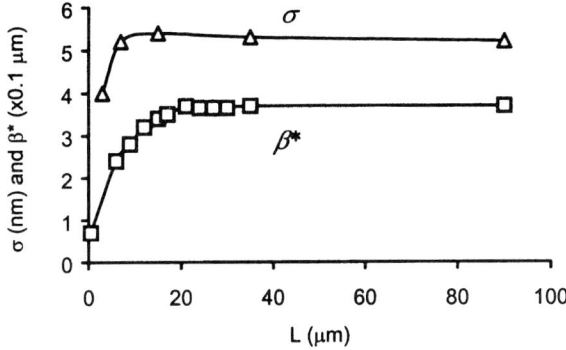

Fig. 2.6. Dependence of measured σ and β^* upon scan size L for a glass disk (based on [268])

length limit) (Fig. 2.6). And understandably so, since the roughness is composed of many wavelengths superimposed upon each other which all affect the cumulative values of σ and β^*, and the wavelengths smaller than the sampling intervals or larger than the scan size cut off and do not contribute to the roughness parameters [268]. Thus, the measured roughness parameters depend upon the short- and long-wavelength limits. This consideration is not only an artifact of the measurement or a result of the measuring devices' limitations. For practical contact problems, asperity may be defined as a roughness detail that participates in the contact and forms a contact spot. Therefore, the size and length of the contact spots are important for the contact of rough surfaces and may provide wavelength limits relevant for the contact problem.

A surface is composed of a large number of length scales of roughness that are superimposed on each other. The variances of surface height and other roughness parameters depend on the resolution of the roughness measurement instrument. As the resolution increases, more small details of the rough profile can be observed. When a rough surface is repeatedly magnified, increasing details of roughness are observed down to nanoscale. The roughness at all magnifications appears quite similar in structure. Such self-affinity can be characterized by fractal geometry.

Archad [12] suggested we present a rough surface as one covered by asperities of a certain size, which have much smaller asperities on the top of them and even smaller asperities on the top of those. He showed that an elastic surface with such a hierarchical structure, which is similar to fractal geometry, leads to an almost linear dependence of the real area of contact with a flat upon the normal force. This, along with the linear proportionality of the friction force to the real area of contact due to the adhesion, could explain the well-known linear proportionality of the friction force to the normal load.

Self-similar curves and surfaces have been studied by mathematicians since the first half of the 20th century. A remarkable property of these curves and surfaces is that they have a fractional "dimension," D, in a sense that when the linear scale is magnified by a certain factor α, the length of the curve or the area of the surface changes proportional to α^D. This is because more fine details are observed with the

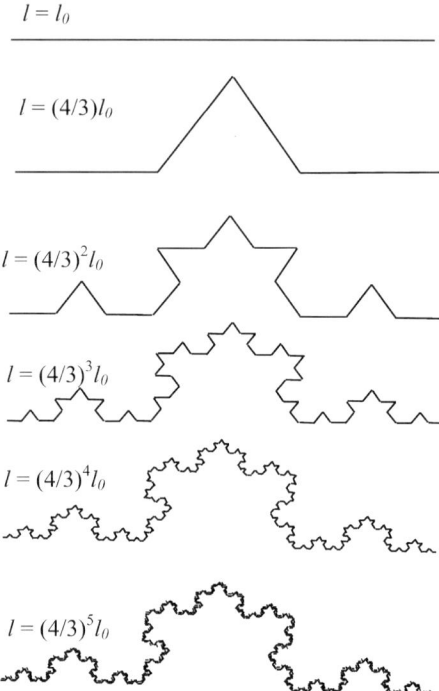

Fig. 2.7. The Koch curve with fractal dimension $D = 1.26$. The curve is built by an iterative procedure, and at each step its length l is increased by the factor $4/3$. If the linear length scale l_0 is increased by 3 times, the total length is increased by $4 = 3^D$

magnification. Thus, when the so-called Koch curve (Fig. 2.7) of length l is magnified by the factor $\alpha = 3$, its length becomes equal to $4l$. Thus, the fractal dimension, $D = \ln(4)/\ln(3) = 1.26$, is between 1 and 2. Unlike most mathematical functions used in engineering, the fractal curves do not have a derivative at any point. Although self-similarity implies equal magnification in all directions, the term self-affinity has a broader meaning and implies that a curve can scale in a certain manner during magnification.

In the 1970s, the term "fractal" was introduced and the concept of self-similar and self-affine objects was widely popularized. It was recognized that fractal geometry could be applied to various physical phenomena, ranging from the coastal line of oceans to the turbulent flow in fluids. The fractals were thought to be a universal tool which could be applied in the situation of noncontinuum behavior that cannot be studied by the continuum functions of traditional calculus.

Since the 1980s, it was suggested that fractal geometry can be applied for the characterization of rough surfaces in tribology (Fig. 2.8) [119, 120, 205, 212, 213, 215]. Majumdar and Bhushan (1990) suggested that the Weierstrass–Mandelbrot self-affine function captures significant features of a self-affine rough profile

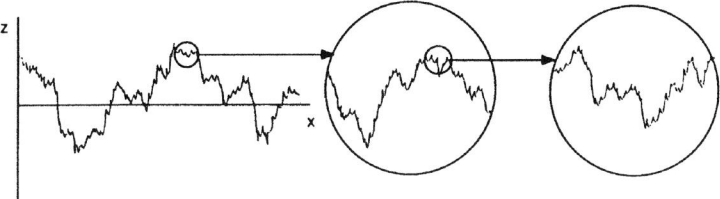

Fig. 2.8. Self-affinity of a surface profile [32]

$$z(x) = G^{(D-1)} \sum_{n=n_j}^{\infty} \frac{\cos 2\pi \gamma^n x}{\gamma^{(2-D)n}}; \quad 1 < D < 2; \; \gamma > 1, \quad (2.12)$$

where D is the fractal dimension, G is a nondimensional scaling constant, and γ^n determines the frequency spectrum of the profile roughness. Nondimensional D and G with the dimension of the length are two parameters that characterize a fractal profile. Ganti and Bhushan [120] extended that analysis and considered the lateral resolution of the measuring instrument as an intrinsic length unit. This generalized analysis allows surface characterization in terms of two fractal parameters—fractal dimension and amplitude coefficient—which, in theory, are instrument independent and unique for each surface. Ganti and Bhushan [120] developed a technique for the simulation of fractal surface profiles. A number of engineered surfaces were measured to validate the generalized fractal analysis, in particular, magnetic tapes, thin-film rigid disks, steel disks, plastic disks, and diamond films, all of varying roughness. For a given surface with varying roughnesses, the fractal dimension essentially remains constant, while the scaling constant varies monotonically with variance of surface heights (σ^2) for a given instrument. Simulated σ shows similar trends in the measured σ for small scan lengths. The coefficient of friction of all surfaces has reasonable correspondence with the scaling constant.

In practice, the profile demonstrates self-affine behavior down to a certain scale length (e.g., of the molecular scale) or a high frequency (short wavelength) limit, ω_h. With a further magnification of the profile, no self-similarity can be found. In a similar manner, there is a low frequency (long wavelength) limit, ω_l, of the fractal behavior [212]. Note that a fractal profile has no characteristic parameters of the length scale. However, short and long wavelength limits effectively provide such parameters of the length dimension, $1/\omega_h$ and $1/\omega_l$. During the contact of two rough surfaces, relevant parameters—such as the number of asperity contacts and the real area of contact—depend upon the short- and long-wavelength limits as power functions with power exponents depending upon D and G.

2.4 Contact of Rough Solid Surfaces

When two rough surfaces come into a mechanical contact, the real area of contact is small in comparison with the nominal area of contact, because the contact takes place only at the tops of the asperities. For two rough surfaces in contact, an equivalent rough surface can be defined for which the values of the local heights, slopes, and local curvature are added to each other. The composite standard deviation of profile heights is related to those of the two rough surfaces, σ_1 and σ_2 as

$$\sigma^2 = \sigma_1^2 + \sigma_2^2. \tag{2.13}$$

The composite correlation length is related to those of the two rough surfaces, β_1^* and β_2^*, as

$$1/\beta^* = 1/\beta_1^* + 1/\beta_2^*. \tag{2.14}$$

Using of the composite rough parameters allows us to effectively reduce the contact problem of two rough surfaces to the contact of a composite rough surface with a flat surface [30, 32].

Two parameters of interest during the elastic and plastic contact of two rough surfaces are the real area of contact, A_r, and the total number of contact spots, N. In most cases, only the highest asperities participate in the contact. This allows us to linearize the dependence of A_r and N upon the roughness parameters during the elastic contact as

$$A_r \propto \frac{W\beta^*}{\sigma E}, \tag{2.15}$$

$$N \propto \frac{1}{\sigma \beta^*}, \tag{2.16}$$

where W is the normal load force and E is the composite elastic modulus. Qualitatively, the higher the asperities, the larger σ is and the smaller A_r is; the wider the asperities, the larger β^* is and smaller A_r is. The larger and wider the asperities, the smaller A_r is [44].

For plastic contact, N, which depends upon the contact topography and thus is independent on whether the contact is elastic or plastic, is still given by (2.16) for a given separation between the surfaces [43], whereas the real contact area is found by dividing the load by the hardness

$$A_r \propto W/H. \tag{2.17}$$

For fractal surfaces, the roughness and contact parameters are related to the high and low frequency limits as [212]

$$\sigma \propto \omega_l^{(D-2)}, \tag{2.18}$$

$$A_r \propto \frac{\omega_l^{(2-D)/2}}{\omega_h^{D/2}}, \tag{2.19}$$

$$N \propto \omega_l^{3(2-D)/2} \omega_h^{D/2}. \tag{2.20}$$

2.5 Surface Modification

As discussed in the preceding sections, surface properties including the topography have significant effect upon the mechanical contact. There are many ways to modify surfaces in order to obtain desirable properties. Two such methods are surface texturing and layer deposition.

2.5.1 Surface Texturing

Since most engineered and natural surfaces are rough, it may be advantageous not to stay with the random roughness, but to texture a surface in a certain manner so that the useful properties of the surface, such as load capacity, low friction, and wear, improve. Surface texturing has became an object of intensive study in the recent decade [28, 191]. Various techniques are used for surface texturing, including machining, ion beam texturing, etching, lithography, and laser texturing. The texturing usually produces a large number of microdimples on a surface that are effective in combination with lubrication. The dimples can serve as microhydrodynamic bearings, reservoirs for lubricant, or traps for wear debris [106]. Surface texturing is commonly used in magnetic storage devices [27, 28] and microelectromechanical systems (MEMS) to prevent adhesion and stiction (sticking of two components to each other due to adhesion) [36, 38]. It is also used in the automotive industry to hone cylinder liners. At this point, most studies in the area of texturing are experimental and concentrate on finding the optimum size and distribution of the dimples. Thus, Hsu and coworkers [335] investigated the effect of dimple size (of the order of dozens of microns) and depth (below one micron) on sliding friction under boundary lubrication conditions. They found that, for a constant dimple surface, smaller and shallower dimples are more advantageous.

Fabrication techniques for creating micro/nanoroughness include lithography (photo, E-beam, X-ray, etc.), etching (plasma, laser, chemical, electrochemical), deformation, deposition, and others.

2.5.2 Layer Deposition

Thin, artificially deposited layers of long-chain molecules can be used to lubricate microdevices. Such monolayers or thin films are commonly produced by the so-called Langmuir–Blodgett method and by chemically grafting the molecules into self-assembled monolayers. In the Langmuir–Blodgett method, a monolayer is formed at a liquid-air interface and then deposited upon the substrate, to which it is bonded by weak van der Waals forces. Self-assembled monolayer (SAM) molecules attach to the substrate by chemical bonds [267].

Besides the SAM method, there are several other techniques of deposition, including adsorption, dip coating, spin coating, anodization, electrochemical deposition, evaporation, plasma, etc.

2.6 Summary

Since fractals have been introduced into surface mechanics [213], the argument continues over whether fractal geometry provides an adequate description of physical phenomena and scaling issues. Interestingly, one of the creators of the classical Greenwood and Williamson [137] statistical model of the surface published an "apology," recognizing that a fractal description is needed instead [138]. Indeed, many rough surfaces demonstrate self-affine properties to a certain extent and at a certain range of scales. Fractals as mathematical objects obviously have a certain beauty and give us a tool to describe noncontinuous phenomena; these features attracted many physicists and other scientists. However, it is questionable whether the fractal description, which ultimately assumes that a rough profile is characterized by only two nondimensional parameters, D and G, can provide more practical information for the analysis of engineering surfaces than traditional statistical characterization. The generalized analysis by Ganti and Bhushan [120] provides an extension of the Majumdar–Bhushan model for tribological applications; however, practical usefulness of fractal analysis in tribology remains the subject of an argument. An ideal fractal surface is composed of roughness at different scales, but it does not possess parameters of length scale. Unlike the ideal surface, a real fractal surface has such parameters, $1/\omega_h$ and $1/\omega_l$. The contact parameters calculated from the fractal models of surfaces, given by (2.19)–(2.20), depend upon ω_h and ω_l, which, in fact, characterize *limits* of fractal behavior, rather than the fractal behavior itself.

It is important to note that the statistical description of a rough surface provides some parameters of the length scale, for example, σ and β^*. However, single length and height parameters do not provide an adequate description of multiscale surfaces that involve several scale lengths. While the roughness parameters provide only a constant scale length, we have observed in this chapter two different types of scale dependence. One is the dependence of the roughness parameters upon the scan size, as shown in Fig. 2.6. This dependence is the measurement artifact and is a result of the measuring equipment's limitations. Another type of scale dependence appears during the contact of rough surfaces. The contact spot's size provides additional length parameters that may interplay with the roughness parameters. For example, if the contact size is smaller than the long wavelength limit of roughness, ω_l, the roughness components with larger wavelengths do not contribute to the roughness and contact parameters, effectively changing the latter.

3

Mechanisms of Dry Friction, Their Scaling and Linear Properties

Abstract Various mechanisms of dry sliding friction of two solids is discussed, including adhesion and adhesion hysteresis, deformation, plastic yield, fracture, the ratchet, cobblestone and third-body mechanisms. It is discussed how all these diverse mechanisms lead to the linear Amontons–Coulomb's empirical law of friction. Various explanations of the linearity of friction are discussed (real area of contact and slope-controlled friction, etc.) and the concept of a "small parameter" responsible for the linearity is suggested.

Dry solid–solid friction is the resistance to sliding and rolling motion. Friction is a universal phenomenon which is observed in a great variety of sliding and rolling situations. Friction is also a complex phenomenon that cannot be reduced to a single mechanism, but rather is a result of a simultaneous action of various mechanisms at different hierarchy and scale levels [30, 32, 63]. In a remarkable way, all these various mechanisms result in a dissipative process, which can often be characterized by only one single parameter, the coefficient of friction that is equal to the ratio of the friction force to the normal load. In this chapter, we will discuss general scaling issues related to solid–solid dry friction, and after that we will consider various mechanisms of friction in order to investigate what they have in common and how they all result in what is observed at the macroscale as the simple process of dry friction.

In this and following chapters, we study friction as a multiscale (hierarchical) phenomenon, showing that the mechanisms of energy dissipation result from the interplay of forces at two or more scale levels. In the following sections, the fundamental mechanisms of solid–solid friction are considered involving heterogeneity, linear, nonlinear, and hierarchical effects. The main mechanisms of dry friction are adhesion, deformation of asperities (plowing), fracture and the so-called ratchet, and third-body mechanisms [30, 32, 63]. For each mechanism, we will identify the "small parameter" that is present because the forces at the interface are smaller than the forces in the bulk. In the following chapters we will show how this small parameter leads to linearity of the friction force as a function of load. We will show that two characteristic scale lengths may be identified for most of these mechanisms. Mapping of dry friction mechanisms using these characteristic scale lengths will be proposed in the next chapter. Then a scale-dependence of these mechanisms is studied, based on the presumption that inhomogeneity at each hierarchy level leads to energy dissipation. Based on this, the second part of the book discusses hierarchical biological surfaces, which are created by nature to decrease or increase solid–solid and solid–liquid adhesion and friction. We show that their hierarchy is a consequence of simultaneously acting physical mechanisms at different scale levels; thus, surface hierarchy is a consequence of the hierarchical nature of friction mechanisms. After discussing in this chapter the well-known manifestation of linearity of friction, the Amontons–Coulomb rule, we will study deviations from linearity in the next chapter. The inherent nonlinearity of friction serves as a basis for creating hierarchical mechanisms and structures.

3.1 Approaches to the Multiscale Nature of Friction

Dry solid–solid friction is a complex and universal phenomenon which is found at various scale sizes from the atomic scale up to the macroscale and at different levels of the hierarchy of a device, from the level of molecules up to surfaces, asperities, components, and systems. Each of these levels is characterized by a different structure and range of scales, and each may have different predominant friction mechanisms (Table 3.1). The atomic scale (on the order of 1 nm or less) is characterized by discrete atoms and quantum-mechanical interactions (chemical bonds), described by the surface energy states. The mesoscale or nanoscale (on the order from 1 nm to 0.1 μm) is characterized by dislocations, surface defects, roughness, and inhomogeneity. Mesoscale description is required in order to provide a link between the

Table 3.1. Dissipation and friction mechanisms corresponding to different hierarchy levels

Ideal situation	Real situation	Mechanism of dissipation leading to friction	Friction mechanism	Hierarchy level
Nonadhesive surfaces	Chemical interaction between surfaces is possible	Breaking chemical adhesive bonds	Adhesion	Molecule
Conservative adhesive forces	Conservative (van der Waals) forces and nonconservative (chemical) bonds	Breaking chemical adhesive bonds	Adhesion	Molecule
Rigid material	Deformable (elastic and plastic) material	Radiation of elastic waves (phonons)	Adhesion	Surface
Smooth surface	Rough surface	Plowing, ratchet mechanism, cobblestone mechanism	Deformation, ratchet, cobblestone mechanisms	Asperity
Homogeneous surface	Inhomogeneous surface	Energy dissipation due to inhomogeneity	Adhesion	Surface

atomic and continuum levels. At the mesoscale, the bulk of the body can be viewed as divided into blocks or domains, so that the quantities which are not defined at the atomic scale, such as the yield strength or the coefficient of friction, can be defined at the macroscale by averaging throughout a mesoscale block or domain.

In order to introduce the mesoscale into friction models, it is instructive to consider the approach of scale-dependent plasticity theories. The scale-dependent yield strength is introduced in this manner by strain-gradient plasticity theories [122]. These theories postulate that the yield strength, which controls the onset of plastic flow, σ_Y, depends not only upon the strain, but also upon spatial strain gradient, $\nabla\varepsilon$, as

$$\sigma_Y = \sigma_{Y0}\sqrt{1 + l\nabla\varepsilon}, \tag{3.1}$$

where σ_{Y0} is the macroscale yield strength and l is a new characteristic length parameter postulated by these theories, which is on the order of micrometer. For two geometrically proportional configurations of different sizes, the strains are the same, but the strain gradient is much greater at a smaller scale configuration [230]. Thus, for submicron-sized systems (those with a typical size greater than l), the value of the yield strength will be considerably greater than the macroscale value, σ_{Y0}. Physically, the yield strength depends on the strain gradient due to the presence of the so-called geometrically necessary dislocations, which are required for strain compatibility, and their density increases with decreasing scale. Figure 3.1(a) shows the randomly distributed statistically stored dislocations during shear and geometrically necessary dislocations during bending that are needed for stress compatibility. Geometrically necessary dislocations during indentation are shown in Fig. 3.1(b). However, in order to introduce this dependence into the theory of plasticity in a strict manner, it is necessary to connect the micron-scale plasticity to the dislocation theories in a multiscale framework, and this is achieved by considering mesoscale blocks (Fig. 3.1(c)) [122, 156]. Bhushan and Nosonovsky [42] showed that such scale dependence of the yield strength and of hardness leads to the scale dependence of the coefficient of friction.

Frictional sliding is a dissipative process, and it is thermodynamically irreversible resulting in an increase of entropy of a system. Friction is not a property of a surface, but rather a system response [30, 32] that results in an increase of the system's disorder and entropy. Friction force is not a fundamental force of nature because it is a result of the action of the electromagnetic and exchange forces between the atoms, which are in principle reversible. For an ideal system of perfectly rigid bodies with potential electric forces acting between the atoms, there would be no energy dissipation and therefore no friction. Real systems, however, are imperfect and involve elastically and plastically deformable as well as brittle bodies; rough, chemically active, and inhomogeneous surfaces; and reversible weak and irreversible strong adhesive bonds. These imperfections result in energy dissipation and frictional resistance to sliding.

It is well known from experiments that the values of the coefficient of friction, when measured at the micro/nanoscale, are different from those at the macroscale, and therefore friction is scale dependent [50, 154, 287]. Various approaches have

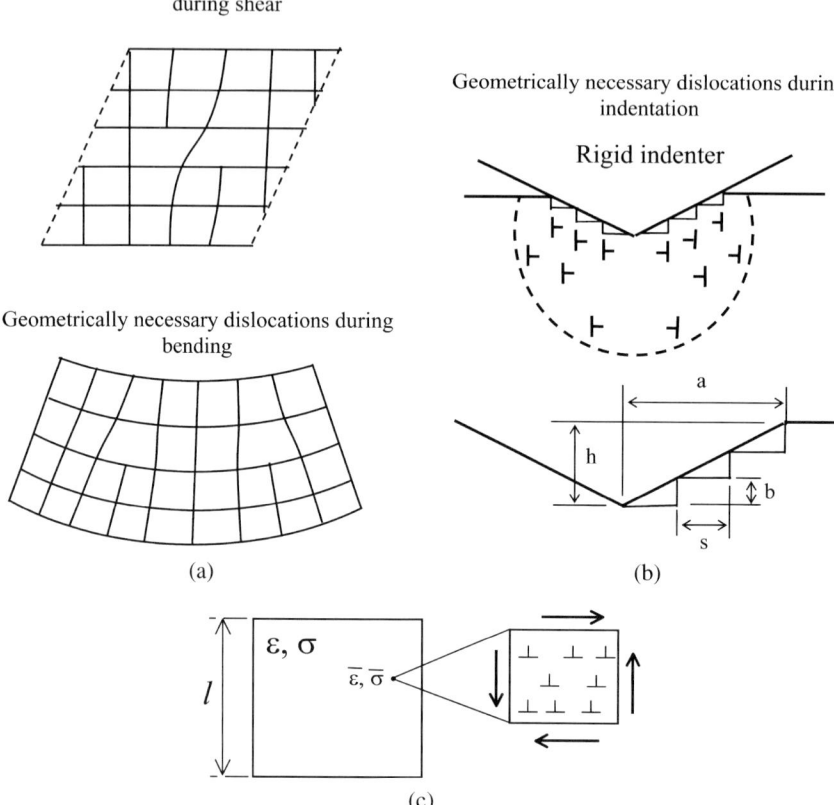

Fig. 3.1. a Statistically stored dislocation during bending and **b** geometrically necessary dislocations during indentation in strain-gradient plasticity [42]. **c** The multiscale framework of the strain gradient plasticity. Dislocation interaction at the microscale is considered via the Taylor relation. The higher-order strain gradient plasticity theory is established on the mesoscale representative cell of size l (based on [156])

been proposed to study and explain the scale-dependence of friction. Many scholars have considered the so-called *scale effect* on friction or *scaling laws* of friction [4, 42–44, 46, 56, 87, 145, 158, 229, 326, 336, 353]. While the origin of the scaling laws is in the geometrical relations, such as surface-to-volume ratios [72, 272], the term "scale effect" implies more general laws dependent upon physical mechanisms, rather than pure geometrical relations. Johnson [168] paid attention to the fact that frictional stress is strongly dependent upon the scale of contact and suggested that gliding dislocations at the surface contribute to the frictional stress. Hurtado and Kim [158] (HK) proposed a model of single-asperity contact with a scale-dependent shear stress. Their model is based on the concept of dislocation-assisted sliding with dislocation loops nucleation at the perimeter of a circular contact zone. The model, however, is limited to the case of commensurate interface between the bodies, which therefore should have the same orientation and spacing of the crystal lattices. This is not a likely situation in most cases. Adams et al. [4] applied the HK model for multiple-asperity elastic contact with a Gaussian statistical distribution of asperity heights and identified parameters responsible for the scale effect.

Bhushan and Nosonovsky [42–44, 46] took a different approach and considered scale-dependent distribution of surface heights combined with scale-dependent frictional stress due to dislocation nucleation from Frank–Read sources (rather than at the perimeter of the contact zone), as well as the strain-gradient plasticity. They later included in their model the effect of asperity and particle deformation, with scale-dependent densities of trapped particles at the interface. Zhang et al. [353] studied scale effects on friction using molecular dynamics (MD) simulation. He and Robbins [145] used MD simulation to study the origin of scale dependence on friction. Deshpande et al. [91] conducted numerical simulation of dislocation motion during frictional plastic deformation and showed that dislocation nucleation from the sources (rather than at the perimeter of the contact zone) results in scale-dependent frictional stress. Kogut and Etsion [189] proposed a model of elastic-plastic frictional contact with scale-independent plasticity, which resulted in the coefficient of friction strongly dependent upon the apparent area of contact, A_a, and normal load, W. Nosonovsky and Bhushan [239] also suggested that the mechanism of load-dependence of friction is similar to that of size dependence. Nosonovsky [235] also studied size, load, and velocity dependence of friction in combination. All these studies investigate some aspects of the scale effect on friction, however, they do not provide us with a general theory of scale dependence of friction.

A different approach is taken by the scholars who try to formulate empirical *friction laws at the nanoscale* rather than the scaling laws of friction [71, 302, 336]. Such friction laws are intended as substitutes for the classical Amontons–Coulomb's empirical laws (better called "rules," because situations in which these rules are not followed do not imply violation of any fundamental laws of nature) of friction, which state that the friction force between two bodies is (1) proportional to the normal load, (2) independent of the nominal contact area between the bodies, and (3) almost independent of the sliding velocity [32]. This approach, however, does not deal with the friction as a universal phenomenon and virtually considers nanoscale and macroscale friction as unrelated.

3.2 Mechanisms of Dry Friction

In this section we discuss major mechanisms of dry friction: adhesion, deformation of asperities, plastic yield, the ratchet, cobblestone, and third body mechanisms.

3.2.1 Adhesive Friction

Adhesion constitutes the most common and best studied mechanism of dry friction, which occurs at a wide range of length scales and conditions.

3.2.1.1 Adhesion between Solid Surfaces

When two surfaces are brought into contact, adhesion or bonding across the interface can occur, and a finite normal force, called the adhesion force, is required to pull apart the two solids [30, 32, 63]. Since the typical range of the adhesion force is in nanometers, the role of adhesion is important at the nanoscale. As we discussed in the preceding sections, for chemically nonactive surfaces, there are two types of interatomic adhesive forces: the strong (chemical) forces, such as covalent, ionic, and metallic bonds, whose rupture corresponds to large absorption of energy (around 400 kJ/mol); and weak forces, such as hydrogen bonds and van der Waals forces (few kJ/mol) [222]. Weak conservative forces act at larger ranges of distance, whereas strong bonds act at short distances.

For macrofriction of nonadhesive surfaces, Bowden and Tabor [63] suggested that the friction force F is directly proportional to the real area of contact A_r and shear strength at the interface τ_f

$$F = \tau_f A_r. \tag{3.2}$$

Every nominally flat surface in reality has roughness. The real area of contact is only a small fraction of the nominal area of contact because the contact takes place only at the summits of the asperities (Fig. 3.2(a)). Various statistical models of contact of rough surfaces show that A_r is almost directly proportional to the applied normal load W, for elastic and plastic surfaces, which explains the empirically observed linear proportionality of F and W (the so-called Amontons–Coulomb's rule), assuming constant τ_f [137]. The physical nature of the surface shear strength τ_f, however, remains a subject of discussion. For the pure interfacial friction, τ_f may be viewed as the shear component of the adhesive force, which is required to move surfaces relative to each other.

Effect of adhesion on elastic contact has been investigated by many researchers [64, 82, 90, 169, 170, 221, 222]. When a smooth sphere comes into contact with elastic half-space, the contact area exceeds that predicted by the Hertzian elastic theory. The difference may be due to adhesion. Two competing models—by Johnson, Kendall and Roberts (JKR) [170] and Derjaguin, Muller, and Toporov (DMT) [90]—have been developed to account for adhesive force during elastic contact. The JKR model assumes that adhesive forces are confined to inside the contact area, whereas the DMT model also considers adhesive forces outside the contact area. Tabor [311] pointed out that these models are valid for different ranges of magnitude of elastic deformation compared to the range of surface forces, with JKR valid for large elastic deformations and DMT in the opposite case [3]. Adhesion of rough elastic surfaces has also been studied in the past years [64, 82, 264, 277, 345].

3.2.1.2 Adhesion Hysteresis

It was recently suggested [211, 284, 310, 351] that nanofriction is not related to adhesion per se, but to adhesion *hysteresis*. The energy needed to separate two surfaces

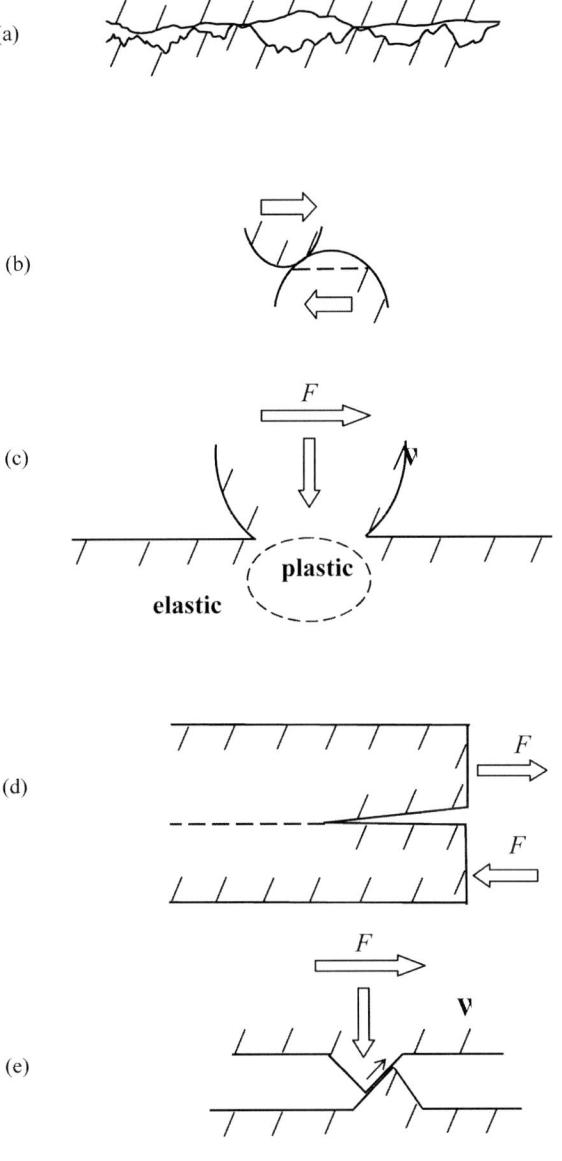

Fig. 3.2. Fundamental mechanisms of friction **a** adhesion between rough surfaces, **b** plowing, **c** the plastic yield, **d** the similarity of a mode II crack propagation and friction, **e** the ratchet mechanism, **f** the third-body mechanism

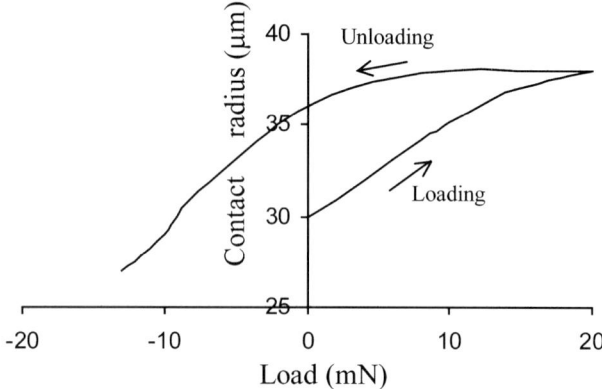

Fig. 3.3. Adhesion hysteresis. Adhesion force is different when surfaces are approaching contact and when separating for polystyrene (based on [211])

is always greater than the energy gained by bringing them together (Fig. 3.3). As a result, the energy is dissipated during the separation process. Adhesion hysteresis, or surface energy hysteresis, can arise even between perfectly smooth and chemically homogeneous surfaces supported by perfectly elastic materials. Adhesion hysteresis exists due to surface roughness and inhomogeneity [211].

The van der Waals force itself is conservative and does not provide a mechanism of energy dissipation. However, adhesion hysteresis due to surface heterogeneity and chemical reactions leads to dissipation [211, 284, 310, 351]. Both sliding and rolling friction involve the creation and consequent destruction of the solid–solid interface. During such a loading-unloading cycle, the amount of energy ΔW is dissipated per unit area.

Since the underlying physical reason of adhesion hysteresis is in surface roughness and chemical heterogeneity, there is a natural way to obtain the hysteresis of a conservative van der Waals force by assuming that the surface is not perfectly rigid, that is, deformable. There are a number of contact models that combine the elastic deformation and adhesion [169], however, these theories do not address the issue of adhesion hysteresis.

Nosonovsky [233] considered a very simple model which, however, can account for adhesion hysteresis. Physically the van der Waals adhesion force and the elastic force are both caused by the atomic interaction. However, at the scale of nanometers, the contacting bodies can still be treated as a continuum, but the effects of adhesion forces are important [169]. The usual approach for the elasto-adhesive problems is to consider the bodies in contact as a continuum media and the interaction between them governed by an adhesive potential. In this section we will study a simple two-dimensional model of solid–solid contact with adhesion. It is expected that the two-dimensional model, while simple, can catch qualitatively the behavior during three-dimensional contact as well.

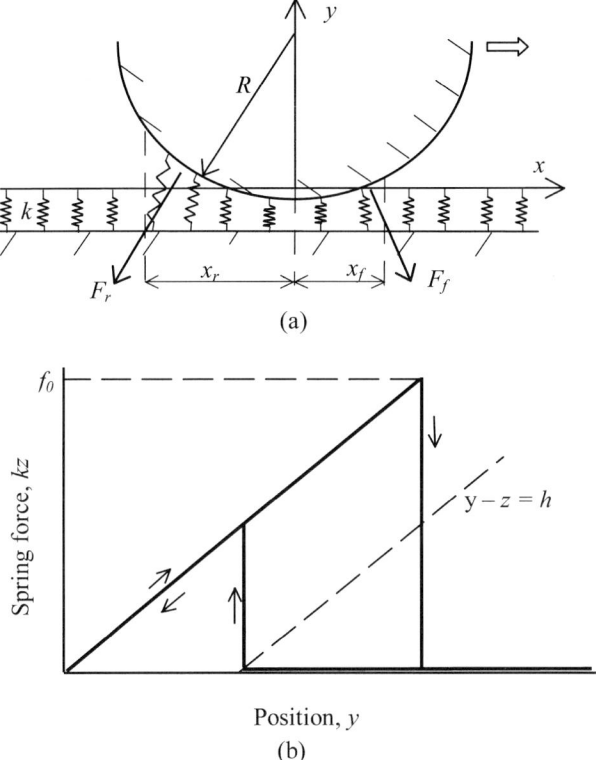

(a)

(b)

Fig. 3.4. a Schematics of a rigid spherical asperity sliding upon a deformable substrate (represented by springs), $z = d - y$, where $d = R - (R^2 - x^2)^{1/2}$, with adhesion force between them. Due to the hysteresis, the position of the springs on approach is different from that at detaching. **b** Dependence of the force, acting upon a spring, on the vertical position of the asperity y during the motion (loading–unloading cycle). **c** The Lennard-Jones, elastic and combined potentials, with the combined potential having two minimums. **d** Normalized energy difference of the two equilibrium states, $\Delta W z_0 / W$, as a function of the normalized elastic modulus, $\alpha = E z_0 / W$ [233]

Consider a solid continuum deformable surface in contact with a rigid cylinder with the van der Waals adhesion force acting between them (Fig. 3.4(a)) and the separation distance z. The cylinder presents an asperity in contact with a substrate. The total energy, T, of a point at the surface is given by [233]

$$T = p(z) + W_{\mathrm{E}}, \tag{3.3}$$

where $p(z)$ is the Lennard-Jones adhesion potential for plane surfaces [169]

$$p_{\mathrm{a}}(z) = -\frac{W}{3}\left[\left(\frac{z_0}{z}\right)^8 - 4\left(\frac{z_0}{z}\right)^2\right], \tag{3.4}$$

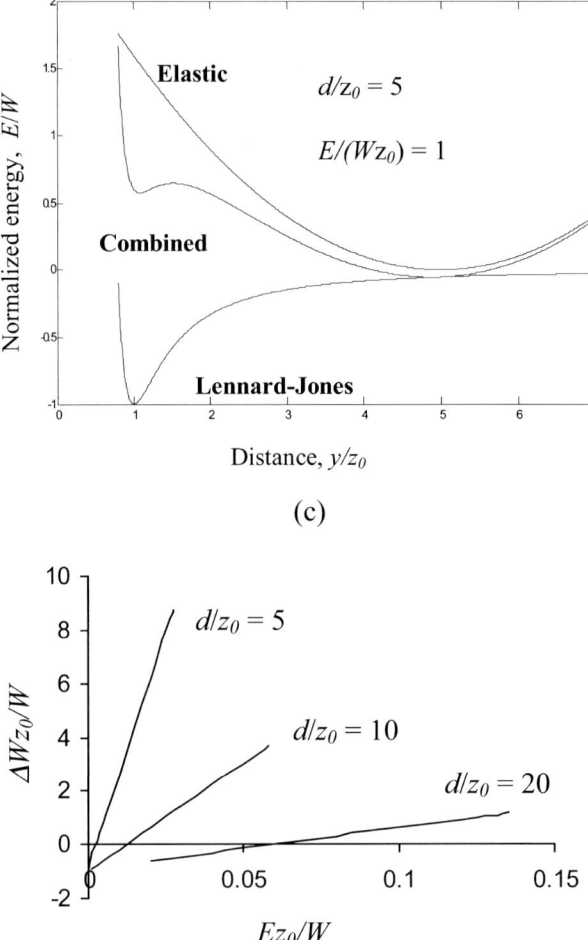

Fig. 3.4. (*Continued*)

and W_{E} is the elastic energy, which can be approximated by

$$W_{\mathrm{E}} = \frac{Ey^2}{2z_0^2},\qquad(3.5)$$

where y is the vertical displacement of the point (individual spring), z is the distance between the point and the rigid asperity, z_0 is the equilibrium distance, and E is the elastic modulus (Fig. 3.4(b)). Equation (3.5) represents a simplified linear elastic law, which may be visualized as a linear spring. Combining (3.3)–(3.5) and noting from Fig. 3.4(a) that $z = d - y$, where $d = R - (R^2 - x^2)^{1/2}$ is a constant distance at a given coordinate x, yields [233]

$$E(z) = -\frac{W}{3}\left[\left(\frac{z_0}{z}\right)^8 - 4\left(\frac{z_0}{z}\right)^2\right] + E\frac{(d-z)^2}{2z_0^2}. \qquad (3.6)$$

As observed from Fig. 3.4(c), the combined potential can have two minimum points which correspond to equilibriums, and thus makes the hysteresis possible. The first equilibrium corresponds to the adhesion forces dominating over the elastic force and is achieved on approach when an element of the deformable surface "jumps to contact" with the rigid asperity. The second equilibrium corresponds to the elastic force dominating over the adhesion and is achieved at separation when an element of the surface detaches from the asperity. The energy barriers between the two states, ΔW, are equal to the hysteresis. Note that even though both the adhesion and elastic forces are reversible, and the energy potential (3.4) is conservative, the hysteresis occurs in the system, which leads to a nonreversible process. The energy is consumed for excitation of elastic vibrations and waves [233].

The normalized energy difference of the two equilibrium states, $\Delta W z_0 / W$, as a function of the normalized elastic modulus, $\alpha = E z_0 / W$, is presented in Fig. 3.4(d) for the values of the normalized distance $d/z_0 = 5, 10$, and 20. For $d/z_0 = 5$, there are two equilibriums when $0.0201 < \alpha < 0.1353$, which correspond to $1.01 < z/z_0 < 1.21$ and $3.75 < z/z_0 < 4.92$. Obviously, the first equilibrium (near $z/z_0 = 1$) corresponds to the substrate, just slightly deformed by the adhesion force, whereas the second equilibrium (near $z/z_0 = d/z_0 = 5$) corresponds to a significant deformation of the substrate, adhered to the asperity. For $d/z_0 = 10$, there are two equilibriums when $0.0012 < \alpha < 0.059$, which correspond to $1.001 < z/z_0 < 1.187$ and $7.99 < z/z_0 < 9.98$. For $d/z_0 = 20$, there are two equilibriums when $0.0001 < \alpha < 0.0273$, which correspond to $1.0013 < z/z_0 < 1.1923$ and $17.51 < z/z_0 < 19.997$. It is observed from Fig. 3.4(d) that the energy difference is almost linearly proportional to the normalized elastic modulus. This is because the energy of the second equilibrium (when the substrate is attached to the asperity) is greater than that of the first equilibrium, and the W_e term, which is proportional to E, dominates [233].

3.2.1.3 Shear Strength Due to Adhesion Hysteresis

Consider a rigid cylinder of radius R and length L rolling along a solid surface with the van der Waals attractive adhesion force between them. From the energy balance, when the cylinder passes the distance d the amount of dissipated energy, $\Delta W A_r$, is equal to the work of the friction force F at the distance d; therefore, the friction force is given by [233]

$$F = A_r \Delta W / d. \qquad (3.7)$$

For a multiasperity contact, the real area of contact, A_r, is only a small fraction of the nominal contact area, which is equal to the surface area covered by the cylinder, Ld.

During frictional sliding of a solid cylinder against a flat surface, the solid–solid interface is created and destroyed in a manner similar to rolling. Based on the adhesion hysteresis approach, the frictional force during sliding is also given by (3.7), and all considerations presented in the preceding section are also valid for the sliding friction.

However, it is well known from the experiments that sliding friction is usually greater than the rolling friction [31, 32]. This is because plowing of asperities takes place during sliding. Even smooth surfaces have nanoasperities, and their interlocking can result in plowing and plastic deformation of the material. Usually, asperities of softer material are deformed by asperities of harder material. The shear strength during plowing is often assumed to be proportional to the average absolute value of the surface slope [31, 32]. It is therefore assumed that in addition to the adhesion hysteresis term, there is another component, H_p, responsible for friction due to surface roughness and plowing [233]

$$F = A_r(\Delta W/d + H_p). \qquad (3.8)$$

The plowing term may be assumed to be proportional to the average absolute value of the surface slope. Note that the normal load is not included in (3.8) directly, however, A_r depends upon the normal load. The right-hand side of (3.8) involves two terms: a term that is proportional to adhesion hysteresis and a term that is proportional to roughness. Nosonovsky [233] pointed out the similarly of (3.8)—that governs energy dissipation during solid–solid friction—to the equations that govern energy dissipation during solid–liquid friction, which will be discussed in the next part of this book.

Summarizing, the adhesive friction provides the mechanism of energy dissipation due to breaking strong adhesive bonds between the contacting surfaces and due to adhesion hysteresis. The value of the force is given by (3.2). In order for adhesive friction to exist, either irreversible adhesion bonds should form or the contacting bodies should be deformable and thus nonideally rigid. The adhesive friction mechanism involves weak short-range adhesive force and strong long-range bulk forces.

3.2.2 Deformation of Asperities

Another important mechanism of friction is deformation of interlocking asperities ([30, 32], as shown in Fig. 3.2(b)). Like adhesion, which may be reversible (weak) and irreversible (strong), deformation may be elastic (i.e., reversible) and plastic (irreversible plowing of asperities). For elastic deformation, a certain amount of energy is dissipated during the loading-unloading cycle due to radiation of elastic waves and viscoelasticity, so an elastic deformation hysteresis exists, similar to adhesion hysteresis. The value of deformational friction force is usually higher than that of adhesive friction and depends on the yield strength and hardness, which trigger a transition to plastic deformation and plowing. The transition from adhesive to deformational friction mechanism depends on load and yield strength of materials and usually results in a significant increase of the friction force [32].

Due to the surface roughness, deformation occurs only at small parts of the nominal contact area, and the friction force is proportional to the real area of contact involving plowing, as given by (3.2). Due to the small size of the real area of contact compared with the nominal area of contact, the plastically deformed regions constitute only a small part of the bulk volume of the contacting bodies.

3.2.3 Plastic Yield

Chang et al. [74] proposed a single-asperity contact model of friction based on plastic yield, which was later modified by Kogut and Etsion [189]. They considered a single-asperity contact of a rigid asperity with an elastic-plastic material. With an increasing normal load, the maximum shear strength grows and the onset of yielding is possible. The maximum shear strength occurs at a certain depth in the bulk of the body (Fig. 3.2(c)). When the load is further increased and the tangential load is applied, the plastic zone grows and reaches the interface. This corresponds to the onset of sliding. Kogut and Etsion [189] calculated the tangential load at the onset of sliding as a function of the normal load using the finite element analysis and found a nonlinear dependence between the shear and tangential forces. This mechanism involves plasticity and implies structural vulnerability of the interface compared to the bulk of the contacting bodies.

3.2.4 Fracture

For brittle material, asperities can break forming wear debris. Therefore, fracture can also contribute to friction. There is also an analogy between mode II crack propagation and the sliding of an asperity [129, 178, 280] (Fig. 3.2(d)). When an asperity slides, the bonds are breaking at the rear, while new bonds are being created at the front. Thus, the rear edge of asperity can be viewed as the tip of a propagating mode II crack, while the front edge can be viewed as a closing crack. Gliding dislocations, emitted from the crack tip, can also lead to the microslip or local relative motion of the two bodies [42]. Calculations have been performed to relate the stress intensity factors with friction parameters [129, 178, 280]. Crack and dislocation propagation along the interface implies that the interface is weak compared to the bulk of the body.

3.2.5 Ratchet and Cobblestone Mechanisms

Interlocking of asperities may result in one asperity climbing upon the other, leading to the so-called ratchet mechanism [30, 32]. In this case, in order to maintain sliding, a horizontal force should be applied which is proportional to the slope of the asperity (Fig. 3.2(e)). At the atomic scale, a similar situation exists when an asperity slides upon a molecularly smooth surface and passes through the tops of molecules and valleys between them. This sliding mechanism is called the "cobblestone mechanism" [161]. This mechanism implies that the strong bonds are acting in the bulk of the body, whereas interface bonds are weak.

3.2.6 "Third Body" Mechanism

During the contact of two solid bodies, wear and contamination particles can be trapped at the interface between the bodies (Fig. 3.2(f)). Along with liquid, which condensates at the interface, these particles form the so-called "third body" which plays a significant role in friction. The trapped particles can significantly increase the coefficient of friction due to plowing. Some particles can also roll and thus serve as rolling bearings, leading to reduced coefficient of friction. However, in most engineering situations, only 10% of the particles roll [30, 32] and thus the third body mechanism leads to an increase in the coefficient of friction. At the atomic scale, adsorbed mobile molecules can constitute the "third body" and lead to significant friction increase [146]. The third body has much weaker bonds to the surface, than those in the bulk of the body.

3.2.7 Discussion

In summary, there are several mechanisms of dry friction. They all are associated with a certain type of heterogeneity or nonideality, including surface roughness, chemical heterogeneity, contamination, and irreversible forces. All these mechanisms are also characterized by the interface forces being small compared to the bulk force. In the following chapters, we will discuss linearity of friction as a result of the presence of a small parameter, nonlinearity of friction, related to heterogeneity and hierarchical structure and multiscale nature of the frictional mechanisms.

3.3 Friction as a Linear Phenomenon

Empirical observations regarding dry friction are summarized in the so-called Amontons–Coulomb's rule, which states that the friction force F is linearly proportional to the normal load W

$$F = \mu W, \tag{3.9}$$

where μ is a constant for any pair of contacting materials, called the coefficient of friction. The coefficient of friction is almost independent of the normal load, nominal size of contact, and sliding velocity. Although there is no underlying physical principle which would require the friction force to be linearly proportional to the normal load, (3.9) is valid for a remarkably large range of conditions and regimes of friction, from macro- to nanoscale, for loads ranging from meganewtons to nanonewtons, and for various material combinations. Two main physical explanations of the linearity of friction have been suggested, based on the friction force proportionality to the real area of contact between the two bodies and to the average slope of a rough surface. These two concepts are considered in the following sections.

3.3.1 Friction, Controlled by Real Area of Contact

Every nominally flat surface is not ideally smooth and has roughness due to small asperities. A contact between the two bodies during friction occurs only at the summits of the asperities. As a result, the real area of contact, A_r, constitutes only a small fraction of the nominal area of contact and depends upon the normal load. For metals at typical loads, the real area of contact constitutes less than 1% of the nominal area of contact. Various statistical models of contacting rough surfaces have been proposed following the pioneering work by Greenwood and Williamson [137]. Using numerical computations, these models conclude that for typical roughness distributions, such as the Gaussian roughness, for both elastic and plastic materials, the real area of contact is almost linearly proportional to the load [3]. For the elastic contact of a smooth surface and a rough surface with the correlation length β^* and standard deviation of profile height σ, the real area of contact is given by

$$A_r \propto \frac{\beta^*}{E^*\sigma} W, \tag{3.10}$$

where E^* is the composite elastic modulus of the two bodies [32]. Note that σ is the vertical and β^* is the horizontal roughness parameters with the dimension of length. The smoother the surface (higher the ratio β^*/σ), the larger A_r. Physically, the almost linear dependence of the real area of contact upon the normal load in this case is a result of the small extent of contact. In other words, it is the consequence of the fact that the real area of contact is a small fraction of the nominal area of contact. With increasing load, as the fraction of the real area of contact grows, or for very elastic materials, such as rubber, the dependence is significantly nonlinear. However, for small real area of contact, with increasing load the area of contact for every individual asperity grows, but the number of asperity contacts also grows, so the average contact area per asperity remains almost constant (Fig. 3.5).

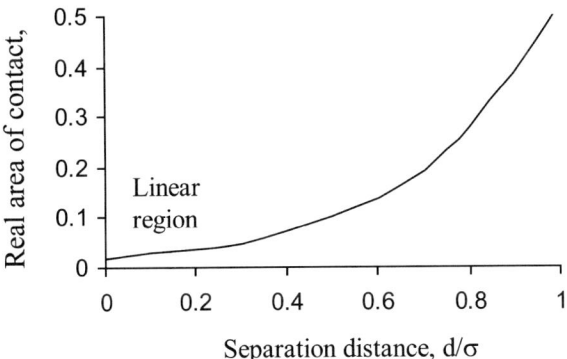

Fig. 3.5. The number of contacts and contact area as a function of separation between the contacting bodies (based on [257]

For plastic contact, the real area of contact is independent of roughness parameters and given by the ratio of the normal load to the hardness of a softer material H_s [32]

$$A_r = \frac{W}{H_s}.$$ (3.11)

Hardness is usually defined in indentation experiments as force divided by the indentation area, so (3.11) naturally follows from this definition. In many cases it may be assumed that the hardness is proportional to the yield strength.

Whether the contact is elastic or plastic may depend upon the roughness parameters, elastic modulus, and hardness. Interestingly, Greenwood and Williamson [137] showed that whether the contact is elastic or plastic does not depend upon the load, but solely upon the so-called plasticity index $\psi = (\sqrt{\sigma/R_p})E^*/H$, where σ is the standard deviation of peak heights and R_p is mean asperity peak radius.

Based on Bowden and Tabor's model (Eq. (3.2)), the friction force due to adhesion is proportional to the real area of contact and adhesive shear strength τ_a. Combining (3.2) and (3.9)–(3.11) yields a linear dependence of F upon W.

Fractal models provide an alternative description of a rough surface. Long before the discovery of fractals by mathematicians, Archard [12] studied multiscale roughness with small asperities on top of bigger asperities, with even smaller asperities on top of those, and so on (Fig. 3.6). According to the Hertzian model, for the contact of an elastic sphere of radius R with an elastic flat with the contact radius a, the contact area $A_r = \pi a^2$ is related to the normal load as

$$A_r = \pi \left(\frac{3RW}{4E^*}\right)^{2/3}.$$ (3.12)

The pressure distribution as a function of the distance from the center of the contact spot, r, is given by

$$p = \left(\frac{6WE^{*2}}{\pi^3 R^2}\right)^{1/3} \sqrt{1 - (r/a)^2}.$$ (3.13)

Let us now assume that the big spherical asperity is covered uniformly by many asperities with a much smaller radius, and these asperities form the contact. For an asperity located at a distance r from the center, the load is proportional to the stress given by (3.13). The area of contact of this small asperity is still given by (3.12)

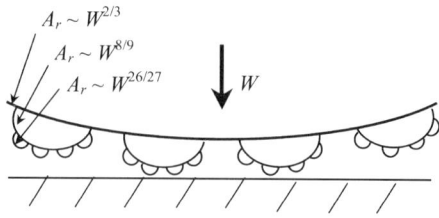

Fig. 3.6. A multiscale rough elastic surface in contact with a flat surface

using the corresponding load. The dependence of total contact area upon W is then given by integration of the individual contact areas by r as [12]

$$A_{\mathrm{r}} \propto \int_0^a \left[W^{(1/3)} \sqrt{1 - r^2/a^2} \right]^{2/3} 2\pi r \, dr$$

$$\propto \int_0^\pi \left[W^{(1/3)} \cos\phi \right]^{2/3} 2\pi (a \sin\phi) a \cos\phi \, d\phi$$

$$\propto W^{(2/9)} a^2 \propto W^{(2/9)} W^{(2/3)} \propto W^{(8/9)}. \tag{3.14}$$

In the above derivation, the variable change $r = a \sin\phi$ and (3.6) were used. The integral of the trigonometric functions can be easily calculated, however, its value is not important for us, because it is independent of a and W.

If the small asperities are covered by the "third-order" asperities of an even smaller radius, the total area of contact can be calculated in a similar way as

$$A_{\mathrm{r}} \propto \int_0^a \left[W^{(1/3)} \sqrt{1 - r^2/a^2} \right]^{8/9} 2\pi r \, dr \propto W^{(8/27)} a^2 \propto W^{(26/27)}. \tag{3.15}$$

For elastic contact, it is found that

$$A_{\mathrm{r}} \propto W^{(3^n - 1)/3^n}, \tag{3.16}$$

where n is the number of orders of asperities, leading to an almost linear dependence of A_{r} upon W with increasing n. Later more sophisticated fractal surface models were introduced, which led to similar results [213].

Thus, both statistical and fractal roughness for elastic and plastic contact, combined with the adhesive friction law (3.2) results in an almost linear dependence of the friction force upon the normal load.

3.3.2 Friction Controlled by Average Surface Slope

Another type of dry friction model is based on the assumption that during sliding asperities climb upon each other (the ratchet mechanism) (Fig. 3.7). From the balance of forces, the horizontal force, which is required to initiate motion, is given by the normal load multiplied by the slope of the asperities

$$F = W \tan\theta, \tag{3.17}$$

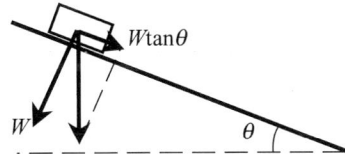

Fig. 3.7. Slope-controlled friction. For a body moving without acceleration upon an inclined surface with slope θ, the shear force, $W \tan\theta$, is proportional to the normal load, W

where θ is the slope angle of the asperities. Comparing (3.9) and (3.17) it may be concluded that, for a rough surface, the coefficient of fiction is equal to the average absolute value of its slope

$$\mu = |\tan\theta|. \tag{3.18}$$

The sign of the absolute value appears in (3.18) because asperities can climb only if the slope is positive. Similar to the ratchet mechanism is the cobblestone mechanism, which is typical for atomic friction.

3.3.3 Other Explanations of the Linearity of Friction

Among other attempts to explain the linearity of the friction force with respect to the load, two modeling approaches are worth mentioning. Sokoloff [301] suggested that the origin of the friction force is in the hardcore atomic repulsion. The vertical component of the repulsion force's vector, which contributes to the normal load, is proportional to the horizontal component of the same vector, which contributes to friction because the vector has a certain average orientation. In a sense, this is still the same slope-controlled mechanism, but considered at the atomic level.

Ying and Hsu [348] suggested an interesting macroscale approach. They noticed that for a spherical asperity of radius R, slightly indented into a substrate, the contact radius, a, is proportional to the second power of the penetration h (Fig. 3.8)

$$a \propto W^{1/3}. \tag{3.19}$$

When such an asperity plows the substrate, the cross-sectional plowing area (or projection of the indented part of the sphere upon a vertical plane) A_p is given by a cubic function of a and thus is proportional to the normal load

$$A_p = \frac{2a^3}{3R} \propto W. \tag{3.20}$$

This is the case of "elastic plowing," the force resisting to sliding is proportional to A_p and, therefore, is linearly proportional to the normal load.

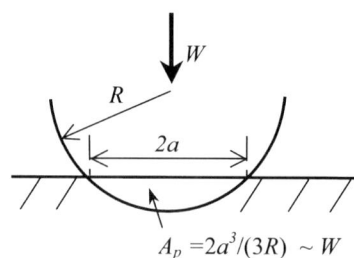

$$A_p = 2a^3/(3R) \sim W$$

Fig. 3.8. "Elastic plowing:" the trans-sectional area of the asperity is linearly proportional to the Hertzian normal load

3.3.4 Linearity and the "Small Parameter"

We have found that several physical mechanisms result in a linear dependence of the friction force upon the normal load. Mathematically, a linear dependence between two parameters usually exists when the domain of a changing parameter is small, and thus a more complicated dependency can be approximated within this domain as a linear function. For example, if the dependency of the friction force upon the normal load is given by

$$F = f(W) \approx f(0) + f'(0)W + \frac{f''(0)}{2}W^2$$
$$= \mu W + \frac{f''(0)}{2}W^2 \qquad (3.21)$$

the dependency can be linearized as $F = \mu W$ if

$$W \ll \frac{2\mu}{f''(0)}. \qquad (3.22)$$

In other words, the ratio of the load W to a corresponding parameter of the system, given by (3.22) (with the dimension of force), is small. That parameter may correspond to the bulk strength of the body.

3.4 Summary

In this chapter we considered several mechanisms of friction that result in a linear dependence of the friction force upon the normal load (Table 3.2). We also discussed the role of the small parameter in the linearity. In more general terms, linearity of the friction is a consequence of the interface forces being small compared to the binding forces acting in the bulk of the body. Since this ratio is small, the ratio of real to nominal areas of contact is also small, which guarantees validity of (3.10) based on the statistical models, as it was explained in the preceding sections. In a similar manner, the small extent of the contact at the interface, compared to the bulk of the material, provides the linear dependencies given by (3.10)–(3.13) and (3.20).

Table 3.2. Mechanisms of friction and linear dependence of the friction force upon the normal load

	Mechanism	Friction force and real area of contact as functions of the normal load
Area-controlled	Elastic hierarchical (Archard)	$F = \tau_a A_r \propto W^{(3^n-1)/3^n}$
	Elastic statistical	$F = \tau_a A_r \propto \frac{\beta^*}{E^* \sigma} W$
	Plastic	$F = \tau_a A_r = \frac{W}{H_s}$
Slope-controlled	Ratchet	$F = W \tan\theta$
Other	Elastic plowing	$F = \tau_a A_p = \frac{2a^3}{3R} \propto W$

4

Friction as a Nonlinear Hierarchical Phenomenon

Abstract Nonlinear effects of dry friction are discussed, including the nonlinearity in the Amontons–Coulomb's laws, velocity dependence, dynamic nonlinearities, stick–slip, interdependence of size-, load-, and velocity dependencies, self-organized criticality. The relation between the nonlinearity and hierarchical organization is discussed. The heterogeneity is studied in relation to the hierarchy and energy dissipation and it is shown how deviation from an "ideal" situation at different hierarchy levels (including elasticity, plasticity, roughness and heterogeneity, reversible and irreversible adhesion) leads to friction. A phase field (mesoscopic functional) approach to the stick–slip is suggested and mapping of various friction mechanisms is presented.

The previous section considered friction as a linear phenomenon, characterized by a small parameter due to a small ratio of interface forces to the bulk forces. This section considers friction as an inherently nonlinear phenomenon and will show that nonlinearity is required for a dissipative process.

4.1 Nonlinear Effects in Dry Friction

Although the Amontons–Coulomb rule states a linear dependence of the friction force on the normal load, dry friction is a nonlinear phenomenon in several important aspects. This section discusses the inherent nonlinearity of the Amontons–Coulomb rule and several important cases of violation of this rule, such as the velocity dependence and stick–slip motion.

4.1.1 Nonlinearity of the Amontons–Coulomb Rule

Despite the linear dependence of the friction force upon the normal load, Amontons–Coulomb's rule is inherently nonlinear. Direction of the friction force depends upon the direction of motion, so that in the vector form the friction force is given by

$$\vec{F} = \frac{\vec{V}}{|\vec{V}|} \mu |\vec{W}|, \tag{4.1}$$

where $\vec{V}$ is sliding velocity. The ratio $\vec{V}/|\vec{V}|$ is nonlinear. This nonlinearity results in some static frictional problems having no solution or a nonunique solution, e.g., the so-called Painlevé [258] paradox. These paradoxes show that the rigid-body dynamics with contact and Coulombian friction is inconsistent. To resolve these problems, the dynamic friction and elastic deformation should be considered [304]. Note also that the coefficient of static friction (for $\vec{V} = 0$) is usually greater than the coefficient of the kinetic friction (for $\vec{V} \neq 0$), which can also be viewed as a manifestation of nonlinearity.

4.1.2 Dynamic Instabilities Associated with the Nonlinearity

The mathematical formulation of quasi-static sliding of two elastic bodies (half-spaces) with a frictional interface, governed by the Amontons–Coulomb's rule, is a classical contact mechanics problem. Interestingly, the stability of such sliding was not investigated until the 1990s, when Adams [2] showed that the steady sliding of two elastic half-spaces is dynamically unstable, even at low sliding speeds. The instability mechanism is essentially one of slip-wave destabilization. Steady-state sliding was shown to give rise to a dynamic instability in the form of self-excited motion. These self-excited oscillations are confined to a region near the sliding interface and can eventually lead to either partial loss of contact or to propagating regions of stick–slip motion (slip waves). The existence of these instabilities depends upon the surfaces' elastic properties, however, it does not depend upon the friction coefficient, nor does it require a nonlinear contact model. The same effect was predicted theoretically by Nosonovsky and Adams [238] for the contact of rough periodic elastic surfaces.

It is well known that two types of elastic waves can propagate in an elastic medium: shear and dilatational waves. In addition, surface elastic waves may exist, and their amplitude decreases exponentially with the distance from the surface. For two slightly dissimilar elastic materials in contact, the interface waves (Rayleigh waves) may exist at the interface zone. Their amplitude decreases exponentially with the distance from the interface.

The above-mentioned instabilities are a consequence of energy being pumped into the interface as a result of the positive work of the driving force (that balances the friction force). As a result, the amplitude of the interface waves grows with time. In a real system, of course, the growth is restricted by the limits of applicability of linear elasticity and linear vibration theory. This type of friction-induced vibration may be, at least partially, responsible for noise and other undesirable effects during friction [238]. These instabilities are a consequence of inherent nonlinearity of the boundary conditions with the Coulombian friction. Whereas the interface waves occur for slightly dissimilar (in the sense of their elastics properties) materials, for very dissimilar materials, waves would be radiated along the interfaces. This provides a different mechanism of pumping energy away from the interface [237].

Adams et al. [5] also demonstrated that dynamic effects lead to new types of frictional paradoxes, in the sense that the assumed direction of sliding used for Coulomb

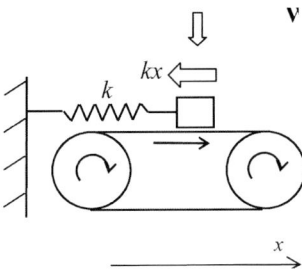

Fig. 4.1. The dynamic friction and the origin of stick–slip motion due to the decrease of the friction force with increasing velocity. When body moves in the same direction as the tape, the friction force increases and the body sticks until the spring force becomes high enough to overcome the static friction and the body can slip in the opposite direction [246]

friction is opposite to that of the resulting slip velocity. In a strict mathematical sense, the Coulomb friction is inconsistent not only with the rigid body dynamics (the Painlevé paradoxes), but also with the dynamics of elastically deformable bodies.

4.1.3 Velocity-Dependence and Dynamic Friction

It is known from experiments that the absolute value of the friction force is not completely independent of the sliding velocity. In fact, it was known to Coulomb, who claimed that for very small velocities friction force grows with increasing velocity, for moderate velocities friction force remains constant, and for high velocities it decreases. Tolstoi [318] was the first who paid attention to the importance of the normal coordinate during dry sliding, showing that separation distance between the sliding bodies grows with increasing velocity. Since there is less time for the asperity contact and, therefore, less time for asperities to deform (e.g. viscoplastically), the body elevates (Fig. 4.1). This usually results in a decrease of the real area of contact and a decrease of the friction force. Various dynamic models have been suggested on the basis of various physical effects, such as time-dependent creep-like relaxation and viscosity [263]. It is usually believed that, for dry friction, increasing velocity results in decreasing friction (the so-called "negative viscosity"), although for some material combinations and friction regimes the opposite trend is observed. Note that the decrease of friction with increasing velocity may lead to a dynamic instability, since decreased frictional resistance will lead to acceleration and further increase of velocity and decrease in friction [220].

To analyze frictional dynamic instabilities, the so-called state-and-rate models of friction were introduced [98, 99, 263]. These models, used at first to study sliding friction for seismic and geophysical applications, showed reasonable agreement with the experimental data. Based on the state-and-rate models, when sliding velocity changes, the friction force first increases and then decreases (due to creep relaxation) to a velocity-dependent steady-state value. According to the Dieterich [98] state-and-rate model, the area of contact is given by

$$A_r = A_{r0}\big[1 + B \ln(\theta/\theta_0)\big], \tag{4.2}$$

where θ is a "state" parameter which is physically equal to the age of contact, B is a constant, and θ_0 is a normalization parameter, such that $\theta = \theta_0$ for $A_r = A_{r0}$. The contact area increases with the age of contact due to the creep. For steady state sliding with a constant velocity V, θ is inversely proportional to the sliding velocity [99].

$$A_r = A_{r0}\big[1 - C \ln(V/V_0)\big], \tag{4.3}$$

where V_0 is a normalization parameter, such that $V = V_0$ for $A_r = A_{r0}$ and C is a constant. Velocity-dependent friction law has been used to resolve some paradoxes associated with dynamic Coulombian friction, such as the "ill-posedness" (a dynamic instability with an unlimited rate of amplitude increase for small wavelengths) of steady frictional sliding of two smooth elastic half-spaces, or dynamic instabilities studied in the preceding section [278].

4.1.4 Interdependence of the Load-, Size-, and Velocity-Dependence of the Coefficient of Friction

The velocity-dependence of the coefficient of friction is related to other deviations from the Amontons–Coulomb's rule [235]. Every nominally flat solid body has small asperities. During sliding, the real area of contact between two bodies, A_r, is smaller than the apparent (or nominal) area of contact, because contact takes place only on the tops of the asperities. The so-called Amontons–Coulomb's empirical laws of friction state that sliding friction force F is (1) proportional to the normal load W, (2) independent of the apparent area of contact between the bodies A_a, and (3) independent of sliding velocity V [31, 32]. In other words, these three rules state that the ratio of the friction force to the normal load, also called the coefficient of kinetic friction, $\mu = F/W$, is a constant, which does not change with changing W, A_a, and V, i.e., independent of W, A_a, and V [235].

The first and second Amontons–Coulomb's rules are related to each other. To illustrate them, let us consider an apparent area of contact A_a, which supports a normal load W resulting in the friction force $F = \mu W$ (Fig. 4.2(a)). A part of the apparent area of contact A_a/c supports the normal load W/c resulting in the friction force $F/c = \mu W/c$, where c is a constant (Fig. 4.2(b)). Let us now increase the normal load from W to cW. According to the first Amontons–Coulomb's rule, which states that F is proportional to W, such an increase results in the friction force F and normal load W acting upon the part of apparent area of contact A_a/c. In other words, for both the whole apparent area of contact A_a and for its fraction A_a/c, the same friction force F corresponds to the load W (Fig. 4.2(c)). Thus, using only the first Amontons–Coulomb's law, it was shown that the ratio F/W is independent of the apparent area of contact, which constitutes the second Amontons–Coulomb's rule. Therefore, the load-dependence of friction is coupled with its dependence on the apparent area or size of contact. In case the load-independence is violated (i.e., the μ is dependent on load), the size-independence will be violated too, i.e. the μ will depend on the apparent area of contact [235].

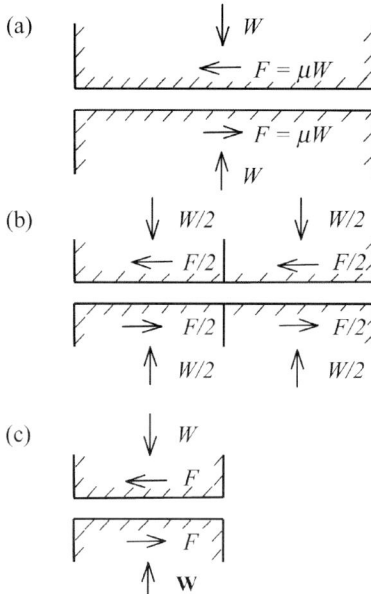

Fig. 4.2. Illustration of coupling of the laws, which state that the coefficient of friction is independent of the normal load and apparent area of contact. **a** The nominal area of contact A_a between two bodies supports the normal load of W and results in friction force $F = \mu W$. **b** Parts of the nominal area of contact, A_a/c (shown for $c = 2$) support the normal load of W/c and results in the friction force $F/c = \mu W/c$. **c** For friction force linearly dependent on the normal load, an increase of the load at the small contact area up to W, results in the friction force equal to F and thus independent of the area of contact. [235]

Amontons–Coulomb's empirical laws of friction are not satisfied in many cases, especially at micro/nanoscale. It is known from experiments that friction is size-dependent and that the coefficient of friction at micro/nanoscale is different from that at the macroscale; load- and velocity-dependence of the coefficient of friction is also well established [42, 70, 71, 154, 225, 287, 302, 336, 353]. Several approaches have been suggested to deal with the laws of friction at micro/nanoscale, which include the formulation of "scaling laws of friction" [42, 229, 239, 336, 353] as well as specific nanofriction laws, which should substitute for the Amontons–Coulomb's laws at the nanoscale [71, 336]. Bhushan and Nosonovsky [42, 46] proposed a model for size effect on friction due to scale-dependent hardness and surface roughness. However, size-, load-, and velocity-effect on friction was investigated separately in these studies, These effects were studied in combination by Nosonovsky [235].

4.1.5 Stick–Slip Motion

Stick–slip motion is an important nonlinear effect similar to the dynamic instability due to decrease of friction force with increasing velocity. Since static friction force is greater than the kinetic friction force, in a situation when a steady force applied to a body via some elastic medium, sticking of two bodies results in a steady increase of applied force. When the applied force exceeds the static friction, resistance instantly drops to the value of kinetic friction and acceleration. This results in a decrease of the applied force and deceleration until the body sticks again, thus alternating the stick and slip phases [263].

Stick–slip motion is found in many macroscale phenomena, such as car brake vibration and noise. However, it is particularly common at the atomic scale. For example, during the contact of an AFM tip with an atomically smooth surface the energy dissipation takes place through stick–slip movement of individual atoms at the contact interface [336].

Another important dynamic effect is the slip waves that can propagate along a frictional interface between two bodies. The slip wave is a propagating stick–slip motion. In a slip wave, a region of slip propagates along the interface, which is otherwise at the stick state. As a result, two bodies shift relative to each other in a "caterpillar" or "carpet" motion. The concept of the slip waves has been applied in areas such as seismology, to study the motion of earth plates, and solid state physics, to investigate gliding of the dislocation at an interface between two bodies [3].

4.1.6 Self-Organized Criticality

Self-organized criticality (SOC) is a concept in the theory of dynamic systems that was introduced in the 1980s [18]. Many conventional physical systems that are studied at equilibrium have a critical point—that is, a point at which a distinction between two phases vanishes and a typical length of fluctuations (referred as the correlation length) tend to grow up to infinity. Simple scaling relationships between various parameters of the system in the vicinity of the critical point can usually be established. These relationships are governed by power laws with certain critical exponents. For example, the critical point of water is at $T_c = 374\,°C$ and $P_c = 218$ atm, and the distinction between liquid and gas water at these conditions disappears. When a small variation of normalized temperature $\tau = (T - T_c)/T_c$ is considered, various physical quantities (e.g., the heat capacity) would change in accordance to the power law, $f(\tau) \sim \tau^\alpha$, where α is the critical exponent [8]. Note that as in the case of linear behavior, the small parameter expansion was used to obtain the power law.

The systems with SOC have a critical point as an attractor, so thus they spontaneously reach the vicinity of the critical point and exhibit power law scaling behavior. SOC is typically observed in slowly driven nonequilibrium nonlinear systems with many degrees of freedom. The main difference between "conventional" systems with phase transitions and SOC systems is that in the conventional equilibrium systems the critical point is reached only by tuning precisely a control parameter (e.g., the temperature tuned to the melting point), while in an SOC system the critical point is as an attractor and thus the transition occurs spontaneously [251].

The best-studied example of SOC is the "sandpile model," representing grains of sand randomly placed into a pile until the slope exceeds a threshold value, transferring sand into the adjacent sites and increasing their slope in turn. Placing a random grain at a particular site may have no effect or it may trigger an avalanche that will affect many sites at the lattice. Thus, the response does not depend on the details of the perturbation. As in the case of a "conventional" phase transition, the correlation length of the system grows to infinity and the power-law scaling behavior is found. Note that the scale of the avalanche is much greater than the scale of the initial perturbation; thus the avalanche belongs to the upper level of the hierarchical organization. Typical external signs of an SOC system are the power-law behavior, the "one-over-frequency" noise distribution signals (fractal in time) and so on [18]. Since SOC allows a system to reach criticality spontaneously and without tuning the controlling parameter, it has been suggested that it plays a major role in spontaneous creation of complexity and hierarchical structures in various natural and social systems [18]. The concept has been applied to such diverse fields as physics, cellular automata theory, biology, economics, sociology, linguistics, and others.

It has been suggested that SOC is responsible for landslides and earthquakes, because it is known that the number of earthquakes and their amplitude are related by a power law. In other words, a number of earthquakes with amplitude greater than a certain level, in a given area during a given period, are related to that level by a power law. During earthquakes, the stress between two plates is accumulated for a long time and released suddenly in a catastrophic event, which is similar to the sandpile avalanche [323].

The similarity between the propagation of slip waves along an interface between dissimilar elastic materials, including earth plates, and microscale slip behavior, such as gliding edge dislocations, is well established [3]. It is therefore not surprising that attempts to identify SOC behavior in other slip systems have been made. Thus, it has been suggested that a transition between the stick and slip phases during dry friction may be associated with SOC, since the slip is triggered in a similar manner to sandpile avalanches and earthquake slides. Zypman et al. [361] showed that in a traditional pin-on-disk experiment, the probability distribution of slip zone sizes follows the power law. In a later work, the same group found nanoscale SOC-like behavior during AFM studies of at least some materials [66, 361]. It is interesting to note that SOC is found only in the two-dimensional sandpile model, whereas most models used for the study of friction are two-dimensional. It should be noted also that fractal surface topography may lead to the power-law distribution of asperity heights, which may be responsible for the power law distribution of slip events. It was suggested that SOC may play a role in wetting behaviors as well [96, 251].

4.2 Nonlinearity and Hierarchy

Different mechanisms of friction acting simultaneously at different scales and hierarchy levels lead to the hierarchical nature of friction. A transition to a higher level is not characterized by a simple statistical averaging of variables but by the emergence of new qualities. Mathematically, the transition to a higher hierarchy level is a result of instability which, in turn, is a consequence of nonlinearity of the system. This is illustrated in Fig. 4.3, which shows energy profile E, a symmetric nonlinear system

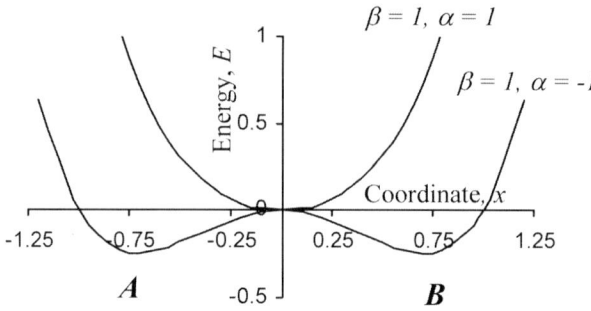

Fig. 4.3. The loss of symmetry due to instability. The energy profile of a non-linear system given by $E = \alpha x^2 + \beta x^4$ is shown for $\beta = 1, \alpha = 1$ and for $\beta = 1, \alpha = -1$. For positive α there is only one equilibrium state, however, when the decreasing parameter α becomes negative, two stable equilibrium states emerge. At the "lower level" the system is described by the coordinate x, whereas at the "upper" level it is described by either position A or B. Due to the instability, it is not possible to predict whether the system will be in the state A or B, so the "lower level" information is not sufficient to deduce the "upper level" description, which, therefore, constitutes a new quality [246]

characterized by a parameter x

$$E = \alpha x^2 + \beta x^4. \tag{4.4}$$

For $\alpha > 0$ and $\beta > 0$, the system has a stable equilibrium at $x = 0$, whereas for $\alpha < 0$ and $\beta > 0$, the equilibrium at $x = 0$ is unstable and two additional stable equilibriums exist at $x = \pm\sqrt{\alpha/(2\beta)}$, marked A and B. When stable equilibrium is destabilized due to decreasing α, the system is transferred from a symmetric to an asymmetric state. At the lower hierarchy level, the system is still symmetric and described by the variable x, whereas at the higher level it is described by either position A or B [228].

The same concept is applicable for transition to a higher hierarchy level for a frictional system. For example, at the atomic level, the contact is characterized by positions and velocities of individual atoms and molecules, having spatially symmetric potentials. The system is discrete, and it is characterized by a large number of variables; it has multiple states of equilibrium and stick–slip motion. At a higher level of description (such as asperity), a small number of variables characterizing position and velocity are used and the motion is interpreted as steady sliding. In a similar manner, interlocking and plowing of asperities with energy sources and sinks is considered at the asperity level, whereas at the surface level it is presented as sliding with a steady energy dissipation.

Dry friction is a result of many mechanisms, acting at different scale and hierarchy levels. The common feature of these mechanisms, on one hand, is the presence of a small parameter due to weak interfacial bonds as compared to the bulk material bonds. This leads to almost-linear friction force as a function of load in most frictional regimes. On the other hand, essential nonlinearity of frictional systems provides hierarchical structures and statistical mechanisms of energy dissipation. The following chapter studies the relation of hierarchy and heterogeneity to dissipation at various hierarchy levels.

4.3 Heterogeneity, Hierarchy and Energy Dissipation

In this section, we will examine the relation of heterogeneity to frictional energy dissipation at various scale and hierarchy levels. First, ideal and real contact situations will be considered, with the real situation characterized by a certain heterogeneity that leads to friction. Then, the relation of heterogeneity to dissipation at various hierarchy levels will be discussed.

4.3.1 Ideal vs. Real Contact Situations

The ideal case of contact between two bodies would involve rigid, chemically inactive, smooth homogeneous surfaces and reversible conservative forces. In such a case, no energy would be dissipated during sliding and no friction would exist. Real surfaces are deformable, rough, and heterogeneous, while forces are not always reversible. This leads to frictional energy dissipation due to increased systems disorder and entropy during the contact. Various friction mechanisms can be activated by different types of inhomogeneity and act at different scale and hierarchy levels (Table 3.1).

In the following sections, we will consider different types of nonideality, identify parameters which characterize inhomogeneity, and relate them to particular friction mechanisms at particular scale and hierarchy levels.

4.3.2 Measure of Inhomogeneity and Dissipation at Various Hierarchy Levels

Several physical models exist for atomic-scale friction, including the Tomlinson [319] model, Frenkel and Kontorova [116] model, and other similar models [225]. These models assume that the substrate atoms form a periodic energy profile with energy barriers that should be overcome during sliding (Fig. 4.4). However, for two atomically smooth noncommensurate bodies in contact, positions of the energy barriers and sinks would not coincide, so virtually no friction is expected. This effect is known as "superlubricity," and it has been argued that superlubricity is in fact observed for graphite [97]. He et al. [146] suggested that chemical inhomogeneity, such as hydrocarbon molecules adsorbed at the surface, provide energy barriers that lead to friction. Sokoloff [300] showed that for static friction to occur, the elastic energy should be higher than the interfacial potential energy barriers, so that the atoms can sink into their interfacial potential minima. Later the model was extended for kinetic friction, and it was shown that even fluctuations in concentration of atomic level defects do not account for static and kinetic friction, but that surface roughness

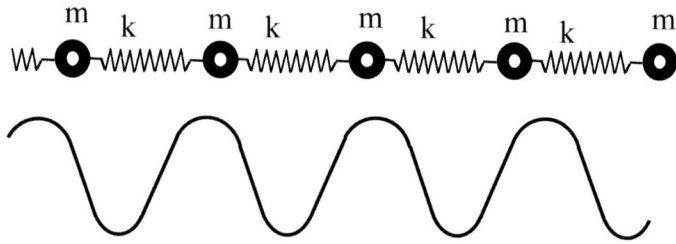

Fig. 4.4. The Tomlinson model of atomic friction, representing the periodical surface forces in the substrate by a periodic potential and elastic forces in the upper body by springs

Table 4.1. Factors responsible for potential energy barriers and elastic energy

	Inhomogeneity, creating potential energy barriers	Elastic energy
Molecule	Surface energy states	
Surface	Atomic roughness and inhomogeneity	Strong atomic bonds
Asperity	Surface roughness	Structural deformation of asperities

(multiple asperity contact) at the micron scale can indeed lead to friction [88]. Thus the atomic-scale energy barriers combined with the micron scale roughness leads to the frictional dissipation.

The same argument can be applied to a more general case of surface heterogeneity, involving surface deformation, roughness, irreversible adhesive bonds, and adhesion hysteresis. We will consider each of these effects at different hierarchy and scale levels. Each level of heterogeneity affects the potential energy profile and increases energy barriers, whereas the deformational energy depends upon scale and hierarchy level (Table 4.1).

4.3.2.1 Rigid vs. Deformable Body

Elastic and plastic deformation is the most obvious reason for energy dissipation. Plastic deformation is irreversible and leads to energy dissipation. However, even reversible elastic deformation combined with reversible adhesion force may lead to energy dissipation due to hysteresis. Consider a smooth rigid cylinder of radius R, representing a two-dimensional asperity, with the profile $y(x)$ given by

$$y = \frac{1}{2R}x^2 + a,\qquad(4.5)$$

where a is the vertical position of the bottom of the cylinder. The cylinder is sliding along a flat deformable surface (Fig. 3.4(a)), which is modeled by springs with the elastic contact k per area, so that the vertical position of a point, z, is given by

$$kz = f,\qquad(4.6)$$

where f is the applied force per area. It is anticipated that this simple model, while it can easily be handled mathematically, captures the main features of the two-dimensional elastic contact of a half-space with a rigid asperity. The Dugdale approximation of the adhesive force is used (Fig. 3.4(c)), according to which the adhesion force-per-unit area (adhesion stress) f_0 is constant in the adhesion zone, which is defined by

$$y - z < h, \tag{4.7}$$

where h is the range of the adhesion force. The Dugdale potential is considered a reasonable approximation for a more complex Lennard-Jones potential [221], because it captures the main features of the latter: attraction at finite distances below a certain critical distance and repulsion at very small distances. It is assumed that the cylinder's radius is much larger than the range of the adhesion force, $R \gg h$, so the quadratic function (4.5) may be used to approximate the cylinder's profile near its contact with the surface.

The dependence of the force acting upon an element of the surface upon the vertical displacement of the element, z, during a loading-unloading cycle is presented in Fig. 3.4(b). First, the element is outside the adhesion zone and displacement is equal to zero. Then, as soon as (4.7) is satisfied, the adhesion stress f_0, which stretches the springs, would be added. The element jumps up and attaches to the cylinder, so that $z = y$. Then z decreases together with y, reaches its minimum, and then increases until the force which acts from the springs is strong enough to overcome the adhesion so that the element detaches from the asperity

$$kz > f_0. \tag{4.8}$$

In the case of $f_0/k > h$, the hysteresis is observed since the position at which the element attaches the asperity is defined by (4.7) and the range of adhesion force h, whereas the position at which the spring detaches from the asperity depends also on the spring constants according to (4.8). The energy loss is due to excitation of oscillation of the springs after the contact. This elastic friction-induced vibration may decline with time due to viscosity (which is not considered in this simple model) and elastic wave (phonon) radiation from the surface.

This simple model [253] predicts that the friction force depends upon the ratio of the adhesion force to elastic modulus of the substrate f_0/k and to the range of adhesion force h

$$F = \int_{-\sqrt{2R(f_0/k-a)}}^{\sqrt{2R(h-a)}} \frac{x}{R} kz \, \mathrm{d}x = \int_{-\sqrt{2R(f_0/k-a)}}^{\sqrt{2R(h-a)}} \frac{x}{R} k \left(\frac{x^2}{2R} + a \right) \mathrm{d}x$$

$$= \frac{k}{2} \left(\frac{f_0^2}{k^2} - h^2 \right). \tag{4.9}$$

We assumed here that the elastic springs are uniformly distributed along the substrate, which leads to a linear energy profile. For discreet springs separated by a certain distance d, the energy profile has minima and barriers (Fig. 3.4(d)), with the height of the barriers dependent on d.

Frictional dissipation as a result of energy deformation is relevant at various scale levels, at hierarchy levels of surface, and single asperity.

4.3.2.2 Smooth vs. Rough Surface

Sokoloff [300] studied atomic friction of two weakly interacting, atomically smooth solids and found, using simple scaling arguments, that neither a smooth surface nor fluctuation of atomic-scale defects provide a friction mechanism. However, surface roughness leading to multiple-asperity contact can provide such a mechanism. For a smooth surface, the atomic-level interfacial potential energy is much weaker than the elastic potential energy, so the contacting surfaces behave as rigid bodies, without being able to reach minimums of interfacial energy. This leads to superlubricity. For microscale asperities, the elastic energy is small due to large distance between the asperities, and multiple equilibrium states with energy barriers between them exist. Thus the ratio of elastic-to-surface energy is the parameter which controls the friction.

4.3.2.3 Homogeneous vs. Heterogeneous Surface

Heterogeneity has a similar effect upon surface energy as roughness [309]. Chemical heterogeneity leads to fluctuations of the surface energy. The molecular dynamics simulation conducted by He et al. [146] showed that adsorbed contamination molecules play a significant role in friction.

As discussed earlier, a criterion for determining the role of inhomogeneity is whether the deformation energy at the range of inhomogeneity is greater than the surface energy barrier ΔW. For the typical distance between energy minimums Δx and length of heterogeneous spots L, the ratio of ΔW to the elastic energy $E \Delta x^2 / (2L)$ is given by the atomic friction parameter

$$r = \frac{2 \Delta W L}{\Delta x E}, \tag{4.10}$$

where E is elastic modulus. For small values of r, it is expected that friction is also small, since elastic forces prevent atoms from occupying surface energy minimum positions.

We mentioned in the preceding sections that friction is linear because surface forces are smaller than the bulk forces. Setting $L = \Delta x$ and calculating r would give a ratio of surface-to-bulk energy at the interatomic scale, which is a small number. In order for friction to exist, inhomogeneity is required at a much larger scale $L \gg \Delta x$, which makes friction a sufficiently multiscale phenomena.

4.3.2.4 Reversible vs. Strong Bonds

Chemical bonds are irreversible, so the energy is dissipated when the bonds are broken. Interface chemical bonds are not different from those in the bulk of the body.

However, the number of bonds is different, and at the interface there are much fewer bonds than in the bulk. As in the preceding cases, the small parameter may be defined which is equal to the ratio of the quantity of bonds at the interface and in the bulk.

4.3.3 Order-Parameter and Mesoscopic Functional

Since friction depends upon heterogeneities, it seems reasonable to introduce the order-parameter method. In the physics of phase transformations, the order-parameter is used in conjunction with the Landau–Ginzburg functional that provides a semi-empirical description of the behavior of a system near the phase-state transition. Although originally this method was applied to the study of phase transitions of the second kind, later it was used to describe a variety of phenomena on the mesoscale such as interfacial phenomena, wetting transitions, near-surface density profiles, formation of microemulsions, and phase separation in polymers [8]. All these phenomena involve coexistence of two phases, of which one may be considered as ordered and the other is disordered. An order-parameter ϕ may be selected in such a manner that for the system in the state of complete disorder, the parameter is equal $\phi = 0$, whereas $\phi = 1$ for complete order. The energy functional is associated with the order-parameter, which involves the term dependent upon ϕ and the term dependent upon the gradient of the order-parameter. The energy functional is written as

$$L = \int \left\{ f(\phi) + \frac{\lambda}{2} |\nabla f(\phi)|^2 \right\} dV, \tag{4.11}$$

where λ is the gradient coefficient, $f(\phi)$ is the energy function, and the integration is performed throughout the whole volume. The gradient term reflects the fact that any heterogeneity, such as creating an interface between the phases, is not energetically profitable. This semi-empirical approach allows us to understand qualitative behavior of near-critical systems without going into details of the interactions in the system.

Einax et al. [104] considered friction as the second-order phase transition using a formulation similar to the Tomlinson model. Their numerical analysis demonstrated that for small interatomic interaction strength, the system is in a frictionless state, but when the strength is above the critical value, there is static friction. They analyzed scaling behavior of the order-parameter near the critical value and calculated corresponding critical exponents. The order-parameter and the energy functional can be powerful mathematical tools to study scaling laws of friction and near-critical behavior.

4.3.4 Kinetics of the Atomic-Scale Friction

A simple kinetic model may be developed for single asperity contact with a periodic substrate that represents the atomic crystal lattice with the period L and energy amplitude B. We'll use the Arrhenius method approach, which is similar to grain growth in a crystal lattice [269]. Suppose that the energy profile is given by

$$W(x) = Fx + B\cos(2\pi x/L), \tag{4.12}$$

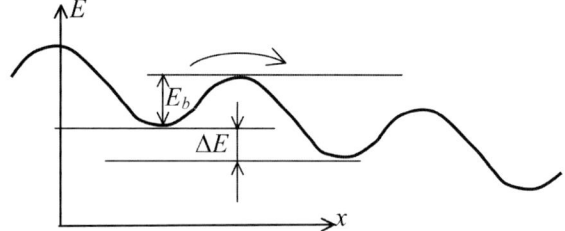

Fig. 4.5. Energy barriers and kinetics of frictional sliding

where the first term represents the driving force F and the second term represents the periodic profile (Fig. 4.5). In the case of $F > 2\pi B/L$, the system now has local energy minima and sliding is expected to occur without any obstacles. However, in the case of $F < 2\pi B/L$, there are local energy minima that correspond to metastable equilibrium states, and the system can drift from one equilibrium state to another due to the thermal drift [269].

Consider two equilibrium states separated with the free energy difference of $\Delta E = FL$ and separated by the activation energy barrier $E_b = B - FL/2$. The probability of a transition from the position with a higher energy to that with a lower energy is given by the probability of a corresponding thermal fluctuation,

$$p_1 = \exp(-E_b/kT). \tag{4.13}$$

The probability of migrating in the opposite direction is given by

$$p_2 = \exp\bigl(-(E_b + \Delta E)/kT\bigr). \tag{4.14}$$

To obtain probabilities per unit of time, the same quantities should be multiplied by the number of vibrations of an atom per second, ν. Assuming that the system is close to equilibrium ($\Delta E/\Delta E_a \ll 1$) the velocity is given by

$$
\begin{aligned}
V &= (\nu/L)\exp(-E_b/kT) - (\nu/L)\exp\bigl(-(E_b + \Delta E)/kT\bigr) \\
&= (\nu/L)\exp(-E_b/kT)\bigl[1 - \exp(-\Delta E/kT)\bigr] \\
&= (\nu/L)\exp(-E_b/kT)\Delta E/kT.
\end{aligned} \tag{4.15}
$$

Or, in terms of F and B,

$$
\begin{aligned}
V &= (\nu/L)\exp\bigl(-(B - FL/2)/kT\bigr) - (\nu/L)\exp\bigl(-(B + FL/2)/kT\bigr) \\
&= (\nu/L)\exp(-B/kT)\bigl[\exp(FL/kT) - \exp(-FL/kT)\bigr] \\
&= \nu\exp(-B/kT)(2F/kT).
\end{aligned} \tag{4.16}
$$

According to (4.16), the sliding velocity for atomic stick–slip motion is proportional to the applied force F. A similar approach can be used for other thermally activated kinetic processes, for example, gliding edge dislocations [269].

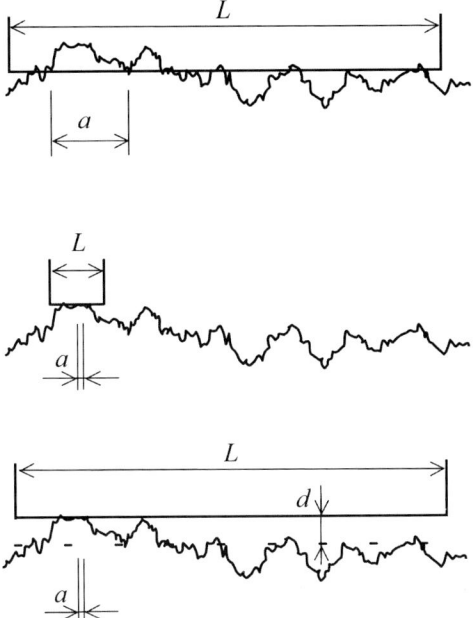

Fig. 4.6. The similarity of the contact extent (vertical separation) and linear scale size of contact [233]

4.4 Mapping of Friction at Various Hierarchy Levels

We noted in the preceding sections that most friction mechanisms involve forces at two scale ranges of size and magnitude, that is, the interface forces and the bulk forces. While the interface forces are defined by the surface inhomogeneity and involve energy barriers associated either with the surface potential energy, surface roughness, contamination, or defects, the bulk forces involve deformation associated with sliding. This may include structural deformation and thus is related to the hierarchical structure of the contacting bodies. Linear scale of the bulk forces, L, is usually greater than the typical linear scale of heterogeneities associated with the energy barriers. Since the amount of bulk forces involved in the contact is related to the extent of contact, another (and more convenient) way to characterize the scale of bulk forces is to use the extent of contact, or the vertical separation between the bodies. The idea is illustrated in Fig. 4.6, which shows a rough solid surface in contact with a smooth surface. The contact is characterized by the apparent size of contact, L, and by an average size of individual contacts, a. The latter is responsible for many physical effects associated with the size of contact. Both changing L and changing the separation d has a similar effect on a, and thus both L and d can serve as parameters characterizing the extent of contact.

A qualitative map of various friction regimes is proposed in Fig. 4.7 as a function of the distance between energy minimums Δx and the extent of contact h (vertical

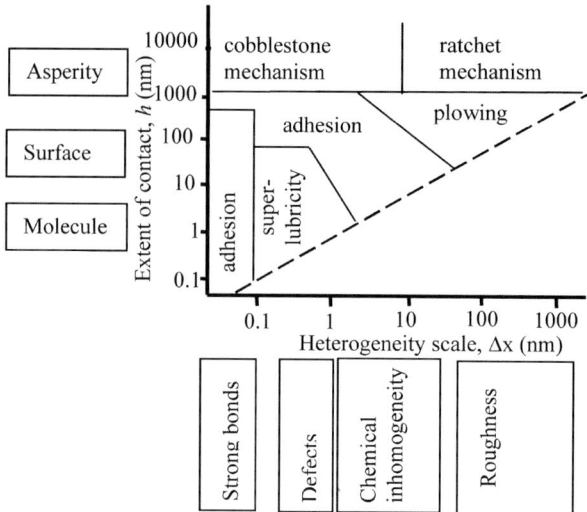

Fig. 4.7. The qualitative map of friction mechanisms presented as a function of the inhomo-geneity scale and the extent of contact [246]

separation or penetration of the contacting bodies). For small h—that is, for the scale of inhomogeneity involved in the contact corresponding to the molecular level— the above-mentioned superlubricity is expected in the case of weak bonds. In the case of strong bonds, however, adhesion can exist. For higher h corresponding to the asperity level of inhomogeneity involved in the contact, the elastic energy is small and adhesion can also exist as well as the cobblestone mechanism. For higher Δx corresponding to surface energy minima due to defects and heterogeneity, an adhesion mechanism can exist. Even higher Δx corresponds to energy minima due to surface roughness, so the plowing and ratchet mechanisms can exist for these values of Δx, with plowing corresponding to higher values of elastic energy. It is noted that the map consists of several domain areas which correspond to different friction mechanisms. Inside each domain the same scaling laws may apply, and a transition from one domain to another might result in a change of friction mechanisms and scaling laws.

4.5 Summary

To summarize, in this section we examined how inhomogeneity and nonideality leads to energy dissipation, which involves an interaction at different scale and hierarchy levels. A qualitative map of friction regimes was suggested based upon this analysis. The next section will study hierarchical surfaces intended for the control of friction.

Part II

Solid–Liquid Friction and Superhydrophobicity

"One who performs his duty without attachment,
surrendering the results unto the Supreme Lord, is unaffected by sinful action,
as the Lotus leaf is untouched by water."
Bhagavad Gita 5.10

5

Solid–Liquid Interaction and Capillary Effects

Abstract Phase transformations are discussed and the classical concepts of the Young, Laplace, and Kelvin equations are introduced. The role of the capillary effects at the nanoscale is further discussed including deviations from the conventional water phase diagram, stability issues associated with capillary effects, etc.

In the preceding chapters we studied multiscale dissipative mechanisms of solid–solid friction. In the second part of the book, we investigate the dissipative mechanisms of solid–liquid friction. In this chapter we introduce basic concepts of the physical chemistry relevant to the wetting of a solid surface by a liquid. We will emphasize the multiscale nature of mechanisms and interactions involved in the wetting process and specific nanoscale mechanisms of wetting.

5.1 Three Phase States of Matter

It is well known that any substance can be in one of the three phase states: solid, liquid, or gas (vapor). There is also the fourth state, plasma; however, it is outside the scope of our consideration. Matter in the solid state has crystalline or amorphous structure with atoms or molecules packed closely together and strongly bonded to each other by the covalent, metallic, or ionic bonds. When examined at the macroscale, a solid body can sustain both compressive and tensile normal stresses and shear stress.

   With increasing temperature or decreasing pressure, a solid can melt and transform into the liquid state. In a liquid, polar molecules still have bonds (hydrogen or van der Waals) with each other, however, these bonds are weaker than the bonds in a solid, and they can be easily ruptured. At the macroscale, a liquid can sustain only normal isotropic stresses (usually compressive and rarely tensile), however, it flows when a shear stress is applied. Usually, molecules of a liquid are packed less closely than those of a solid, so the solid is denser than the liquid. A notorious exception is water, which at 0 °C is denser than ice. This property of water is known as the "water anomaly," and because of it ice flows at the water surface.

With further increase of the temperature or decrease of the pressure, a liquid can transform into vapor (gas). The characteristic feature of a gas is that it tends to expand and occupy all available space. The density of a gas is much lower than that of liquid and solid. The distance between gas molecules is large, and it can be assumed in many cases that there is no interaction between the gas molecules, except for the hardcore repulsion during their collisions. The model of the "ideal gas" is based on this assumption [283, 6].

Transitions between solid, liquid, and vapor states are known as the "phase transitions of the first kind," as opposed to the "phase transitions of the second kind." In general, phase transitions are characterized by an abruptly increased (or decreased) order in the system. The phase transitions of the first kind are also characterized by a significant amount of energy consumed or released during the transition. For example, in order to completely transfer one liter of boiling water at 100 °C into vapor, the amount of energy is needed is several times greater than that needed to heat one liter of water from 0 °C to 100 °C. The energy is consumed for breaking the bonds between the molecules. An example of phase transitions of the second kind is the transition to the superconductive state or transitions between different magnetic states in a metal (paramagnetic and ferromagnetic phases).

The transition from solid to liquid is called melting, while the opposite transition is called freezing. The transition from liquid to vapor is called evaporation or boiling, while the opposite transition is called condensation. The direct transition from solid to vapor (without the liquid phase stage) is known as sublimation, while the opposite transition is called deposition [6, 283].

At a given temperature and pressure, after a sufficient period of time required for the phase transition processes, the system reaches a thermodynamic equilibrium. The state of the system at equilibrium is a function of only temperature and pressure. Therefore, it is convenient to use a phase diagram that shows the phase state of the substance as a function of temperature and pressure (Fig. 5.1). The boundaries or equilibrium lines between the solid, liquid, and vapor states are shown in the phase diagram as curves, corresponding to pressure as functions of tempera-

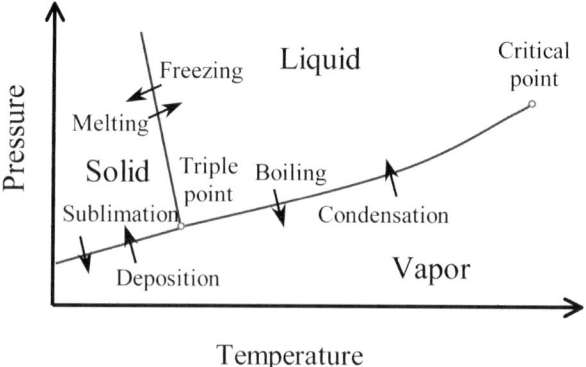

Fig. 5.1. Schematic of phase diagram of water

ture, $P(T)$. Note that for water, the solid–liquid state equilibrium line has a positive slope, whereas for most other substances the slope is negative. This is because the water anomaly, the transition from ice to liquid water at a given temperature, corresponds to increasing pressure (and, therefore, decreasing density).

The unique combination of temperature and pressure at which all three phases can coexist at equilibrium is called the triple point. The liquid–vapor equilibrium line goes from the triple point to another characteristic point in the phase diagram, the critical point. At the critical point, the energy barriers associated with the liquid–vapor phase transition vanish and the distinct boundary between the two states disappears. Instead, a continuous change of density may lead to the gradual liquid–vapor transformation above the critical point [6, 283].

5.2 Phase Equilibrium and Stability

When a phase transition line in the phase diagram is crossed, it is normally expected that the substance will change its phase state. However, such a change would require additional energy input for nucleation of seeds of the new phase. For example, the liquid–vapor transition requires nucleation of vapor bubbles (this process is called cavitation), while the liquid–solid transition requires nucleation of ice crystals. If special measures are taken to prevent nucleation of the seeds of the new phase, it is possible to postpone the transition to the equilibrium state phase indefinitely [321]. In this case, for example, water can remain liquid at temperatures below $0\,°C$ (supercooled water) or above $100\,°C$ (superheated water). Such a state is metastable and therefore fragile. A metastable equilibrium can be destroyed easily with a small energy input due to a fluctuation. At the stable equilibrium state, a metastable state corresponds to a local energy minimum, however, the energy barrier separating the metastable state from an unstable state is very small (Fig. 5.2).

An interesting and important example of a metastable phase state is "stretched" water, i.e., water under tensile stress or negative pressure. When liquid pressure is reduced below the liquid–vapor equilibrium line for a given temperature, it is expected

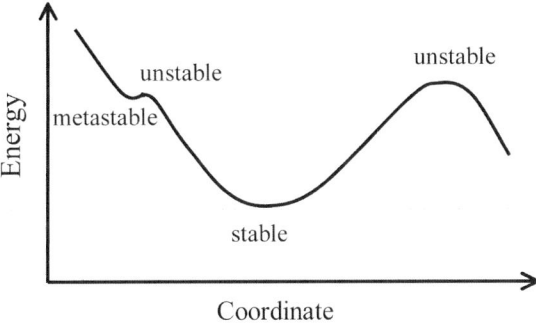

Fig. 5.2. Schematic of an energy profile showing stable, unstable, and metastable equilibriums. The metastable equilibrium is separated by a very small energy barrier

to transform into the vapor state. However, such a transition requires the formation of a new liquid–vapor interface, usually in the form of vapor bubbles, which needs additional energy input and, therefore, creates energy barriers. As a result, the liquid can remain in a metastable liquid state at low pressure, even when the pressure is negative [321].

There are two factors that act upon a vapor bubble inside liquid: the pressure and the interfacial tension. While the negative pressure (tensile stress) is acting to expand the size of the bubble, the interfacial tension is acting to collapse it. For a small bubble, the interfacial stress dominates, however, for a large bubble, the pressure dominates. Therefore, there is a critical radius of the bubble, and a bubble with a radius larger than the critical radius would grow, whereas one with a smaller radius would collapse. The value of the critical radius is given by

$$R_c = \frac{2\gamma_{LV}}{P_{sat} - P},$$
(5.1)

where γ_{LV} is the liquid–vapor surface tension, P_{sat} is the saturated vapor pressure, and P is the actual liquid pressure [150]. The corresponding energy barrier is given by

$$E_b = \frac{16\pi (\gamma_{LV})^3}{3(P_{sat} - P)^2}.$$
(5.2)

Cavitation (bubble formation) becomes likely when the thermal fluctuations have energy comparable with E_b.

With further decrease of pressure in the negative region, the so-called spinodal limit can be reached (Fig. 5.3). At that pressure, the critical cavitation radius becomes of the same order as the thickness of the liquid–vapor interface. In this case, there is a lower energy path of nucleation connecting the liquid to the gas phase by

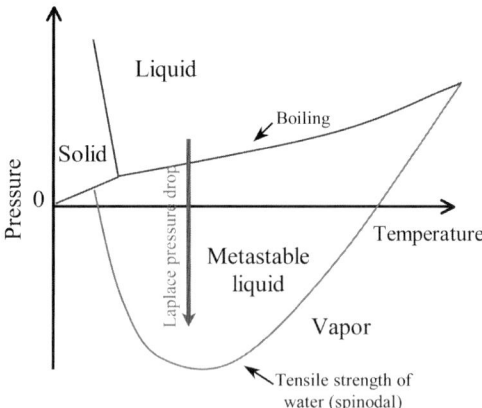

Fig. 5.3. Schematic of phase diagram of water showing the metastable liquid region. While the energy barriers separating the metastable states are small, at the nanoscale these barriers are very significant

expansion of a smoothly varying density profile [150]. In the phase diagram, the line that corresponds to the spinodal limit is expected to go all the way to the critical point. This critical spinodal pressure at a given temperature effectively constitutes the tensile strength of liquid water. Various theoretical considerations, experimental observations, and molecular dynamic simulation results have been used to determine the value of the spinodal limit. At room temperature, the spinodal pressure is between −150 and −250 MPa [186].

Moderate negative water pressures were obtained in the 19th century using a tube sealed with a piston [321]. However, it is very difficult to obtain deeply negative pressure at the macroscale because of nucleation. The values that have been reported constitute −19 MPa with the isochoric cooling method [149], −17.6 MPa with the modified centrifugal method [11], and −25 MPa with the acoustic method [136]. The situation is different at the micro- and nanoscale. A pressure of −140 MPa in the microscopic aqueous inclusions in quartz crystals was reported [7]. Tas et al. [313] showed that water plugs in hydrophilic nanochannels can be at a significant absolute negative pressure due to tensile capillary forces.

An interesting example of negative pressure in nature is in tall trees, such as the California redwood (*Sequoia sempervirens*) (Fig. 5.4(a)). Water can climb to the top of the tree due to the capillary effect, and if a tree is tall enough, a negative pressure may be required to supply water to the top [188, 324]. A height difference of more than 10 meters would correspond to the pressure difference of more than one atmosphere. Cavitation damage is also a significant technical problem, in particular, for boat propellers, which reduce water pressure due to high speed (Fig. 5.4(b)).

5.3 Water Phase Diagram at the Nanoscale

The phase diagram shows phase states of water at equilibrium at a given pressure and temperature. It does not take into account the effects of the interface energy, which may lead to energy barriers and metastable states. However, with a physical system's decreasing size, the surface-to-volume ratio grows, and surface and interfacial effects become increasingly important. This is why at the nanoscale the system may be found at a state which is far from the macroscale equilibrium at the given temperature and pressure. Therefore, the water phase diagram does not always reflect the situation at the nanoscale.

One particularly important example of that, which has a direct relevance to nanoscale friction, is in condensed water capillary bridges (menisci) between very small asperities and, in particular, between an AFM tip (with a radius on the order of 10 nm) and a flat sample (Fig. 5.5). When the contact takes place in air, such a capillary bridge is almost always present due to the condensation of the water vapor from air. The menisci have negative curvature (which can be estimated from the Kelvin equation, as will be explained later) and the pressure inside them is reduced compared to the pressure outside. This reduced pressure leads to the attractive meniscus force between the tip and the sample. This meniscus force (or capillary force) is a

(a)

(b)

Fig. 5.4. Effect of negative water pressure in everyday life. **a** California Redwood (Sequoia) is an example of a tall tree which needs to develop a negative pressure in order to suck water from the roots up to the top. **b** Cavitation damage of a boat propeller. A rapid rotation of the propeller leads to water pressure decrease and cavitation (bubbling) that can damage to propeller over a long period of time

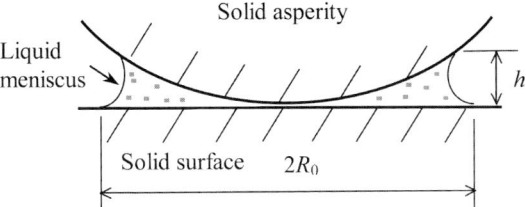

Fig. 5.5. Schematic of a condensed water capillary bridge (meniscus) between a solid spherical asperity (e.g., an AFM tip) and a flat solid sample

significant (and often the dominant) part of the adhesion force, and it can be measured with the AFM. Recent results show that the pressure inside these small bridges can be deeply negative, approaching the spinodal pressure [346]. The negative pressure in the capillary bridges is possible because the size of the bridges is smaller than the critical cavitation radius given by (5.1) and, therefore, no cavitation occurs.

It has been suggested also that water in the nanoscale capillary bridges may include ice or be a mixture of liquid water with ice [166]. Such a mixture would have high viscosity and may act like glue, demonstrating elastic response. The capillary adhesive force, F_{cap}, will lead to the component of the friction force, μF_{cap}, that is present even when no external load is applied to the asperity.

There are several ways to estimate the capillary force between an asperity of radius R and a flat substrate. An approximate value is given by

$$F_{cap} = 4\pi R \gamma_{LV} \cos \theta, \qquad (5.3)$$

where θ is the contact angle between water and the substrate material. Note that the capillary bridges are usually formed between solid asperities only if material of the asperities is hydrophilic, in other words, if $\theta < 90°$, so that $F_{cap} > 0$. Equation (5.3) is based on the assumption that the meniscus radius is small compared with R and that the meniscus has an almost cylindrical shape. A more accurate calculation of the capillary force should involve exact calculation of the meniscus shape and pressure inside the capillary bridge.

Equation (5.3) states that F_{cap} is independent of the relative humidity and it provides no information about the pressure inside the capillary bridge. However, it is known from the experiments that F_{cap} depends upon the relative humidity (Fig. 5.6). Most experiments show that the capillary force first increases with increasing relative humidity up to approximately 30% and decreases with a further increase of the relative humidity after that [14, 147]. The reasons for this trend are still not completely clear. It has been suggested that water on mica can form a monolayer of ice-like phase. With decreasing humidity the ice-like water monolayer breaks into islands and thus the pull-off force decreases. Water often forms adsorbed layers at the surface of many materials. These layers may have an ice-like structure [14]. The layers are connected to the capillary bridge.

We have seen that water at the nanoscale can remain liquid at very low pressures. There is evidence of ice presence at the ambient temperature and pressure in the

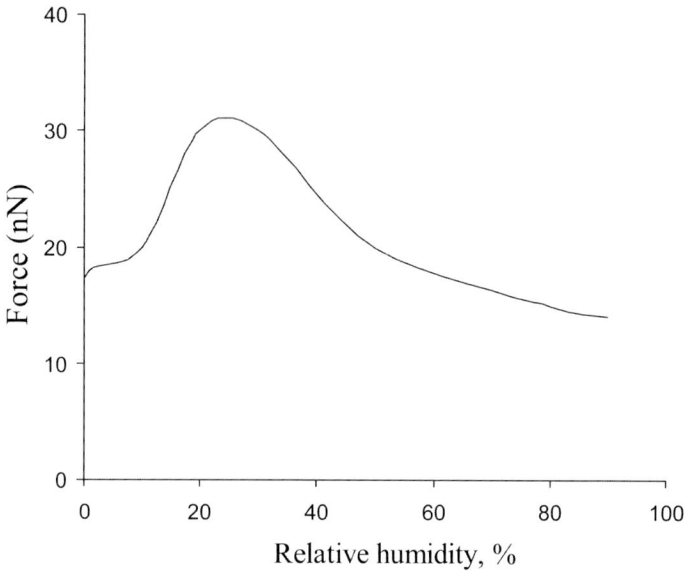

Fig. 5.6. A typical dependence of the pull-off adhesion force between an asperity and a tip upon the relative humidity measured by an AFM. The dependence has a maximum at relative humidity close to 30%

adsorbed water layers. In addition, the thermal fluctuations can play a significant role at the nanoscale. Therefore, the macroscale phase diagram goes not always reflect nanoscale behavior adequately.

5.4 Surface Free Energy and the Laplace Equation

Atoms and molecules near the surface have bonds with a smaller number of neighboring atoms and molecules than those in the bulk. As a result, the atoms and molecules at the surface have higher energy. This excess energy is called free surface energy, γ, and it is measured in Joules per area units. Since atoms at the edge of the body have even fewer bonds than at the surface, their energy is even higher. These considerations lead to the conclusion that, in general, the surface energy grows with decreasing radius of curvature, R, of the surface

$$\gamma(R) = \gamma_0 \left(1 - \frac{2\delta}{R} + \cdots \right), \tag{5.4}$$

where δ is the so-called Tolman's length of the molecular scale and γ_0 is the surface energy for a flat surface [9]. However, this curvature-dependence is significant only for very small curvature radii comparable with the molecular length, so for most practical considerations the surface energy of a flat surface can be taken for calculations, $\gamma = \gamma_0$.

The free surface energy is defined as the energy needed to create a new interface with a unit area and it quantifies the disruption of chemical bonds that occurs when a surface is created. Surfaces are less energetically favorable than the bulk of a material; otherwise there would be a driving force for surfaces to be created, and surface is all there would be. Cutting a solid body into pieces disrupts its bonds, and therefore consumes energy. For an interface between two materials or phase states of the same material, the interface energy can be defined in a similar manner, e.g., γ_{SL}, γ_{SV}, and γ_{LV} for solid–liquid, solid–vapor, and liquid–vapor free interface energies, respectively.

Expanding the interface between solid, liquid, and vapor phases is usually energetically unfavorable. For a vapor bubble of radius R, expanding the size of the bubble for a small amount dR would lead to the surface area change of $8\pi R dR$ and volume change of $4\pi R^2 dR$. If pressure inside the bubble extends the outside pressure by ΔP, the work of the pressure is given by $\Delta P 4\pi R^2 dR$, whereas the change of the surface energy is given by $\gamma_{LV} 8\pi R dR$. It is concluded from the earlier discussion that the droplet is at equilibrium if these energy changes are equal, that is, $\Delta P = 2\gamma_{LV}/R$. In the general case, the surface is not necessarily spherical, and the pressure change along the curved interface is given by the Laplace equation (sometimes called the Young–Laplace equation [6, 283])

$$\Delta P = \gamma_{LV}\left(\frac{1}{R_1} + \frac{1}{R_2}\right), \tag{5.5}$$

where R_1 and R_2 are the principal radii of curvature of the interface at a given point [6, 283]. It follows immediately from (5.5) that a liquid–vapor interface has the constant mean curvature $1/R_1 + 1/R_2$ at any point. It also follows that pressure inside a gas bubble or liquid droplet is larger than the pressure outside for the amount given by the Laplace equation. The pressure calculated from (5.5) is called the Laplace pressure [283].

For some interfaces, the mean curvature is negative. An important example of an interface with a negative mean curvature is the condensed water meniscus between an asperity and a flat surface, which was discussed earlier in this chapter. The pressure inside such a meniscus is reduced compared to the atmospheric pressure outside, and if the negative mean curvature is low enough, the pressure will be below zero. Taking the typical liquid–air interface energy $\gamma_{LV} = 0.072$ N/m and the atmospheric pressure $\Delta P = 10^5$ Pa will result in the mean curvature radius of 720 nm corresponding to the pressure drop of one Atmosphere.

5.5 Contact Angle and the Young Equation

When a liquid front comes in contact with a flat solid surface under the angle θ (Fig. 5.7), propagation of the liquid front for a small distance dt results in a net energy change of $dt (\gamma_{SL} - \gamma_{SV} + \gamma_{LV} \cos\theta)$. Therefore, for the liquid front to be at equilibrium, the Young equation should be satisfied [283]

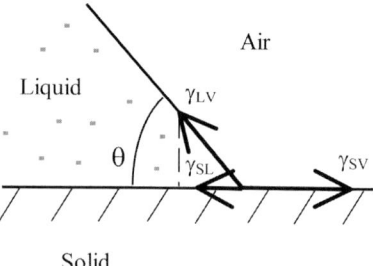

Fig. 5.7. A water–vapor surface coming to the solid surface at the contact angle of θ. From the balance of the tension forces, the Young equation should be satisfied, $\gamma_{LV}\cos\theta = \gamma_{SV} - \gamma_{SL}$

$$\gamma_{LV}\cos\theta = \gamma_{SV} - \gamma_{SL}. \tag{5.6}$$

It is clear from (5.6) that three situations are possible. If $(\gamma_{SV} - \gamma_{SL})/\gamma_{LV} > 1$, complete wetting takes place with the liquid fully adsorbed by the solid surface ($\theta = 0$). If $(\gamma_{SV} - \gamma_{SL})/\gamma_{LV} < -1$, the complete rejection of the liquid by the solid surface takes place ($\theta = 180°$). The most common is the intermediate situation of partial wetting ($-1 < (\gamma_{SV} - \gamma_{SL})/\gamma_{LV} < 1$, $0 < \theta < 180°$). A liquid that has the contact angle $\theta < 90°$ is often referred to as a "wetting liquid," while that with $\theta > 90°$ is "a nonwetting liquid." Corresponding surfaces are called, in the case of water, "hydrophilic" and "hydrophobic," respectively.

An alternative form of the equation for the contact angle θ is given by the so-called Young–Dupré equation which involves the work of cohesion of the solid with liquid, W,

$$\gamma_{LV}(1 + \cos\theta) = W. \tag{5.7}$$

Since W is the work of cohesion, which is equal to the energy required to create a solid–liquid interface of the unit area while the solid–vapor and liquid–vapor interfaces are being destroyed, the work of cohesion is given by

$$W = \gamma_{SV} - \gamma_{SL} + \gamma_{LV}. \tag{5.8}$$

Substituting (5.8) into (5.7) results in (5.6). According to the Young equation, a liquid–vapor interface always comes in contact with the solid surface under the same contact angle θ.

The free interface energies can also be viewed as surface tension forces. These forces are applied to the three-phase contact line (the triple line) and directed toward the corresponding interface. The surface tensions are measured in force units per length of the contact line, N/m, the same units as the interface energy, J/m². If a section of the triple line of length L moved for the distance of dx in the direction perpendicular to the line, the new surface area $Ld x$ is created, which requires the energy of $\gamma Ld x$. Since the work is equal to the force times the distance, the corresponding surface tension force is given by γL. From this simple consideration it is seen that the surface tensions and surface free energies are essentially the same. Historically, the Young equation was formulated in early 1800s in terms of forces,

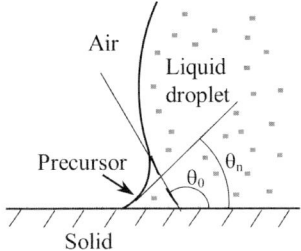

Fig. 5.8. Scale dependence of the contact angle. The apparent contact angle θ_0 is observed at the macroscale, while the nanoscale value, θ_n, may be significantly different due to the presence of a precursor or film

after the thermodynamic concept of free energy in general and free surface energy in particular was introduced by Helmholtz, Gibbs and other scientists in the second half of the 19th century [283].

Young's equation was originally formulated for the horizontal component of the tension force. Assuming three tension forces act upon the triple line from the directions of the three phases (Fig. 5.7), the balance of the horizontal projection of these forces leads to (5.6). As for the vertical component of the force, it is balanced by the elastic response of the solid surface. This is a valid assumption for the thermodynamic equilibrium (the quasi-thermodynamic approximation). If the system is not at equilibrium, the tensile stress, caused by the vertical component, would increase locally the chemical potential of the system, so that the material can dissolve and change its shape [289, 344].

Young's equation does not take into account a number of factors, which can significantly affect the contact angle at the micro- and nanoscale. It is emphasized that the contact angle provided by Young's equation is a macroscale parameter, so it is called sometimes "the apparent contact angle." The actual angle under which the liquid–vapor interface comes in contact with the solid surface at the micro- and nanoscale can be different. There are several reasons for that. First, water molecules tend to form a thin layer upon the surfaces of many materials. This is because of a long-distance van der Waals adhesion force that creates the so-called disjoining pressure [89]. This pressure is dependent upon the liquid layer thickness and may lead to the formation of stable thin films or precursors (Fig. 5.8). In this case, the shape of the droplet near the triple line transforms gradually from a spherical surface into a precursor layer, and thus the nanoscale contact angle is much smaller than the apparent contact angle. In addition, adsorbed water monolayers and multilayers are common for many materials.

Second, even carefully prepared atomically smooth surfaces exhibit certain roughness and chemical heterogeneity. Water tends to first cover the hydrophilic spots with high surface energy and low contact angle [75]. The tilt angle due to roughness can also contribute to the apparent contact angle.

Third, Young's equation provides the value of the so-called static contact angle, that is, it ignores any dynamic effects related to the change of the droplet's shape. The very concept of the static contact angle is not well defined. For practical purposes, the contact angle, which is formed after a droplet is gently placed upon a surface and stops propagating, is considered the static contact angle. However, depositing the droplet involves adding liquid while leaving it may involve evaporation, so it is difficult to avoid dynamic effects.

Fourth, for a small droplet and curved triple lines, the effect of the contact line tension may be significant. Molecules at the surface of a liquid or solid phase have higher energy because they are bonded to fewer molecules than those in the bulk. This leads to surface tension and surface energy. In a similar manner, molecules at the edge have fewer bonds than those at the surface, which leads to line tension and the curvature dependence of the surface energy. This effect becomes important when the radius of curvature is comparable with the Tolman's length [9]. However, the triple line at the nanoscale can be bending, and the radius of curvature can be very small, so that the line tension effects become important [266]. Thus, while the contact angle is a convenient macroscale parameter, wetting is governed by interactions at the micro- and nanoscale, which determine the contact angle hysteresis and other wetting properties.

5.6 Kelvin's Equation

Due to evaporation, a certain amount of water vapor is always present in the air. The evaporation and condensation may have different rates; however, they reach equilibrium at a certain pressure of the vapor. Partial pressure of vapor at which it is at equilibrium with liquid water is called saturation pressure, p_s. The actual partial vapor pressure, p, may be smaller than the saturation pressure. The relative humidity is the ratio of p/p_s. If a liquid-gas interface with negative mean curvature is present—for example, a water meniscus—the Laplace pressure, calculated from (5.5) is reduced as compared to the atmospheric pressure. In this case, the equilibrium of the water in the meniscus and vapor in air can be reached at a much lower pressure than p_s. The relation between the relative humidity and the mean curvature of the meniscus at a given temperature T is given by the Kelvin equation [283, 6]

$$\gamma_{LV}\left(\frac{1}{R_1} + \frac{1}{R_2}\right) = \frac{RT}{V} \ln\left(\frac{p}{p_s}\right),\qquad(5.9)$$

where $R = 8.314\,\text{J/(Kmol)}$ is the universal gas constant and V is the molar volume of air. At standard conditions ($T = 273\,\text{K}$, $P = 10^5\,\text{Pa}$), $V = 0.023\,\text{m}^3/\text{mol}$. The radius of curvature obtained by the Kelvin equation is sometimes called the Kelvin radius, $1/R_k = 1/R_1 + 1/R_2$. The Kelvin equation can be written as

$$R_k = \frac{\gamma_{LV} V}{RT} \ln\left(\frac{p_s}{p}\right).\qquad(5.10)$$

Another way to look at the Kelvin equation is to say that it predicts—for a given relative humidity, p/p_s, as soon as the condensation and evaporation processes reach a thermodynamic equilibrium—a meniscus with the mean curvature R_k given by (5.10) can form. The meniscus should also satisfy the Young equation (5.6) at the triple line. Since the mean curvature of the meniscus is negative, this condition can usually be satisfied at the points where a solid surface is concave, for example, near inside corners of a vessel or at the points of asperity contacts. Therefore, a meniscus forms near asperity contacts with the Laplace pressure drop given by the combination of (5.5) and (5.10)

$$\Delta P = \frac{RT}{V} \ln\left(\frac{p}{p_s}\right).$$

(5.11)

It is observed from (5.11) that the Laplace pressure drop inside the meniscus is expected to grow indefinitely with decreasing relative humidity. The value of the Laplace capillary force can be calculated by multiplying ΔP by the area of the foundation of the meniscus

$$F_{cap} = \Delta P \pi R_0^2,$$

(5.12)

where R_0 is the radius of the foundation of the meniscus. Since the Kelvin radius decreases quickly with decreasing relative humidity, the area of the foundation will decrease too. Assuming a conical asperity in contact with a flat surface, there will be a linear proportionality of the two radii

$$R_0 \propto R_k.$$

(5.13)

Combining (5.10)–(5.13) yields

$$F_{cap} \propto -\ln\left(\frac{p}{p_s}\right).$$

(5.14)

That is, with reduced relative humidity the capillary adhesion force is expected to grow. For a more complicated shape of an asperity (e.g., spherical rather than conical), the dependence of R_0 upon R_k would be more complicated than (5.13) [305]; however, it may be expected that (5.14) still provides a general trend.

It is indeed observed from the experiments, that the capillary force grows with decreasing relative pressure [49]. However, this trend is observed only for relative humidity greater than 30% (Fig. 5.6). At that level of relative humidity, the Kelvin radius is on the order of nanometers [305], and the height of a capillary bridge can constitute only several molecules and is comparable with the thickness of the liquid–vapor interface. In this situation, with the further decrease of the Kelvin radius there is just not enough molecules to sustain the liquid phase and cause the attractive capillary force, so the Kelvin equation breaks down. As a result, the Laplace force cannot have unlimited growth with the decreasing relative humidity.

5.7 Capillary Effects and Stability Issues

The significant role of the capillary force at the nanoscale raises the question of its stability. In classical mechanics, stability analysis of a solution plays a major role

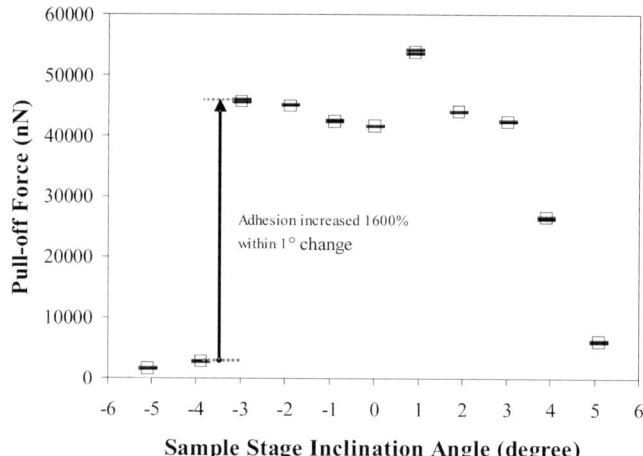

Fig. 5.9. Variation of adhesion force between a rough spherical colloidal probe in contact with a tilted flat sample as a function of tilt angle. Changing the tilt angle results in the changing contact spots due to roughness and thus leads to a very significant change in the adhesion force [254]

both in the statics and in dynamics. It is well known that in order for a solution to exist physically, it should be at a stable equilibrium and correspond to a local minimum of the potential energy. Since the 1960s it has been recognized that nonequilibrium processes play a very significant role in many physical phenomena, such as the hierarchical and self-organizing structures, and thus the instability and nonstable equilibrium are important to analyze. At the nanoscale, mechanical instabilities are inherent in many processes. For example, mechanical hysteresis of an AFM cantilever is caused by the destabilization of the mechanical equilibrium of the cantilever. Adhesion hysteresis that leads to energy dissipation during loading–unloading cycle with adhesion is a specific nanoscale example.

Capillary effects constitute another important area of nanomechanics that involves instabilities. The stability of the capillary force with respect to roughness details has not received enough attention from the scientists. The classical models of contact with adhesion, such as the Johnson–Kendall–Roberts (JKR) and Derjaguin–Muller–Toporov (DMT) models [222], do not address stability, although experiments show that the capillary force is very sensitive to small changes in roughness. During the multiple-asperity contact of two rough solid surfaces, multiple menisci can form between the bodies. There are only two restrictions upon the geometry of the meniscus at the thermodynamic equilibrium: (1) the curvature of the menisci should satisfy (5.3), and (2) the contact angle should be constant at the triple line (solid–liquid–air contact line). These two requirements may lead to nonunique shapes of the menisci for a given topography of rough surfaces. Furthermore, a slight change of topography may lead to a significant change of the shape of the meniscus and, therefore, of the adhesion force, as shown by Yang et al. [345]. They measured the adhesion force between a rough glass sphere attached to the AFM cantilever and a tilted smooth silicon sample. The response was not stable with respect to the tilt

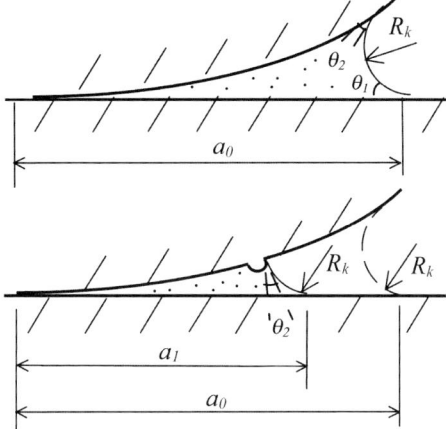

Fig. 5.10. Small roughness (a bump) upon the asperity may result in a big change of the radius of the meniscus, a_0 [254]

angle, that is, a small change of the tilt angle resulted in a large (sometimes 300–400%) change in the pull-off adhesion force (Fig. 5.9). While conventional models of contact with adhesion, such as the JKR and DMT, overlook this effect and assume that meniscus is stable, it has been reported that the pull-off adhesion force between rough surfaces is very sensitive with respect to small changes in roughness. Fig. 5.10 shows how superimposing a tiny roughness detail upon a 2D asperity can dramatically change the size of the meniscus and, therefore, the capillary force.

5.8 Summary

In this chapter, we introduced and discussed the fundamental equations that govern wetting and capillary effects: the Laplace, Young, and Kelvin equations. Phase transitions and equilibrium as well as the phase diagrams at the macro and nanoscale have been examined as well.

6

Roughness-Induced Superhydrophobicity

Abstract The concept of roughness-induced superhydrophobicity is introduced and discussed, including the Wenzel and Cassie equations, contact angle hysteresis and theoretical calculation of the contact angle for common types of surfaces. The effect of the triple line pinning vs. adhesion hysteresis is discussed as well as multiscale mechanisms of dissipation leading to contact angle hysteresis.

In the preceding chapter we considered general mechanisms and equations related to wetting. In this chapter we study the effect of surface roughness upon wetting and, in particular, the phenomenon of superhydrophobicity, induced by the surface roughness.

6.1 The Phenomenon of Superhydrophobicity

Numerous micro/nanotribological and micro/nanomechanical applications, such as in micro/nanoelectromechanical systems (MEMS/NEMS) require surfaces with low adhesion and stiction [29, 33–36, 38, 50]. As the size of these devices decreases, the surface forces tend to dominate over the volume forces, and adhesion and stiction constitute a challenging problem for proper operation of these devices. This makes the development of nonadhesive surfaces crucial for many of these emerging applications. It has been suggested that extremely water-repellent (superhydrophobic) surfaces produced by applying a micropatterned roughness combined with hydrophobic coatings may satisfy the need for the nonadhesive surfaces [53, 239–243, 245, 248]. Wetting may lead to formation of menisci at the interface between solid bodies during sliding contact, which increases adhesion/friction. As a result of this, the wet friction force is greater than the dry friction force, which is usually undesirable [30, 32, 34]. On the other hand, high adhesion is desirable in some applications, such as adhesive tapes and adhesion of cells to biomaterial surfaces, therefore, enhanced wetting by changing roughness would be desirable in these applications [239–241].

The primary parameter that characterizes wetting is the static contact angle, which is defined as the measurable angle that a liquid makes with a solid. The contact angle depends on several factors, such as roughness and the manner of surface preparation, and its cleanliness [6, 161]. If the liquid wets the surface (referred to as wetting liquid or hydrophilic surface), the value of the static contact angle is $0 \leq \theta \leq 90°$, whereas if the liquid does not wet the surface (referred to as nonwetting liquid or hydrophobic surface), the value of the contact angle is $90° < \theta \leq 180°$. The term hydrophobic/philic, which was originally applied only to water ("hydro-" means "water" in Greek), is often used to describe the contact of a solid surface with any liquid. The term "oleophobic/philic" is used sometimes to refer to wetting by oil. The terms "superphobic/philic" are also sometimes used. Surfaces with high energy, formed by polar molecules, tend to be hydrophilic, whereas those with low energy and built of nonpolar molecules tend to be hydrophilic.

Surfaces with a contact angle between 150° and 180° are called superhydrophobic. For liquid flow and other applications requiring low solid–liquid friction, in addition to high contact angle, superhydrophobic surfaces should also have very low water contact angle hysteresis. Contact angle hysteresis is the difference between the advancing and receding contact angles, which are two stable values. If additional liquid is added to a sessile drop, the contact line advances, and each time motion ceases, the drop exhibits an advancing contact angle. Alternatively, if liquid is removed from the drop, the contact angle decreases to a receding value before the contact retreats. For a droplet moving along a solid surface (e.g., if the surface is tilted) there is another definition. The contact angle at the front of the droplet (advancing contact angle) is greater than that at the back of the droplet (receding contact angle) due to roughness, resulting in contact angle hysteresis (Fig. 6.1(a)). It has been disputed that the two definitions are equivalent [192]; however, in many cases the two definitions have the same meaning. Surfaces with low contact angle hysteresis have a very low water roll-off angle that denotes the angle to which a surface must be tilted for roll off of water drops [108, 180].

One of the ways to increase the hydrophobic or hydrophilic properties of a surface is to increase surface roughness, so roughness-induced hydrophobicity has become the subject of extensive investigation. Wenzel [337] found that the contact angle of a liquid with a rough surface is different from that with a smooth surface. Cassie and Baxter [73] showed that air (or gas) pockets may be trapped in the cavities of a rough surface, resulting in a composite solid–liquid–air interface, as opposed to the homogeneous solid–liquid interface. Shuttleworth and Bailey [297] studied the spreading of a liquid over a rough solid surface and found that the contact angle at the absolute minimum of surface energy corresponds to the values predicted by Wenzel [337] or Cassie and Baxter [73]. Johnson and Dettre [171] showed that the homogeneous and composite interfaces correspond to the two metastable equilibrium states of a droplet. Bico et al. [55]), Marmur [216, 217], Lafuma and Quèrè [197], Patankar [260, 261], He et al. [148], and other authors investigated the metastability of artificial superhydrophobic surfaces and showed that whether the interface is homogeneous or composite may depend on the history of the system (in particular, whether the liquid was applied from the top or from the bottom). Extrand [108] pointed out that whether the interface is homogeneous or composite depends on droplet size, due to the gravity. It was also suggested that the so-called two-tiered (or double)

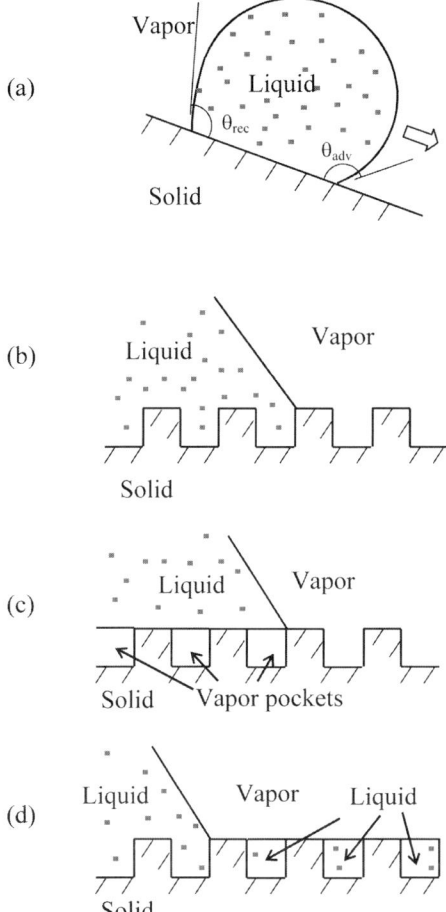

Fig. 6.1. a Schematics of a droplet on a tilted substrate showing advancing (θ_{adv}) and receding (θ_{rec}) contact angles. The difference between these angles constitutes the contact angle hysteresis. Configurations described by **b** the Wenzel equation for the homogeneous interface, **c** Cassie–Baxter equation for the composite interface with air pockets, and **d** the Cassie equation for the homogeneous interface

roughness, composed by superposition of two roughness patterns at different length-scale [151, 262, 308], and fractal roughness [293]) may lead to superhydrophobicity. Herminghaus [151] showed that certain self-affine profiles may result in superhydrophobic surfaces even for wetting liquids, in the case where the local equilibrium condition for the triple line (line of contact between solid, liquid and air) is satisfied. Nosonovsky and Bhushan [240, 241] pointed out that such configurations, although formally possible, are likely to be unstable. Nosonovsky and Bhushan [241, 242] proposed a stochastic model for the wetting of rough surfaces with a certain probability associated with every equilibrium state. According to their model, the overall contact angle with a two-dimensional rough profile is calculated by assuming that the overall configuration of a droplet is a result of superposition of numerous metastable states. The probability-based concept is consistent with the experimental data [197], which suggests that transition between the composite and homogeneous interfaces is gradual, rather than instant.

It has been demonstrated experimentally that roughness changes contact angle in accordance with the Wenzel model. Yost el al. [350]) found that roughness enhances the wetting of a copper surface with Sn–Pb eutectic solder, which has a contact angle of 15–20° for a smooth surface. Shibuchi et al. [293] measured the contact angle of various liquids (mixtures of water and 1,4-dioxane) on alkylketen dimmer (AKD) substrate (contact angle not larger than 109° for smooth surface). They found that for wetting liquids the contact angle decreases with increasing roughness, whereas for nonwetting liquids it increases. Semal et al. [288] investigated the effect of surface roughness on contact angle hysteresis by studying a sessile droplet of squalane spreading dynamically on multilayer substrates (behenic acid on glass) and found that an increase in microroughness slows the rate of droplet spreading. Erbil et al. [105] measured the contact angle of polypropylene (contact angle of 104° for smooth surface) and found that the contact angle increases with increasing roughness. Burton and Bhushan [67] measured the contact angle with roughness of patterned surfaces and found that, in the case of hydrophilic surfaces, it decreases with increasing roughness; for hydrophobic surfaces, it increases with increasing roughness. Bhushan and Jung [39–41] and Jung and Bhushan [172–174] studied wetting properties of hydrophobic and hydrophilic leaves and patterned surfaces and found similar trends.

In the last decade, material scientists paid attention to natural surfaces which are extremely hydrophobic. Among them are leaves of water-repellent plants such as *Nelumbo nucifera* (lotus) and *Colocasia esculenta*, which have high contact angles with water [24, 227, 330]. First, the surface of the leaves is usually covered with a range of different waxes made from a mixture of hydrocarbon compounds that have a strong phobia of being wet. Second, the surface is very rough due to so-called papillose epidermal cells, which form asperities or papillae. In addition to the microscale roughness of the leaf due to the papillae, the surface of the papillae is also rough with submicron sized asperities composed of the wax [330]. Thus, they have hierarchical micro- and nanosized structures, which were studied extensively by Burton and Bhushan [68] and Bhushan and Jung [39]. Water droplets on these surfaces readily sit on the apex of the nanostructures because air bubbles fill in the valleys of the structure under the droplet. Therefore, these leaves exhibit considerable superhydrophobicity. The water droplets on the leaves remove any contaminant particles from their surfaces when they roll off, leading to self-cleaning ability referred to as the lotus-effect. Other examples of biological surfaces include duck feathers and butterfly wings. Their corrugated surfaces provide air pockets that prevent water from completely touching the surface. Study and simulation of biological objects with desired properties is referred to as "biomimetics," which comes from a Greek word "biomimesis" meaning to mimic life.

As far as the realization of strongly water-repellent artificial surfaces is concerned, they can be constructed by chemically treating surfaces with low-surface-energy substances, such as polytetrafluoroethylene, silicon, or wax, or by fabricating extremely rough hydrophobic surfaces directly [148, 180, 293]. Sun et al. [308] studied an artificial poly(dimethylsiloxane) (PDMS) replica of a lotus leaf surface and found a water contact angle of 160° for the rough surface, whereas for the smooth PDMS surface it is about 105°.

As stated earlier, when two solids come in contact in the presence of a wetting liquid, a meniscus is often formed [30, 32–34]. Meniscus results in the normal meniscus force which, in turn, results in an increase in the tangential friction force. The magnitude of the meniscus force depends on the number of asperity contacts and asperity radii, which depend on roughness, and on surface tension of the liquid and the contact angle. The contact angle, as stated earlier, depends on surface roughness, and thus roughness affects the wet friction force [240].

6.2 Contact Angle Analysis

In this section, we consider the dependence of the contact angle on the surface tension for a liquid in contact with a smooth and a rough solid surface, forming a homogeneous interface. The surface atoms or molecules of liquids or solids have energy above that of similar atoms and molecules in the interior, which results in surface tension or free surface energy being an important surface property. This property is characterized quantitatively by the surface tension or free surface energy γ, which is equal to work, that is required to create a unit area of the surface at constant volume and temperature. The units of γ are J/m^2 or N/m and it can be interpreted either as energy per unit surface area or as tension force per unit length of a line at the surface. When a solid is in contact with liquid, the molecular attraction will reduce the energy of the system below that for the two separated surfaces. This may be expressed by the Dupré equation

$$W_{SL} = \gamma_{SA} + \gamma_{LA} - \gamma_{SL}, \tag{6.1}$$

where W_{SL} is the work of cohesion per unit area between two surfaces, γ_{SA} and γ_{SL} are the surface energies (surface tensions) of the solid against air and liquid, and γ_{LA} is the surface energy (surface tension) of liquid against air [161].

If a droplet of liquid is placed on a solid surface, the liquid and solid surfaces come together under equilibrium at a characteristic angle called the static contact angle θ_0. This contact angle can be determined from the condition of the net free surface energy of the system being minimized [6, 161]. The total energy E_{tot} is given by

$$E_{tot} = \gamma_{LA}(A_{LA} + A_{SL}) - W_{SL}A_{SL}, \tag{6.2}$$

where A_{LA} and A_{SL} are the contact areas of the liquid with air and the solid with liquid, respectively. It is assumed that the droplet is small enough that the gravitational potential energy can be neglected. It is also assumed that the volume and pressure are constant, so that the volumetric energy does not change. At the equilibrium

$dE_{\text{tot}} = 0$, which yields

$$\gamma_{\text{LA}}(dA_{\text{LA}} + dA_{\text{SL}}) - W_{\text{SL}}dA_{\text{SL}} = 0. \tag{6.3}$$

For a droplet of constant volume, it is easy to show using geometrical considerations that

$$dA_{\text{LA}}/dA_{\text{SL}} = \cos\theta_0. \tag{6.4}$$

Combining (6.1), (6.3), and (6.4), the well-known Young equation for the contact angle is obtained

$$\cos\theta_0 = \frac{\gamma_{\text{SA}} - \gamma_{\text{SL}}}{\gamma_{\text{LA}}}. \tag{6.5}$$

Equation (6.5) provides the value of the static contact angle for given surface tensions. Note that although we use the term "air," the analysis does not change in the case of another gas, such as a liquid vapor.

6.3 Heterogeneous Surfaces and Wenzel and Cassie Equations

In this section, we discuss the so-called heterogeneous interface and introduce the equations that govern the contact angle for the heterogeneous interface.

6.3.1 Contact Angle with a Rough and Heterogeneous Surfaces

The Wenzel [337] equation, which was derived using the surface force balance and empirical considerations, relates the contact angle of a water droplet upon a rough solid surface, θ, with the contact angle upon a smooth surface, θ_0 (Fig. 6.1(b)), though the nondimensional surface roughness factor, R_f, is equal to the ratio of the surface area to its flat projected area

$$\cos\theta = \frac{dA_{\text{LA}}}{dA_{\text{F}}} = \frac{dA_{\text{SL}}}{dA_{\text{F}}}\frac{dA_{\text{LA}}}{dA_{\text{SL}}} = R_f\cos\theta_0, \tag{6.6}$$

$$R_f = \frac{A_{\text{SL}}}{A_{\text{F}}}. \tag{6.7}$$

The dependence of the contact angle on the roughness factor is presented in Fig. 6.2 for different values of θ_0. The Wenzel model predicts that a hydrophobic surface ($\theta_0 > 90°$) becomes more hydrophobic with an increase in R_f, while a hydrophilic surface ($\theta_0 < 90°$) becomes more hydrophilic with an increase in R_f [172, 240].

In a similar manner, for a surface composed of two fractions, one with a fractional area f_1 and the contact angle θ_1 and the other with f_2 and θ_2, respectively (so that $f_1 + f_2 = 1$), the contact angle is given by the Cassie equation

$$\cos\theta = f_1\cos\theta_1 + f_2\cos\theta_2. \tag{6.8}$$

Effect of roughness

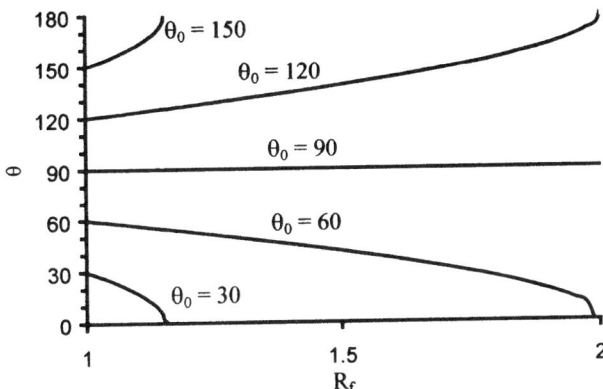

Fig. 6.2. Contact angle for rough surface (θ) as a function of the roughness factor (R_f) for various contact angles for smooth surface (θ_0) [240]

For the case of a composite interface (Fig. 6.1(c)), consisting of the solid–liquid fraction ($f_1 = f_{SL}, \theta_1 = \theta_0$) and liquid–air fraction ($f_2 = f_{LA} = 1 - f_{SL}, \cos\theta_2 = -1$), combining (6.7) and (6.8) yields the Cassie–Baxter equation

$$\cos\theta = R_f f_{SL}\cos\theta_0 - 1 + f_{SL} \text{ or } \cos\theta = R_f\cos\theta_0 - f_{LA}(R_f\cos\theta_0 + 1). \quad (6.9)$$

The opposite limiting case of $\cos\theta_2 = 1$ ($\theta_2 = 0°$ corresponds to the water-on-water contact) yields

$$\cos\theta = 1 + f_{SL}(\cos\theta_0 - 1). \quad (6.10)$$

Equation (6.10) is sometimes used [94] for the homogeneous interface instead of (6.6), if the rough surface is covered by holes filled with water (Fig. 6.1(d)).

6.3.2 The Cassie–Baxter Equation

Two situations in wetting of a rough surface should be distinguished: the homogeneous interface without any air pockets (sometimes called the Wenzel interface, since the contact angle is given by the Wenzel equation or by (6.6)) and the composite interface, with air pockets trapped between the rough details as shown in Fig. 6.3(a) (sometimes called the Cassie or Cassie–Baxter interface, since the contact angle is given by (6.9)).

While (6.9) for the composite interface was derived using (6.6) and (6.8), it could also be obtained independently. For this purpose, two sets of interfaces are considered: a liquid–air interface with the ambient and a flat composite interface under the droplet, which itself involves solid–liquid, liquid–air, and solid–air interfaces. For fractional flat geometrical areas of the solid–liquid and liquid–air interfaces under

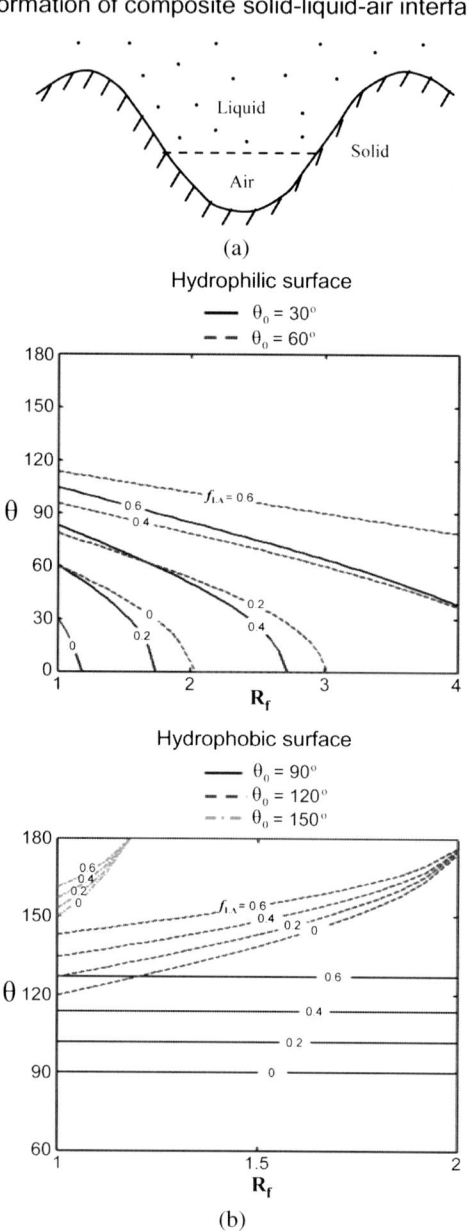

Fig. 6.3. a Formation of a composite solid–liquid–air interface for rough surface, **b** contact angle for rough surface (θ) as a function of the roughness factor (R_f) for various f_{LA} values on the hydrophilic surface and the hydrophobic surface, and **c** f_{LA} requirement for a hydrophilic surface to be hydrophobic as a function of the roughness factor (R_f and θ_0) [172]

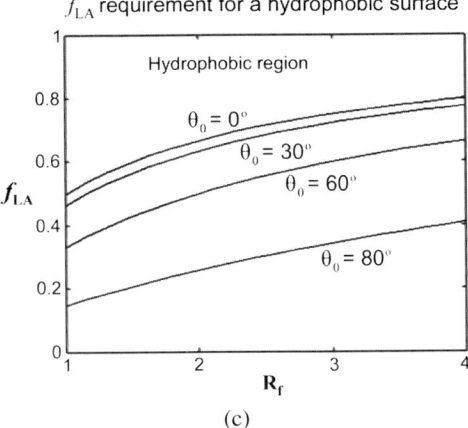

(c)

Fig. 6.3. (*Continued*)

the droplet, f_{SL} and f_{LA}, the flat area of the composite interface is

$$A_C = f_{SL}A_C + f_{LA}A_C = R_f A_{SL} + f_{LA}A_C. \tag{6.11}$$

In order to calculate the contact angle in a manner similar to the derivation of (6.6), the differential area of the liquid–air interface under the droplet, $f_{LA}dA_C$, should be subtracted from the differential of the total liquid–air area dA_{LA}, which yields

$$\cos\theta = \frac{dA_{LA} - f_{LA}dA_C}{dA_C} = \frac{dA_{SL}}{dA_F}\frac{dA_F}{dA_C}\frac{dA_{LA}}{dA_{SL}} - f_{LA}$$
$$= R_f f_{SL}\cos\theta_0 - f_{LA}$$
$$= R_f\cos\theta_0 - f_{LA}(R_f\cos\theta_0 + 1). \tag{6.12}$$

The dependence of the contact angle on the roughness factor for hydrophilic and hydrophobic surfaces is presented in Fig. 6.3(b).

According to (6.12), even for a hydrophilic surface, the contact angle increases with an increase of f_{LA}. At a high value of f_{LA}, a surface can become hydrophobic; however, the value required may be unachievable or the formation of air pockets may become unstable. Using the Cassie–Baxter equation, the value of f_{LA} at which a hydrophilic surface could turn into a hydrophobic one, is given by [172]

$$f_{LA} \geq \frac{R_f\cos\theta_0}{R_f\cos\theta_0 + 1} \quad \text{for } \theta_0 < 90°. \tag{6.13}$$

Figure 6.3(c) shows the value of f_{LA} requirement as a function of R_f for four surfaces with different contact angles θ_0. Hydrophobic surfaces can be achieved above certain f_{LA} values as predicted by (6.13). The upper part of each contact angle line is the hydrophobic region. For the hydrophobic surface, the contact angle increases with an increase in f_{LA} for both smooth and rough surfaces.

Shuttleworth and Bailey [297] studied spreading of a liquid over a rough solid surface and found that the contact angle at the absolute minimum of surface energy corresponds to the values given by (6.6) (for the homogeneous interface) or (6.12) (for composite interface). According to their analysis, spreading of a liquid continues until simultaneously (6.5) (the Young equation) is satisfied locally at the triple line and the minimal surface condition is satisfied over the entire liquid–air interface. The minimal surface condition states that the sum of inversed principal radii of curvature, R_1 and R_2 (mean curvature), is constant at any point, and thus governs the shape of the liquid–air interface.

$$\frac{1}{R_1} + \frac{1}{R_2} = \text{const.} \tag{6.14}$$

The same condition is also the consequence of the Laplace equation, which relates pressure change through an interface to its mean curvature.

Johnson and Detre [171] showed that the homogeneous and composite interfaces correspond to the two stable or metastable states of a droplet. Even though it may be geometrically possible for the system to become composite, it may be energetically profitable for the liquid to penetrate into valleys between asperities and to form the homogeneous interface. Marmur [216] formulated geometrical conditions for a surface under which the energy of the system has a local minimum and the composite interface may exist. Patankar [261] pointed out that whether the homogeneous or composite interface exists depends on the systems history, i.e., on whether the droplet was formed at the surface or deposited. However, the above-mentioned analyses do not provide an answer to our question: Which of the two possible configurations, homogeneous or composite, will actually form?

6.3.3 Limitations of the Wenzel and Cassie Equations

The Cassie equation (6.8) is based on the assumption that the heterogeneous surface is composed of well-separated distinct patches of different material, so that the free surface energy can be averaged. It has also been argued that when the size of the chemical heterogeneities is very small (of atomic or molecular dimensions), the quantity that should be averaged is not the energy, but the dipole moment of a macromolecule [162], and (6.8) may have to be replaced by

$$(1 + \cos \theta)^2 = f_1(1 + \cos \theta_1)^2 + f_2(1 + \cos \theta_2)^2. \tag{6.15}$$

Experimental studies of polymers with different functional groups showed a good agreement with (6.15), and the dependence on the dipole moment is shown only in the case of polymers and may be due to the nature of the polymeric molecular chains [320].

Later investigations put the Wenzel and Cassie equations into a thermodynamic framework, however, they showed also that there is no one single value of the contact angle for a rough or heterogeneous surface [171, 204, 216]. The contact angle can be in a range of values between the so-called receding contact angle, θ_{rec}, and the advancing contact angle, θ_{adv}. The system tends to achieve the receding contact angle when liquid is removed (e.g., at the rear end of a moving droplet), whereas the advancing contact angle is achieved when the liquid is added (e.g., at the front end of a moving droplet). When the liquid is neither added nor removed, the system tends to have a static or "most stable" contact angle, which is given approximately by (6.5)–(6.10). The difference between θ_{adv} and θ_{rec} is known as the "contact angle hysteresis," and it reflects a fundamental asymmetry of wetting and dewetting and the irreversibility of the wetting/dewetting cycle. Although for surfaces with the roughness carefully controlled on the molecular scale it is possible to achieve contact angle hysteresis as low as $< 1°$ [139], it cannot be eliminated completely, since even the atomically smooth surfaces have a certain roughness and heterogeneity. The contact angle hysteresis is a measure of energy dissipation during the flow of a droplet along a solid surface. A water-repellent surface should have a low contact angle hysteresis to allow water to flow easily along the surface.

We emphasize that the contact angle provided by (6.5)–(6.10) is a macroscale parameter, so it is called sometimes "the apparent contact angle." The actual angle under which the liquid–vapor interface comes in contact with the solid surface at the micro- and nanoscale can be different. There are several reasons for that. First, water molecules tend to form a thin layer upon the surfaces of many materials. This is because of a long-distance van der Waals adhesion force that creates the so-called disjoining pressure [89]. This pressure is dependent upon the liquid layer thickness and may lead to the formation of stable thin films. In this case, the shape of the droplet near the triple line transforms gradually from the spherical surface into a precursor layer, and thus the nanoscale contact angle is much smaller than the apparent contact angle. In addition, adsorbed water monolayers and multilayers are common for many materials. Second, even carefully prepared atomically smooth surfaces exhibit certain roughness and chemical heterogeneity. Water tends to cover at first the hydrophilic spots with high surface energy and low contact angle [75]. The tilt angle due to the roughness can also contribute to the apparent contact angle. Third, the very concept of the static contact angle is not well defined. For practical purposes, the contact angle which is formed after a droplet is gently placed upon a surface and stops propagating is considered the static contact angle. However, depositing the droplet involves adding liquid while leaving it may involve evaporation, so it is difficult to avoid dynamic effects. Fourth, for small droplet and curved triple lines, the effect of the contact line tension may be significant. Molecules at the surface of a liquid or solid phase have higher energy because they are bonded to fewer molecules than those in the bulk. This leads to surface tension and surface energy. In a similar manner, molecules at the edge have fewer bonds than those at the surface, which leads to the line tension and the curvature dependence of the surface energy. This effect becomes important when the radius of curvature is comparable with the so-called Tolman's length, normally of the molecular size [9]. However, the triple line at the nanoscale can be curved so that line tension effects become important [266]. The contact angle that accounts for the contact line effect for a droplet with radius R is given by $\cos \theta = \cos \theta_0 + 2\tau/(R\gamma_{LV})$, where τ is the contact line tension and is the value given by the Young equation [62]. Thus, while the contact angle is a convenient macroscale parameter, wetting is governed by interactions at the micro- and nanoscale, which determine the contact angle hysteresis and other wetting properties (Table 6.1).

Table 6.1. Wetting of a superhydrophobic surface as a multiscale process [245]

Scale level	Characteristic length	Parameters	Phenomena	Interface
Macroscale	Droplet radius (mm)	Contact angle, droplet radius	Contact angle hysteresis	2D
Microscale detail (μm)	Roughness	Shape of the droplet, position of the (h) liquid–vapor interface	Kinetic effects	3D solid surface, 2D liquid surface
Nanoscale	Molecular heterogeneity (nm)	Molecular description	Thermodynamic and dynamic effects	3D

6.3.4 Range of Applicability of the Wenzel and Cassie Equations

Gao and McCarthy [125] showed experimentally that the contact angle of a droplet is defined by the triple line and does not depend upon the roughness under the bulk of the droplet. A similar result for chemically heterogeneous surfaces was obtained by Extrand [109]. Gao and McCarthy [125] concluded that the Wenzel and Cassie–Baxter equations "should be used with the knowledge of their fault." The questions remained, however: Under what circumstances can the Wenzel and Cassie–Baxter equations be safely used and under what circumstances do they become irrelevant?

For a liquid front propagating along a rough two-dimensional profile (Fig. 6.4(a)–(b)), the derivative of the free surface energy (per liquid front length), W, by the profile length, t, yields the surface tension force $\sigma = dW/dt = \gamma_{SL} - \gamma_{SV}$. The quantity of practical interest is the component of the tension force that corresponds to the advancing of the liquid front in the horizontal direction for dx. This component is given by $dW/dx = (dW/dt)(dt/dx) = (\gamma_{SL} - \gamma_{SV})dt/dx$. It is noted that the derivative $R_f = dt/dx$ is equal to Wenzel's roughness factor in the case when the roughness factor is constant throughout the surface. Therefore, the Young equation, which relates the contact angle with solid, liquid, and vapor interface tensions, $\gamma_{LV} \cos\theta = \gamma_{SV} - \gamma_{SL}$, is modified as [234]

$$\gamma_{LV} \cos\theta = R_f(\gamma_{SV} - \gamma_{SL}). \tag{6.16}$$

The empirical Wenzel equation (6.1) is a consequence of (6.7) combined with the Young equation.

Nosonovsky [234] showed that for a more complicated case of a nonuniform roughness, given by the profile $z(x)$, the local value of $r(x) = dt/dx = (1 + (dz/dx)^2)^{1/2}$ matters. In the cases that were studied experimentally by Gao and McCarthy [125] and Extrand [109], the roughness was present ($r > 1$) under the bulk of the droplet, but there was no roughness ($r = 0$) at the triple line, and the contact angle was given by (6.6) (Fig. 6.4(c)). In the general case of a 3D rough surface $z(x, y)$, the roughness factor can be defined as a function of the coordinates $r(x, y) = (1 + (dz/dx)^2 + (dz/dy)^2)^{1/2}$.

Whereas the Wenzel equation (6.6) is valid for uniformly rough surfaces—that is, surfaces with $r =$ const—for nonuniformly rough surfaces the generalized Wenzel

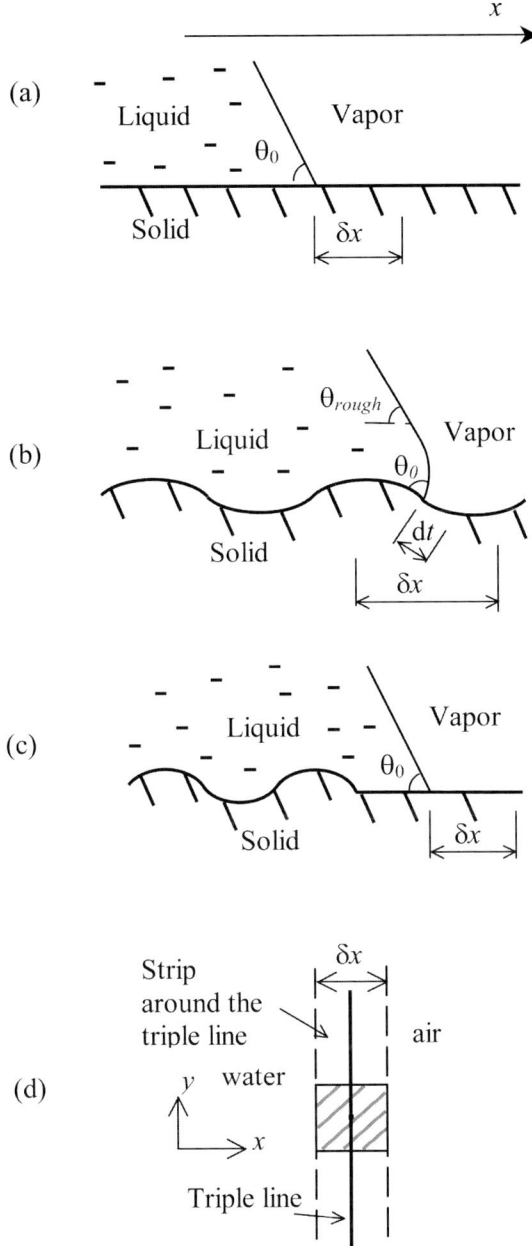

Fig. 6.4. Liquid front in contact with a **a** smooth solid surface, **b** rough solid surface, prop-agation for a distance dt along the curved surface corresponds to the distance dx along the horizontal surface. **c** Surface roughness under the bulk of the droplet does not affect the con-tact angle. **d** Averaged value of $\Theta(x_0, y_0)$ at the triple is obtained by integration along the square area (dashed) with the side δx [250]

Table 6.2. Summary of experimental results for uniform and nonuniform rough and chemically heterogeneous surfaces. For nonuniform surfaces, the results are shown for droplets larger than the islands of nonuniformity. Detailed quantitative values of the contact angle in various sets of experiments can be found in the referred sources [234]

Experiment	Roughness / hydrophobicity at the triple line and at the rest of the surface	Roughness at the bulk (under the droplet)	Experimental contact angle (compared with that at the rest of the surface)	Theoretical contact angle, Wenzel/Cassie equations	Theoretical contact angle, generalized Wenzel–Cassie (6.17)–(6.18)
Gao and McCarthy [125]	Hydrophobic Rough Smooth	Hydrophilic Smooth Rough	Not changed Not changed Not changed	Decreased Decreased Increased	Not changed Not changed Not changed
Extrand [109]	Hydrophilic Hydrophobic	Hydrophobic Hydrophilic	Not changed Not changed	Increased Decreased	Not changed Not changed
Bhushan et al. [53]	Rough	Rough	Increased	Increased	Increased
Barbieri et al. [22]	Rough	Rough	Increased	Increased	Increased

equation is formulated to determine the local contact angle (a function of x and y) with a rough surface at the triple line [234]

$$\cos \theta = r(x, y) \cos \theta_0. \tag{6.17}$$

Equation (6.17) is consistent with the experimental results of the scholars, who showed that roughness beneath the droplet does not affect the contact angle [109, 125], since it predicts that only roughness at the triple line matters. It is consistent also with the results of the researchers who confirmed the Wenzel equation (for the case of the uniform roughness) and of those who reported that only the triple line matters (for nonuniform roughness) (Table 6.2). The main difference between the Wenzel equation (6.6) and the Nosonovsky equation (6.17) is that the latter takes into account only the roughness in the vicinity of the triple line.

The Cassie equation for the composite surface can be generalized in a similar manner introducing the spatial dependence of the local densities, f_1 and f_2, of the solid–liquid interface with the contact angle, as a function of x and y, given by

$$\cos \theta_{\text{composite}} = f_1(x, y) \cos \theta_1 + f_2(x, y) \cos \theta_2, \tag{6.18}$$

where $f_1 + f_2 = 1$, θ_1 and θ_2 are contact angles of the two components [234].

While (6.17)–(6.18) can be used in the case when $r(x, y)$, $f_1(x, y)$, and $f_2(x, y)$ are constant at the triple line, in a more general case an integration along the triple line should be performed.

Another way to view the generalized equations is to consider the transition in (6.8) from two components to a big number of surface components and, in the limit, to the continuously changing local contact angle

$$\cos\theta_{\text{composite}} = \sum f_n \cos\theta_n = \iint_A \cos\theta \, dx \, dy = \oint_T \Theta(x, y) \, dt, \qquad (6.19)$$

where the first integration is performed by the area A of the strip with the thickness δx along the triple line (Fig. 6.4(d)), the path integration is performed along the triple line T, and the locally averaged value of $\cos\theta$ is given by

$$\Theta(x_0, y_0) = \int_{x_0-\delta x/2}^{x_0+\delta x/2} \int_{y_0-\delta x/2}^{y_0+\delta x/2} \cos\theta \, dx \, dy. \qquad (6.20)$$

If the local value of the cosine is given by $\cos\theta = f_1(x_0, y_0)\cos\theta_1 + f_2(x_0, y_0) \cdot \cos\theta_2$, then

$$\Theta(x_0, y_0) = \cos\theta_1 \int_{x_0-\delta x/2}^{x_0+\delta x/2} \int_{y_0-\delta x/2}^{y_0+\delta x/2} f_1(x, y) \, dx \, dy$$

$$+ \cos\theta_2 \int_{x_0-\delta x/2}^{x_0+\delta x/2} \int_{y_0-\delta x/2}^{y_0+\delta x/2} f_2(x, y) \, dx \, dy, \qquad (6.21)$$

and (6.19) yields (6.18). Note that the difference between (6.19) and the conventional Cassie equation (6.8) is that only the area A of the strip next to the triple line is considered and not the entire solid–liquid area. The apparent contact angle given by (6.19) can be observed at a large distance from the solid surface, comparing with the length of T, so that the effect of local heterogeneities smoothens. This is possible, for example, if the roughness or heterogeneity has an axisymmetric distribution or constant along the triple line.

Equations (6.17)–(6.18) are useful only if the average values of r, f_1, and f_2 are constant at the triple line, thus providing a unique value of the apparent contact angle. An example of such a situation is the two-dimensional configuration, which was discussed earlier. Another example is if the roughness/heterogeneity has an axisymmetric distribution with the droplets sitting at the center, as considered by Nosonovsky and Bhushan [250].

The important question remains: What should be the typical size of roughness/heterogeneity details in order for the generalized Wenzel and Cassie equations (6.17)–(6.18) to be valid? Some scholars have suggested that roughness/heterogeneity details should be comparable to the thickness of the liquid–vapor interface and thus "the roughness would have to be of molecular dimensions to alter the equilibrium conditions" [23], whereas others have claimed that roughness/heterogeneity details should be small compared with the linear size of the droplet [22, 46, 171, 204]. The interface in our analysis is an idealized 2D object, which has no thickness. In reality, the triple line zone has two characteristic dimensions: the thickness (of the order of molecular dimensions) and the length (of the order of the droplet size).

The apparent contact angle, given by (6.17)–(6.18), may be viewed as the result of averaging the local contact angle at the triple line by its length, and thus the size of the roughness/heterogeneity details should be small compared to the length (and not the thickness) of the triple line. A rigorous definition of the generalized equation requires the consideration of several scale lengths. The length dx needed for averaging the energy gives the length over which the averaging is performed to obtain $r(x, y)$. This length should be larger than the roughness details. However, it is still smaller than the droplet size and the length scale at which the apparent contact angle is observed (at which local variations of the contact angle level out). Since of these three lengths (the roughness size, dx, the droplet size) the first and the last are of practical importance, we conclude that the roughness details should be smaller than the droplet size. When the liquid–vapor interface is studied at the length scale of roughness/heterogeneity details, the local contact angle, θ_0, is given by the Young equation. The liquid–vapor interface at that scale has perturbations, caused by the roughness/heterogeneity, and the scale of the perturbations is the same as the scale of the roughness/heterogeneity details. However, when the same interface is studied at a larger scale, the effect of the perturbation vanishes, and the apparent contact angle is given by (6.17)–(6.18) (Fig. 6.4(d)). This apparent contact angle is defined at the scale length for which the small perturbations of the liquid–vapor interface vanish, and the interface can be treated as a smooth surface. The values of $r(x, y)$, $f_1(x, y)$, $f_2(x, y)$ in (6.17)–(6.18) are average values by area (x, y) with a size larger than a typical roughness/heterogeneity detail size. Therefore, the generalized Wenzel and Cassie equations can be used at the scale at which the effect of the interface perturbations vanish or, in other words, when the size of the solid surface roughness/heterogeneity details is small compared with the size of the liquid–vapor interface, which is of the same order as the size of the droplet.

We used the surface energy approach to find the validity domain of the Wenzel and Cassie equations (uniformly rough surfaces) and generalized it for a more complicated case of nonuniform surfaces. The generalized equations explain a wide range of existing experimental data which could not be explained by the original Wenzel and Cassie equations.

6.4 Calculation of the Contact Angle for Selected Surfaces

In this section, we calculate the contact angle of a liquid with a number of rough surfaces. The model presented in the preceding sections combines the effect of surface area, possibility of formation of composite interface, and the effect of sharp edges. Several selected rough surfaces are considered, as shown in Fig. 6.5. First, two-dimensional surface profiles are analyzed, followed by more complex three-dimensional surfaces. Based on the analysis, roughness optimization for contact angles was conducted by Nosonovsky and Bhushan [240].

6.4.1 Two-Dimensional Periodic Profiles

6.4.1.1 Sawtooth Periodic Profile

Let us consider a surface with a sawtooth profile with the tooth angle (or the absolute value of slope) of α (Fig. 6.5). Using (6.6), the roughness factor is calculated as

Sawtooth periodic profile

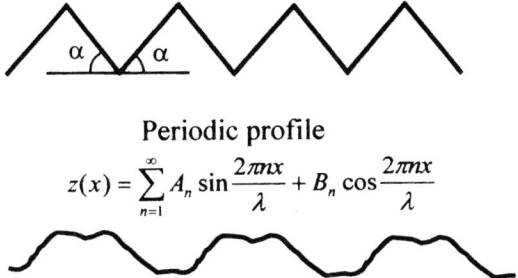

Periodic profile

$$z(x) = \sum_{n=1}^{\infty} A_n \sin\frac{2\pi nx}{\lambda} + B_n \cos\frac{2\pi nx}{\lambda}$$

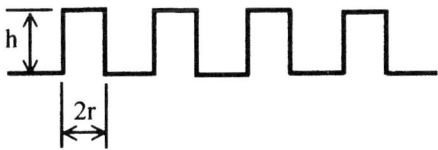

A surface with rectangular asperities

A surface with hemispherically topped
cylindrical asperities

A surface with conical or pyramidal asperities

Random Gaussian surface

$\sigma, \ \beta^{\bullet}$

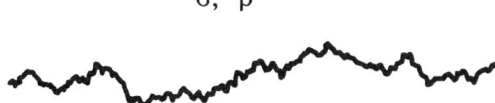

Fig. 6.5. Various rough surfaces [240]

$$R_f = \frac{A_{SL}}{A_F} = (\cos \alpha)^{-1}. \tag{6.22}$$

Using (6.6), the contact angle is given as

$$\cos \theta = \frac{\cos \theta_0}{\cos \alpha}. \tag{6.23}$$

An increase of α above $\alpha_0 = 180 - \theta_0$ will result in a transition from a complete solid–liquid contact to a composite solid–liquid–air interface, and (6.23) cannot be used any further. Substituting the value of slope $\alpha = \alpha_0$ into (6.23), the value of θ, which corresponds to α_0, can be obtained, which gives that the critical value α_0 corresponds to the contact angle $\theta = 180°$. This means that by increasing the tooth angle toward the critical value, a surface with the contact angle approaching $180°$ can be produced for a given θ_0. However, sharp edges, which may lead to pinning of the triple line, make the sawtooth profile undesirable. In addition to this, the sawtooth profile provides roughness only in the direction perpendicular to the grooves, which may act as open capillaries to reinforce wetting, which is also undesirable [240].

6.4.1.2 General Periodic Profile

For a general form of the surface $z(x, y)$, the solid–liquid area of contact is equal to

$$A_{SL} = A_F \iint_{A_F} \sqrt{1 + \left(\frac{\partial z}{\partial x}\right)^2 + \left(\frac{\partial z}{\partial y}\right)^2} \, dx \, dy. \tag{6.24}$$

A periodic two-dimensional surface profile with the period λ can be presented as a Fourier series

$$z(x) = \sum_{n=1}^{\infty} A_n \sin \frac{2\pi n x}{\lambda} + B_n \cos \frac{2\pi n x}{\lambda}. \tag{6.25}$$

The derivatives of $z(x)$ are given as

$$\frac{dz}{dx} = \frac{2\pi}{\lambda} \sum_{n=1}^{\infty} A_n n \cos \frac{2\pi n x}{\lambda} - B_n n \sin \frac{2\pi n x}{\lambda},$$

$$\frac{dz}{dy} = 0. \tag{6.26}$$

Substituting (6.24) and (6.26) into (6.7) provides an expression for the roughness factor of a periodic profile [240]

$$R_f = \frac{1}{\lambda} \int_0^\lambda \sqrt{1 + \frac{4\pi^2}{\lambda^2} \left(\sum_{n=1}^{\infty} A_n n \cos \frac{2\pi n x}{\lambda} - B_n n \sin \frac{2\pi n x}{\lambda}\right)^2} \, dx. \tag{6.27}$$

It is possible to determine whether the composite interface is possible by considering the slope of the profile. In order for the composite interface to form, the absolute value of the slope must exceed the critical angle α_0 at any point

$$\left| \frac{2\pi}{\lambda} \sum_{n=1}^{\infty} A_n n \cos \frac{2\pi n x}{\lambda} - B_n n \sin \frac{2\pi n x}{\lambda} \right| > \tan(\alpha_0) = \tan(-\theta_0). \tag{6.28}$$

As an example, let us consider a sinusoidal profile

$$z(x) = A_1 \sin \frac{2\pi x}{\lambda}. \tag{6.29}$$

By substituting (6.29) into (6.27) and integrating, a closed-form solution can be obtained [240]

$$
\begin{aligned}
R_f &= \frac{1}{\lambda} \int_0^\lambda \sqrt{1 + (2\pi A_1/\lambda)^2 \cos^2(2\pi x/\lambda)} \, dx \\
&= \frac{1}{2\pi} \int_0^{2\pi} \sqrt{1 + (2\pi A_1/\lambda)^2 \cos^2 x} \, dx \\
&= \frac{2}{\pi} \sqrt{1 + (2\pi A_1/\lambda)^2} \int_0^{\pi/2} \sqrt{1 - \frac{(2\pi A_1/\lambda)^2}{1 + (2\pi A_1/\lambda)^2} \sin^2 x} \, dx \\
&= \frac{2}{\pi} \sqrt{1 + (2\pi A_1/\lambda)^2} E\left(\frac{(2\pi A_1/\lambda)}{\sqrt{1 + (2\pi A_1/\lambda)^2}} \right),
\end{aligned}
\tag{6.30}
$$

where $E(x)$ is the so-called elliptical integral of the second kind, the values of which are tabulated in the handbooks

$$E(k) = \int_0^{\pi/2} \sqrt{1 - k^2 \sin^2 x} \, dx. \tag{6.31}$$

The maximum absolute value of the slope of the sinusoidal profile (6.29) is achieved at $x = 0$ and equal to $2\pi A_1/\lambda$. With an increase of A_1/λ, the slope increases, and a composite interface may be formed. For the composite interface to form, the slope at some points should exceed the critical value α_0. By using (6.28) and setting $x = 0$ and using (6.25), the condition for existence of the composite interface is found as

$$\frac{2\pi A_1}{\lambda} > \tan(-\theta_0). \tag{6.32}$$

The contact angle can be calculated by substituting R_f from (6.30) into (6.6). The dependence of the contact angle on amplitude for the sinusoidal profile is presented in Fig. 6.6(a). It is observed that lower values of θ correspond to lower values of θ_0 at the transition to the composite interface, unlike in the case of the sawtooth surface, which has critical values corresponding to $\theta_0 = 180°$. For $\theta_0 = 100°$ the critical value of $R_f = 5.67$ ($\theta = 131°$), for $\theta_0 = 120°$ the critical value of $R_f = 1.73$ ($\theta = 140°$), for $\theta_0 = 150°$ the critical value of $R_f = 0.58$ ($\theta = 159°$). Further increase of A_1/λ may lead to a corresponding increase of R_f and θ. However, we will discussed in the following, the composite interface can be destabilized. Therefore, the sinusoidal interface is not recommended for producing superhydrophobic surfaces. In addition to this, the sinusoidal profile provides roughness only in the direction perpendicular to the grooves, which may act as open capillaries to reinforce wetting, which is also undesirable [240].

6.4.2 Three-Dimensional Surfaces

The analysis of profiles provides critical values of the roughness parameters when the contact line is parallel to the grooves. Three-dimensional surfaces, which constitute a more general case, with various typical shapes of asperities are considered in this section.

6.4.2.1 Array of Asperities of Identical Shape and Size

Let us consider a rough surface with rectangular asperities which have a square foundation with side $2r$ and height h (Fig. 6.5). For each asperity, the area of surface is given by

$$A_{\text{asp}} = 8rh + 4r^2, \tag{6.33}$$

whereas the flat projection area is $4r^2$. Assuming the asperities are randomly distributed throughout the surface with the density of η asperities per unit area, the total contact surface area is given by

$$A_{\text{SL}} = A_{\text{F}} + A_{\text{F}}\eta\left(8rh + 4r^2\right) - A_{\text{F}}4\eta r^2 = A_{\text{F}}\left(1 + 8\eta r^2\right). \tag{6.34}$$

The roughness factor is found using (6.7) and (6.34)

$$R_{\text{f}} = 1 + 8\eta rh = 1 + 2p^2 h/r, \tag{6.35}$$

where p is a packing parameter, which characterizes asperities packing, for asperities with a square foundation, $p = 2r\sqrt{\eta}$. The packing parameter is equal to the fraction of the surface area which is covered by asperities.

In a similar manner, R_{f} can be calculated for asperities with cylindrical foundation of height h and hemispherical top of radius r (Fig. 6.5). For each asperity, the area of surface is given by

$$A_{\text{asp}} = 2\pi r^2(1 + h/r), \tag{6.36}$$

whereas the flat projection area is given by πr^2. Assuming that asperities are randomly distributed throughout the surface with the density of η asperities per unit area, the total contact surface area is given by

$$A_{\text{SL}} = A_{\text{F}} + A_{\text{F}}\eta 2\pi r^2(1 + h/r) - A_{\text{F}}\eta\pi r^2 = A_F\left[1 + \eta\pi r^2(1 + 2h/r)\right]. \tag{6.37}$$

The roughness factor is found using (6.7) and (6.37)

$$R_{\text{f}} = 1 + \eta\pi r^2(1 + 2h/r) = 1 + p^2(1 + 2h/r), \tag{6.38}$$

where the packing parameter for asperities with a circular foundation is $p = r\sqrt{\pi\eta}$ [240].

For conical asperities of height h, radius r, and side length $L = \sqrt{h^2 + r^2}$, it can be obtained in a similar manner

$$A_{\text{asp}} = \pi r^2 (1 + L/r), \tag{6.39}$$

$$A_{\text{SL}} = A_{\text{F}} + A_{\text{F}} \eta \pi r^2 (1 + L/r) - A_{\text{F}} \eta \pi r^2 = A_{\text{F}} (1 + \eta \pi r L)$$

$$= A_{\text{F}} \left(1 + \eta \pi r^2 \sqrt{1 + (h/r)^2} \right) \tag{6.40}$$

and

$$R_{\text{f}} = 1 + \eta \pi r L = 1 + \eta \pi r^2 \sqrt{1 + (h/r)^2} = 1 + p^2 \sqrt{1 + (h/r)^2}, \tag{6.41}$$

where the packing parameter for asperities with a circular foundation is $p = r\sqrt{\pi\eta}$ [240].

For pyramidal asperities with a square foundation of width $2a$ and height h, the corresponding quantities are given as

$$A_{\text{asp}} = 4r^2 \left(1 + \sqrt{1 + (h/r)^2} \right), \tag{6.42}$$

$$A_{\text{SL}} = A_{\text{F}} + 4A_{F}\eta r^2 \left(1 + \sqrt{1 + (h/r)^2} \right) - 4A_{\text{F}}\eta r^2$$

$$= A_{\text{F}} \left(1 + 4\eta r^2 \sqrt{1 + (h/r)^2} \right) \tag{6.43}$$

and

$$R_{\text{f}} = 1 + 4\eta r^2 \sqrt{1 + (h/r)^2} = 1 + p^2 \sqrt{1 + (h/r)^2}, \tag{6.44}$$

where the packing parameter for asperities with a square foundation is $p = 2r\sqrt{\eta}$ [240].

The dependence of the contact angle on the normalized radius of the asperities (taken as p) for $\theta_0 = 120°$ and for different ratios of h/r is presented in Fig. 6.6(a), on the basis of (6.6), (6.35), (6.38), (6.42), and (6.44), for the rectangular, hemispherically topped, conical and pyramidal asperities. It is observed that with an increase of p the value of the contact angle increases and reaches $180°$. For higher aspect ratios, the increase of θ is faster.

In order to determine the critical values of roughness parameters which correspond to the transition of the homogeneous interface to the composite interface, it should be analyzed whether the local slope can exceed the critical value α_0 and whether the composite interface is likely to remain stable. It is difficult to conduct such an analysis due to its complexity; however, an estimate can be made using the fact that with increasing average absolute value of the slope of the surface, both the local slope increases and the destabilization of the composite interface becomes less likely, since the surface is less smooth. Based on this, we assume here that, in a manner similar to the two-dimensional profiles, the absolute value of the surface slope is responsible for transition to the composite liquid–solid–air interface, and consider an average absolute value of the slope. For rectangular, hemispherically topped, conical, and pyramidal asperities, the mean absolute value of the slope, m, is equal to the density of the asperities and the flat projection area times the average absolute value of the slope (equal to twice the aspect ratio),

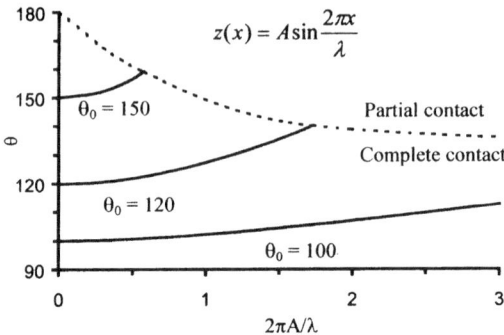

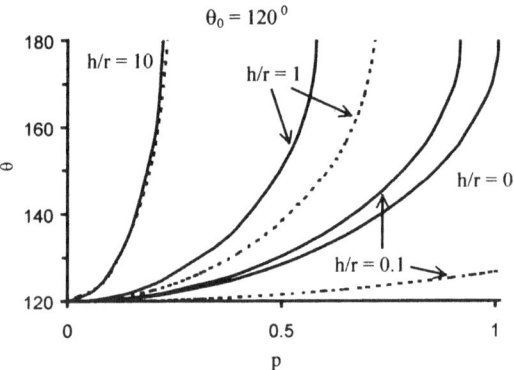

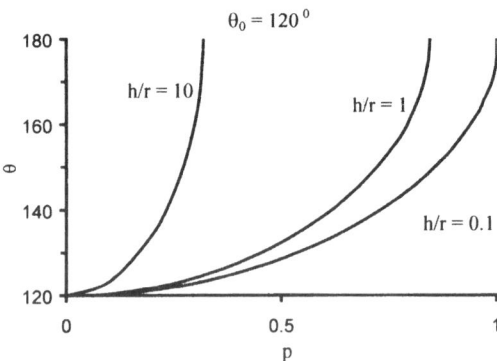

Fig. 6.6. Contact angle for **a** rough surface (θ) as a function of surface parameters for the surface with sinusoidal profile, rectangular (dotted line)/hemispherically-topped cylindrical (solid line) and conical/pyramidal asperities dependence of the roughness factor (R_f), and **b** contact angle (θ) as a function of roughness parameters for a Gaussian surface [240]

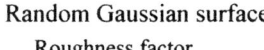

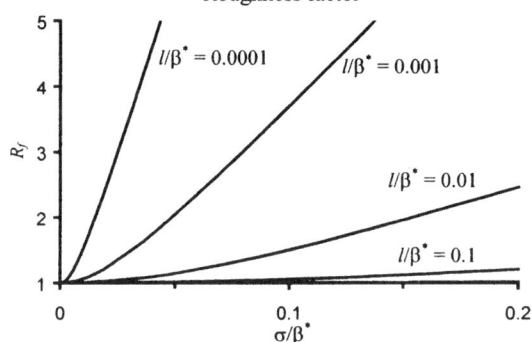

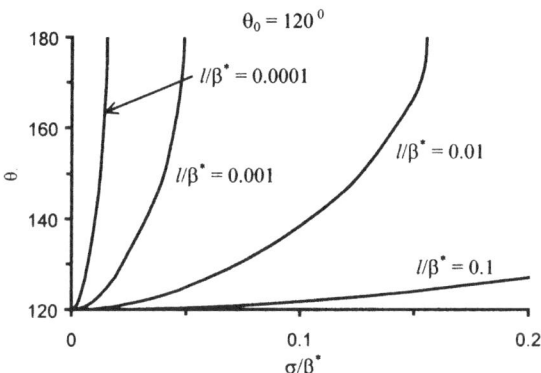

Fig. 6.6. (*Continued*)

$$m = \eta \pi r^2 (h/r) = \eta \pi hr. \qquad (6.45)$$

The critical value can be found using a similar approach as in the derivation of (6.28)

$$m_0 = \eta \pi hr = \tan(180 - \theta_0) = \tan(-\theta_0). \qquad (6.46)$$

Based on (6.35), (6.38), (6.41), (6.44), and (6.46), it may be shown that for the selected value of $\theta_0 = 120°$, for rectangular, hemispherically topped, conical, and pyramidal asperities, the contact angle may approach 180° before the critical value of roughness is reached for the values of h/r shown in Fig. 6.6(a) [240].

The equations developed here are used to calculate the contact angle for a lotus leaf and compare it with measured data. The lotus leaf has almost hemispherically topped asperities (papillae), which are covered with wax [227, 330]. The static contact angle for a water droplet against a wax surface was reported by Craig et al. [86] as 104° and by Kamusewitz et al. [175] as 103°. Based on the data reported by Wagner et al. [330], the number of asperities (papillae) can be estimated for the lotus leaf as 3400 per mm^2 ($\eta = 0.0034\ \mu m^{-2}$), average radius of hemispherically topped asperities $r = 10\ \mu m$, and aspect ratio $h/r \sim 1$. Based on (6.7), these values correspond to the roughness factor $R_f \sim 4$ and the contact angle $\theta = 165°$ (using $\theta_0 = 104°$ for wax). During the measurements conducted by Burton and Bhushan [68], the value of the static contact angle for deionized water on a lotus leaf was found to be $156° \pm 2°$. Neinhuis and Barthlott [227] reported the contact angle value of $162°$ for a water droplet on the lotus leaf.

6.4.2.2 Random Rough Surface

A nominally flat random rough surface can be considered as a superposition of a flat plane and a two-dimensional random process, which is characterized by a height distribution and an autocorrelation function. Many engineered and natural rough surfaces can be characterized by a Gaussian height distribution and exponential autocorrelation function [30, 32]. In this case, a rough surface is described by only two parameters: the standard deviation of asperity heights, σ, and correlation length, β^*. The correlation length β^* is a spatial parameter, and it can be viewed as a measure of randomness. It is responsible for the horizontal scale of the surface, whereas σ is responsible for the vertical scale of the surface. Measured roughness is dependent on the short- and long-wavelength limit of measurement [30, 32].

The absolute value of slope of a Gaussian surface also has a Gaussian distribution with the mean

$$m = \frac{\sigma}{L}\sqrt{\frac{1 - [\exp(-\beta^*/L)]^2}{\pi}}, \tag{6.47}$$

where L is the sampling interval or short-wavelength limit, which is a distance between data points during a measurement [338]. For a surface, the sampling interval is given by a low-wavelength limit of the Gaussian roughness, and is comparable with the atomic dimensions [240].

An element of the area of a surface with slopes of $\partial z/\partial x$ and $\partial z/\partial y$ in x- and y-directions is given by

$$dA = \sqrt{1 + (\partial z/\partial x)^2 + (\partial z/\partial y)^2}\, dx\, dy. \tag{6.48}$$

The distribution of $\sqrt{1 + (\partial z/\partial x)^2 + (\partial z/\partial y)^2}$ is not Gaussian in general, but in most applications the slope is small and the mean value of slope m can be taken to calculate the mean value of $\sqrt{1 + (\partial z/\partial x)^2 + (\partial z/\partial y)^2}$. It can also be assumed that slopes in x- and y-directions are the same. Using (6.25) and substituting m, given by (6.47) for the slope into (6.48) and integrating, the roughness factor can be calculated [240] as

$$R_f = \frac{1}{A_{SL}} \iint_{A_f} \sqrt{1 + (\partial z/\partial x)^2 + (\partial z/\partial y)^2}\, dx\, dy = \sqrt{1 + 2m^2}$$

$$= \sqrt{1 + 2\frac{\sigma^2}{l^2}\frac{1 - \exp(-l/\beta^*)^2}{\pi}}. \tag{6.49}$$

For small l/β^*

$$R_f \approx \sqrt{1 + 2\left(\frac{\sigma}{\beta^*}\right)^2 \frac{2}{\pi(l/\beta^*)}}.$$ (6.50)

Furthermore, for small values of $\sigma^2/(l\beta^*)$

$$R_f \approx 1 + \frac{2}{\pi}\left(\frac{\sigma}{\beta^*}\right)^2\left(\frac{\beta^*}{l}\right).$$ (6.51)

In order to estimate the critical value of roughness parameters, we assume, as in the previous section, that average absolute value of the surface slope is responsible for transition to the composite solid–liquid–air interface. The absolute value of slope is given by (6.47) so, in a manner similar to the derivation of (6.46), the critical values of the Gaussian surface roughness parameters are

$$m_0 = \left(\frac{\sigma}{l}\sqrt{\frac{1 - [\exp(-\beta^*/l)]^2}{\pi}}\right)_0 = \tan(-\theta_0).$$ (6.52)

The dependence of the roughness factor on σ/β^* is presented in Fig. 6.6(b) based on (6.49). Using the roughness factor, the dependence of the contact angle on σ/β^* is presented in Fig. 6.6(b). It is observed that both the roughness factor and the contact angle increase with increasing σ/β^*. Based on (6.52), it may be shown that for the selected value of $\theta_0 = 120°$, the contact angle may approach $180°$ before the critical values of roughness parameters are reached. It is noted that for most natural and engineered Gaussian surfaces, the ratio $\sigma/\beta^* \ll 0.1$, and the average value of slope is small ($m \ll 1$). Therefore, although the roughness is below the critical value, it is difficult to achieve high contact angles with Gaussian random surfaces with a realistic value of σ/β^* [240].

6.4.3 Surface Optimization for Maximum Contact Angle

Among the several types of surfaces considered in the preceding sections, the highest contact angles are achieved with the sawtooth profile and rectangular/hemispherically topped/conical/pyramidal asperities. As it was stated earlier, the sawtooth profile is undesirable due to its sharp edges, which may pin the triple line, and because the grooves may reinforce wetting. Therefore, the rectangular, hemispherically topped, conical, and pyramidal asperities should be considered as the most appropriate for producing the highest contact angles. In order to reduce contact angle hysteresis, it is desirable to avoid asperities with sharp edges which may cause pinning of the triple line. Therefore, hemispherically topped asperities are the most appropriate. A case also will be made later for pyramidal asperities.

Two-tiered roughness involving two wavelengths has been considered by some authors (e.g. [151]) to decrease wetting. However, it is more likely to involve sharp edges, which are undesirable, and lead to unstable composite solid–liquid–air interface.

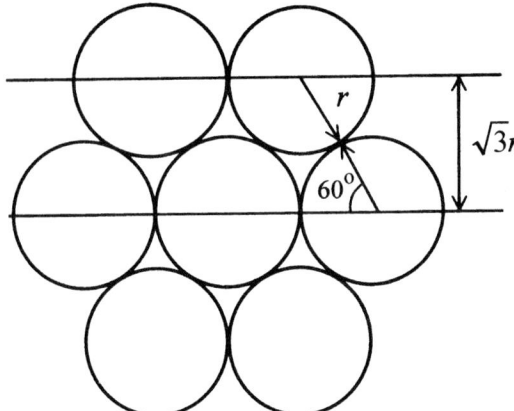

Fig. 6.7. Hexagonal (honeycomb) pattern of packing of circular asperities for highest packing density [240]

Based on (6.28), (6.31), and (6.44), and on the results shown in Fig. 6.6(a), the maximum contact angle can be achieved by increasing the aspect ratio h/r and the packing parameter p. The maximum aspect ratio may be achieved by increasing asperity height. The maximum packing parameter may be achieved by packing the asperities as tight as possible. The square of the packing parameter p^2 is equal to the ratio of the foundation area of the asperities to the total surface area, therefore, the higher value of p corresponds to higher packing density. For asperities with a circular foundation, the square pattern of asperities distribution results in packing of $1/(2r)$ rows per unit area with $1/(2r)$ asperities per unit length in the row. The higher density of asperities packing can be achieved by the hexagonal (honeycomb) distribution of asperities (Fig. 6.7). This distribution pattern results in packing of $1/(\sqrt{3}r)$ rows of asperities per unit length with $1/(2r)$ asperities per unit length in the row, or $\eta = 1/(2\sqrt{3}r^2)$, which yields

$$p = r\sqrt{\pi\eta} = \sqrt{\frac{\pi}{2\sqrt{3}}} \approx 0.952. \qquad (6.53)$$

Therefore, the recommendation for surface optimization is to take hexagonally packed hemispherically topped asperities with a high aspect ratio (needle-like). It is noted that certain leaves tend to have the distribution of the papillae close to the hexagonal [240].

An alternative shape, which provides packing density $p = 1$, is given by pyramidal asperities with a square foundation. In order to avoid pinning due to sharp edges, the tops may be rounded with the hemispheres. Rectangular asperities do not provide space for liquid to penetrate, therefore, in the case of asperities with square foundation, the pyramidal shape should be used. It should be noted that valleys with rounded edges have the same effect on contact angle as asperities do [240].

The foundation radius of individual asperities, r (for circular foundation) or foundation side length $2r$ (for square foundation) should be small as compared to typical droplets. The upper limit of droplet size may be estimated based on the requirement that the gravity effect is small compared to the surface tension (a bigger droplet is likely to be divided into several small droplets). The gravitational energy of the droplet is given by its density ρ, multiplied by the volume, gravitational constant $g = 9.81$ m/s^2, and radius,

$$W_g = \frac{4}{3}\pi r^3 \rho g r,\tag{6.54}$$

whereas the energy due to the surface tension can be estimated by droplet surface area multiplied by the surface tension

$$W_g = 4\pi r^2 \gamma_{LA}.\tag{6.55}$$

Based on $W_g \ll W_s$, we find the maximum droplet radius is smaller than the capillary length

$$r_{max} \ll \sqrt{\frac{3\gamma_{LA}}{\rho g}}.\tag{6.56}$$

Typical quantities for water, $\rho = 1000$ kg/m^3 and $\gamma_{LA} = 72$ mJ/m^2 result in $r_{max} \ll 4.7$ mm. Although the small droplets will tend to unite into bigger ones, the minimum droplet radius is limited only by molecular scale, so it is desirable to have r as small as possible.

To summarize, the highest possible contact angle and lowest contact angle hysteresis, which is desirable in applications, may be achieved by using hemispherically topped asperities with hexagonal packing pattern or by pyramidal asperities with a rounded top. These recommendations can be used for producing superhydrophobic surfaces [240].

For wetting liquids, roughness results in a decreased contact angle in accordance with (6.6). Therefore, in order to create a superhydrophobic surface using the effect of roughness, a hydrophobic film is required. Hydrophobic coating is a well-known method of increasing water-repellency of a material [285].

6.5 Contact Angle Hysteresis

6.5.1 Origin of the Contact Angle Hysteresis

If liquid is added to a sessile droplet, the contact line advances and each time motion ceases, the drop exhibits an advancing contact angle. Alternatively, if liquid is removed from the drop, the contact angle decreases to a receding value before the contact retreats (Fig. 6.8(a)). For a droplet moving along the solid surface (e.g., if the surface is tilted) there is another definition (Fig. 6.8(b)). The contact angle at the front of the droplet (advancing contact angle) is greater than that at the back of the droplet (receding contact angle) due to roughness, resulting in contact angle hysteresis (Fig. 6.1(b)). It is arguable whether the two definitions (for added/removed liquid and for a moving droplet at a tilted surface) are equivalent [192]. However, in many cases the two definitions have the same meaning [243].

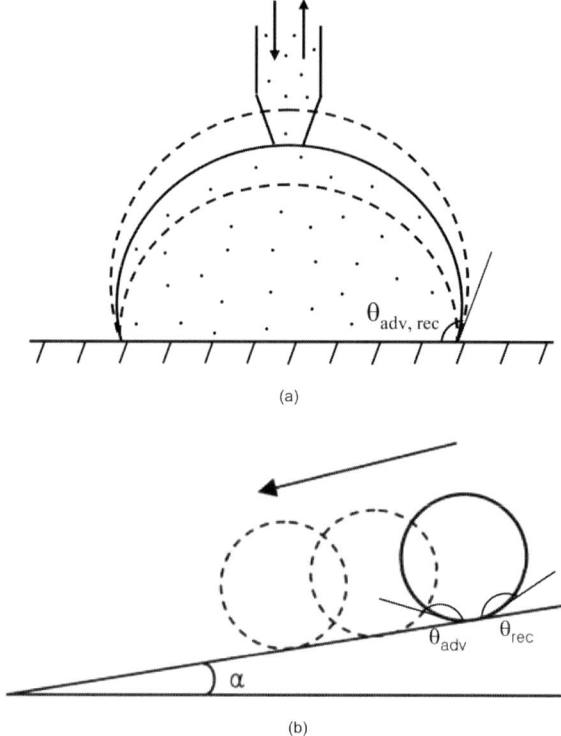

(a)

(b)

Fig. 6.8. a Liquid droplet in contact with rough surface (advancing and receding contact angles are θ_{adv} and θ_{rec}, respectively), and **b** tilted surface profile (the tilt angle is α) with a liquid droplet

The contact angle hysteresis is the measure of energy dissipation during the flow of a droplet. The exact reasons for contact angle hysteresis are not known, although it is clear that they are related to surface heterogeneity and roughness at various scales [167, 171, 243]. The dissipation that leads to contact angle hysteresis may occur either (1) in the bulk volume of the droplet, (2) at the solid–liquid contact area, or (3) near the solid–liquid–air contact line (the triple line). While the bulk interactions involving liquid viscosity are eliminated in the quasi-static case (very low flow velocity), both the contact area interactions (due to so-called adhesion hysteresis) and triple line interactions (due to pinning of the triple line by roughness details) remain significant even at the limit of zero flowing velocity. We will show in Sect. 8.2 that both the contact area interactions and the triple line interactions are equally important for contact angle hysteresis analysis.

Nosonovsky and Bhushan [251] suggested that contact angle hysteresis, as well as the wetting regime transition, involve self-organized criticality. Adding liquid to a droplet is similar to adding grains to a sandpile. The triple line tends to be pinned at the "critical locations," such as the edges of the microstructures, which serve as attractors. After that, the triple line suddenly advances to its new location.

6.5.2 Pinning of the Triple Line

A sharp edge can pin the line of contact of the solid, liquid, and air (also known as the "triple line") at a position far from stable equilibrium, i.e., at contact angles different from θ_0 [107]. This effect is illustrated in the bottom sketch of Fig. 6.9,

Smooth surface

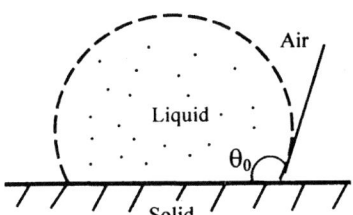

Effect of roughness

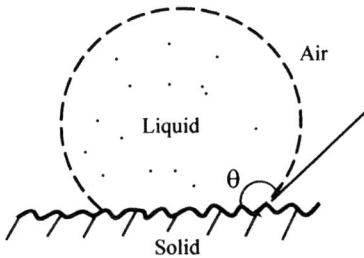

Effect of sharp edges

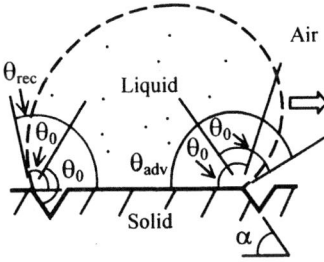

Fig. 6.9. Droplet of liquid in contact with a solid surface–smooth surface, contact angle θ_0; rough surface, contact angle θ; and a surface with sharp edges. For a droplet moving from left to right on a sharp edge (shown by arrow), the contact angle at a sharp edge may be any value between the contact angle with the horizontal plane and with the inclined plane. This effect results in difference of advancing ($\theta_{adv} = \theta_0 + \alpha$) and receding ($\theta_{rec} = \theta_0 - \alpha$) contact angles [240]

which shows a droplet propagating along a solid surface with grooves. At the edge point, the contact angle is not defined and can have any value between the values corresponding to the contact with the horizontal and inclined surfaces. For a droplet moving from left to right, the triple line will be pinned at the edge point until it will be able to proceed to the inclined plane. As observed from Fig. 6.9, the change of the surface slope (α) at the edge causes the pinning. Because of the pinning, the value of the contact angle at the front of the droplet (dynamic maximum advancing contact angle or $\theta_{adv} = \theta_0 + \alpha$) is greater than θ_0, whereas the value of the contact angle at the back of the droplet (dynamic minimum receding contact angle or $\theta_{rec} = \theta_0 - \alpha$) is smaller than θ_0, This phenomenon is known as contact angle hysteresis [107, 161, 171]. A hysteresis domain of the dynamic contact angle is thus defined by the difference $\theta_{adv} - \theta_{rec}$. The liquid can travel easily along the surface if the contact angle hysteresis is small. It is noted that the static contact angle lies within the hysteresis domain. Therefore, increasing the static contact angle up to the values of a superhydrophobic surface (approaching 180°) will also result in reduction of the contact angle hysteresis. In a similar manner, the contact angle hysteresis can also exist even if the surface slope changes smoothly, without sharp edges.

6.5.3 Contact Angle Hysteresis and the Adhesion Hysteresis

The contact angle hysteresis is related to the more general phenomenon known as adhesion hysteresis, which is also observed during solid-solid contact [76, 349]. When two solid surfaces come in contact, the energy required to separate them is always greater than the energy gained by bringing them together, and thus the loading-unloading cycle is a thermodynamically irreversible dissipative process [76, 211, 284, 349, 351]. It was argued that for adhesive dry friction, the frictional shear stress is related to the adhesion hysteresis, rather than the adhesion per se. However, currently there is no quantitative theory relating the adhesion hysteresis with friction in a manner consistent with the experimental data.

When liquid comes in contact with a solid, the solid–liquid interface is created while solid–vapor and liquid–vapor interfaces are destroyed. The work of adhesion between the liquid and the solid per unit area is given by the Dupré equation

$$W = \gamma_{SV} + \gamma_{LV} - \gamma_{SL} = \gamma_{LV}(1 + \cos\theta). \tag{6.57}$$

As stated earlier, the energy gained for surfaces coming to contact is greater than the energy required for their separation (or the work of adhesion) by the quantity ΔW, which constitutes the adhesion hysteresis. For a smooth surface, the difference between the two values of the interface energy (measured during loading and unloading) is given by ΔW_0. These two values are related to the advancing contact angle, θ_{adv0}, and receding contact angle, θ_{rec0}, of the smooth surface, assuming that for a smooth surface, the adhesion hysteresis is the main contributor into the contact angle hysteresis

$$\cos\theta_{adv0} - \cos\theta_{rec0} = \frac{\Delta W_0}{\gamma_{LV}}. \tag{6.58}$$

For a composite interface with a micropatterned surface built of flat-top columns ($R_f = 1$), the fraction of the solid–liquid area is given by f_{SL}, so the adhesion hysteresis of a rough surface, ΔW, is related to that of a smooth surface, ΔW_0, as $\Delta W = f_{SL}\Delta W_0$, while the term that includes the surface roughness effect, H_r, should be added. The contact angle hysteresis is then given by

$$\cos\theta_{adv} - \cos\theta_{rec} = \frac{\Delta W}{\gamma_{LV}} + H_r = \frac{f_{SL}\Delta W_0}{\gamma_{LV}} + H_r$$
$$= f_{SL}(\cos\theta_{adv0} - \cos\theta_{rec0}) + H_r, \qquad (6.59)$$

where θ_{adv} and θ_{rec} are the advancing and receding contact angles for a rough surface [53, 233]. It is assumed that for a rough surface, the contact angle hysteresis involves two terms, ΔW corresponding to the adhesion hysteresis, and the roughness parameter H_r, corresponding to the surface roughness. It is observed from (6.59) that small values of f_{SL} provide both a high contact angle and low contact angle hysteresis. Thus the effect of adhesion hysteresis is due to the change of the solid liquid area, f_{SL}. The first term in the right-hand part of (6.59), which corresponds to the inherent contact angle hysteresis of a smooth surface, is proportional to the fraction of the solid–liquid contact area, f_{SL}. The second term, H_r, may be assumed to be proportional to the length density of the pillar edges or, in other words, to the length density of the triple line [53]. Thus (6.59) involves both the term proportional to the solid–liquid interface area and to the triple line length.

Now let us consider the term corresponding to the roughness, H_r. During motion, the droplet passes from one metastable state to another, and these states are separated by energy barriers. For an exact theoretical calculation of the contact angle hysteresis, a thermodynamic analysis of energy barriers for a moving droplet would be required, which is a complicated problem in three-dimensional geometry. For many practical applications, microfabricated patterned surfaces with small three-dimensional pillars uniformly distributed along the surface are especially important. The main contribution of roughness is expected to be from the sharp edges of the pillars, which may pin a moving droplet. Therefore, the surface roughness term is assumed to be proportional to the density of the edges, which is equal to the perimeter of a pillar, πD, times number of pillars per unit area, $1/P^2$

$$H_r \propto \frac{D}{P^2}, \qquad (6.60)$$

where D is the diameter of the pillars, and P is the pitch distance between them. It is convenient to introduce a nondimensional parameter proportionality constant c, and thus (6.60) is written as

$$H_r = cS_f^2 = (cD)\frac{D}{P^2}, \qquad (6.61)$$

where $S_f = D/P$ is the nondimensional spacing factor [243, 244].

In the preceding chapters we studied solid-solid friction and the effect of adhesion hysteresis upon it. The solid-solid friction involves two major mechanisms: the adhesion (including adhesion hysteresis) and roughness-dependent mechanisms,

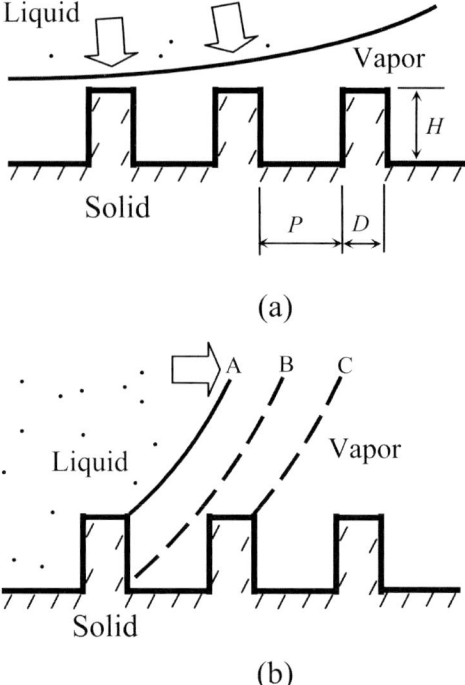

Fig. 6.10. Schematics showing the two modes of the solid–liquid friction ("rolling" and "sliding"). **a** A droplet rolls upon a patterned surface and the contact area at the top of the pillars is the dominant mechanism. **b** A droplet propagates steadily and the interactions at the triple line dominate in the dissipation [233]

such as the deformation. In a somewhat similar manner, the solid–liquid dissipation, characterized by the contact angle hysteresis, involves two mechanisms: the adhesion hysteresis and pinning of the triple line. While the first mechanism is "two-dimensional" in the sense that the dissipation is proportional to the contact area, the second is "one-dimensional" in the sense that the dissipation is proportional to the triple line length. One mechanism involves the "rolling" of a droplet while the other one involves its "sliding" (Fig. 6.10).

6.6 Summary

In this section we presented the theory of roughness-induced superhydrophobicity. The Wenzel and Cassie–Baxter equations provide the contact angle with rough and heterogeneous surfaces. We discussed the range of applicability of these equations. We also discussed contact angle hysteresis and found that it is governed by two factors: the adhesion hysteresis which is inherently present at any surface due to the nanoscale roughness or heterogeneity, and the kinetic effects related to pinning of the triple line. Two wetting regimes are possible: the homogeneous (Wenzel) regime and the composite (Cassie) regime with air bubbles trapped between the solid and liquid. For practical applications, the composite regime is required because it results in a low contact angle hysteresis and, therefore, low rates of dissipation and low adhesion, as well as in high contact angle. The transition between the Cassie and Wenzel regimes for micropatterned superhydrophobic surfaces was discussed. While the exact micro- and nanoscale mechanism of this transition is still not clear, the experimental data suggest that simple microscale geometrical parameters control this transition.

Wetting of a micropatterned surface is complicated, and involves processes at several scale levels. While the macroscale parameters, such as the contact angle and contact angle hysteresis, may be determined approximately by macroscale equations, such as (6.5)–(6.10), these equations do not provide a complete description of the macroscale behavior of the system. The contact angle hysteresis is dependent upon micro- and nanoscale effects that control energy dissipation due to the adhesion, kinetic effects, and the fine structure of the triple line. As it will be shown in the following chapter, the Cassie–Wenzel wetting state transition is also governed by these micro- and nanoscale effects as well as by dynamic effects such as the capillary waves. Furthermore, the very concept of the contact angle is relevant only at the macro- and, to some extent, at the microscale, while the lower scale is dominated by effects such as layer and precursor formation, disjoining pressure, surface heterogeneity, contact line tension, and a finite thickness of the liquid–vapor interface. Therefore, despite its apparent simplicity, a droplet upon a rough surface constitutes a multiscale system. In order to control wetting, it is necessary to control parameters at different scale levels. It is not surprising that biological superhydrophobic surfaces have roughness at different scale lengths.

7

Stability of the Composite Interface, Roughness Optimization and Meniscus Force

Abstract Stability of the composite (Cassie) interface is one of the crucial issues for a successful creation of superhydrophobic surfaces. Destabilizing factors, such as surface waves and capillarity are discussed and a probabilistic model is presented for 2D and 3D roughness. Various related issues, such as the metastability of the Cassie state, similarity of the bubbles and droplets, effect of surface roughness on the capillary adhesion force, roughness optimization and hierarchical roughness are also discussed.

In the preceding chapter we discussed roughness-induced hydrophobicity in general. In this chapter we will study the stability of the composite interface, which is required for low contact angle hysteresis. We will discuss the possibility of partial destabilization and consider two-dimensional and three-dimensional models that imply such destabilization. We will also discuss the effect of roughness upon the meniscus adhesion force.

7.1 Destabilization of the Composite Interface

The composite interface is preferred for superhydrophobicity; however, composite interface can be fragile, since several effects, such as the surface capillary waves (Fig. 7.1(a)) or condensation of small droplets (Fig. 7.1(b)), can destroy it. Therefore, it is important to study stability of the composite interface. An interface can satisfy the equilibrium conditions given by the Young equation; however, it may be unstable, so that stability of a homogeneous interface should be analyzed. Mathematically, this means that in addition to satisfying the equilibrium condition for the net energy

$$dE = 0 \qquad (7.1)$$

the stable configuration should satisfy the minimum net energy condition

$$d^2 E > 0. \qquad (7.2)$$

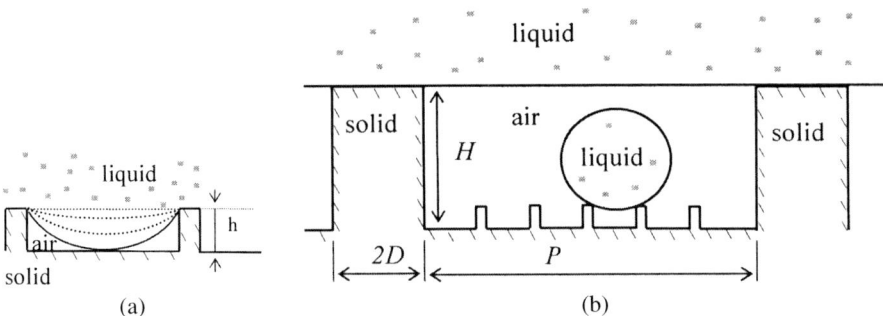

Fig. 7.1. Destabilization of the composite interface by **a** increasing amplitude of a liquid–air interface capillary wave. If wave amplitude is high enough so that it touches the valley between asperities, the liquid can fill the valley. **b** Condensation of a small droplet upon a rough surface with pillars. Smaller scale roughness is required to support small size droplets [244]

The interface may be destabilized due to small perturbations caused by various external influences and effects, for example, by the capillary or gravitational waves. Furthermore, the configuration may have many equilibrium states (metastability) with a certain probability to find the system at a given state. These phenomena are considered in the present section.

7.1.1 Destabilization Due to Capillary and Gravitational Waves

A wave may form at the liquid–air interface due to the gravitational or capillary forces

$$z = A\cos(kx - \omega t), \tag{7.3}$$

where z is vertical displacement, k and ω are the wavenumber and frequency, which are related to each other as

$$\omega^2 = gk + \frac{\gamma_{LA}}{\rho}k^3, \tag{7.4}$$

where g is the gravity constant, ρ is the liquid density, and γ_{LA} is the liquid–air interface energy [200]. For most micro/nanoscale applications, the effect of gravity is small and the frequency is given by

$$\omega = \sqrt{\frac{\gamma_{LA}k^3}{\rho}}. \tag{7.5}$$

The capillary waves may lead to composite interface destabilization, as will be shown in the following [241].

It is assumed that the interface energy γ_{LA} is a constant for given materials and that it is size independent. Generally speaking, this is not true for very small thickness of liquid comparable with the range of intermolecular forces. However, here we are assuming that the relevant size of the surface roughness, as well as the thickness of

Composite solid-liquid-air interface

Smooth liquid-air interface

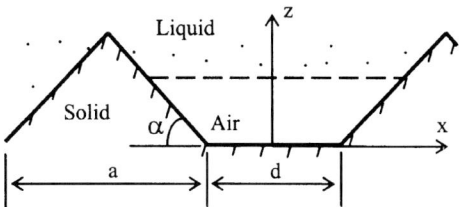

Wavy liquid-air interface

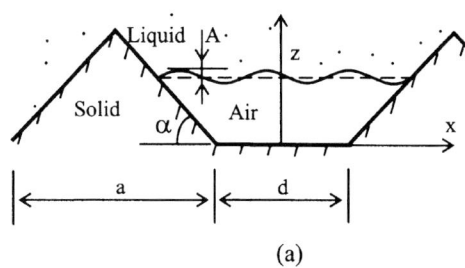

(a)

Stochastic distribution of air pockets

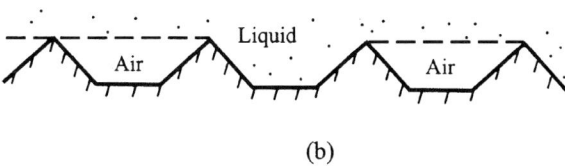

(b)

Fig. 7.2. Sawtooth profile **a** with smooth liquid–air interface and with wavy liquid–air interface, and **b** with stochastic distribution of air pockets [241]. **c** Formation of a composite solid–liquid–air interface for sawtooth and smooth profiles, and **d** destabilization of the composite interface for the sawtooth and smooth profiles due to dynamic effects. Dynamic contact angle $\theta_d > \theta_0$ corresponds to advancing liquid–air interface, whereas $\theta_d < \theta_0$ corresponds to receding interface [240]

the liquid layer is greater than the range of the intermolecular forces and therefore that γ_{LA} is constant.

Consider a sawtooth profile (Fig. 7.2(a)), with teeth height $a \tan \alpha / 2$ and distance between teeth d. The teeth represent asperities of a rough surface. It is assumed that

Formation of composite solid-liquid-air interface

Sawtooth profile

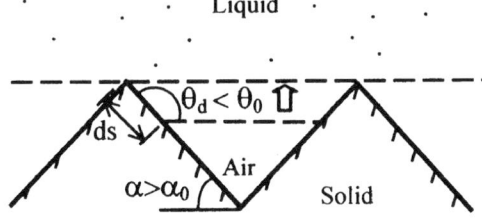

Smooth profile

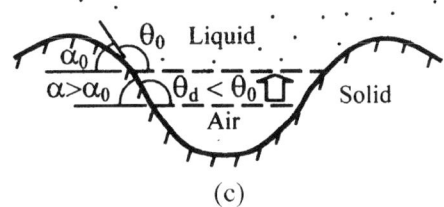

(c)

Destabilization of composite interface due to dynamic effects

Sawtooth profile

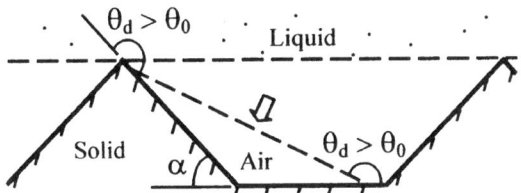

Smooth profile

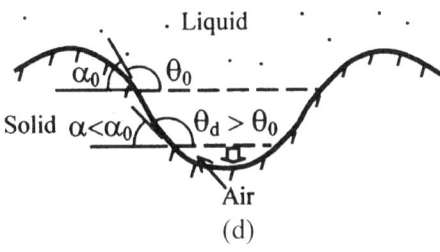

(d)

Fig. 7.2. (*Continued*)

the model of the sawtooth profile can capture the important features of more complicated rough surfaces. The horizontal liquid–air interface is located at a distance z
from the valley and has small waves of amplitude A and wave number k. The total
change of the energy of the system, from the energy of the homogeneous solid–liquid
interface, is given by the sum of surface changes throughout the inclined and horizontal portions of the surface and corresponding liquid–air parts, plus the waves'
energy. The changes of surface energy at inclined and horizontal portions of the surfaces and corresponding liquid–air parts are given by corresponding surfaces' lengths
times the corresponding interface energies. The length of the inclined portion of the
interface is $z/\sin\alpha$, and the length of the corresponding section of the wavy surface
is $(z/\tan\alpha)(kL_0/2\pi)$, where L_0 is the length of the liquid–air interface per wave
period, given by the integral

$$L_0 = \int_0^{2\pi/k} \sqrt{1 + (Ak)^2 \sin^2(xk)}\, dx = 4\sqrt{1 + (Ak)^2}\, E\left(\frac{Ak}{\sqrt{1 + (Ak)^2}}\right), \quad (7.6)$$

where $E(x)$ is the elliptical integral of the second kind [241]. The length of the
horizontal portion of the interface is d, and the length of the corresponding section
of the wavy surface is $dkL_0/(2\pi)$. The energy change is given by

$$
\begin{aligned}
U(z) &= -\frac{2z}{\sin\alpha}(\gamma_{SL} - \gamma_{SA}) + \frac{2z}{\tan\alpha}\frac{kL_0}{2\pi}\gamma_{LA} \\
&\quad - d\left[(\gamma_{SL} - \gamma_{SA}) - \gamma_{LA}\frac{kL_0}{2\pi}\right]H(z) + E_0 \\
&= \frac{2z\gamma_{LA}}{\sin\alpha}\left(\cos\theta_0 + \cos\alpha\frac{kL_0}{2\pi}\right) + d\gamma_{LA}\left[\cos\theta_0 + \frac{kL_0}{2\pi}\right]H(z) + E_0, \quad (7.7)
\end{aligned}
$$

and E_0 is the energy of the waves, γ_{SL} and γ_{SA} are interface energies for the solid–
liquid and solid–air interfaces, correspondingly, and $H(z)$ is the step function, such
that $H(z) = 0$ for $z \leq 0$ and $H(z) = 1$ for $z > 0$. It is assumed in (7.7) that $z > A$
and the Young equation is used.

In the limiting case of a flat liquid–air interface ($A = 0$), the surface energy is
given by [241]

$$
\begin{aligned}
U(z) &= -\frac{2z}{\sin\alpha}(\gamma_{SL} - \gamma_{SA}) + \frac{2z}{\tan\alpha}\gamma_{LA} - d\left[(\gamma_{SL} - \gamma_{SA}) - \gamma_{LA}\right]H(z) \\
&= \frac{2z\gamma_{LA}}{\sin\alpha}(\cos\theta_0 + \cos\alpha) + d\gamma_{LA}(\cos\theta_0 + 1)H(z), \quad A < z. \quad (7.8)
\end{aligned}
$$

For $z > 0$, the energy may increase or decrease with increasing z, depending on the
sign of $(\cos\theta_0 + \cos\alpha)$, since both γ_{LA} and $\sin\alpha$ are positive. In particular, if $180° -
\alpha > \theta_0$, the energy increases with z, whereas otherwise it decreases. The stable
position corresponds to the minimum value of the energy, which is $z = a/(2\tan\alpha)$
(liquid staying at the tops of the asperities) for $180° - \alpha < \theta_0$ and $z = 0$ for $180° -
\alpha > \theta_0$ (homogeneous solid–liquid interface).

For the wavy liquid–air interface based on (7.7) the energy may increase with increasing z, if $\cos\theta_0 + \cos\alpha kL_0/(2\pi) > 0$, for $z > A$. However, for $z < A$, the waves

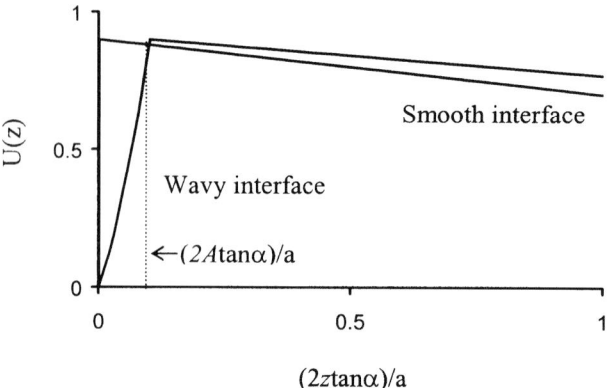

Fig. 7.3. Energy change as a function of interface position for smooth and wavy liquid–air interfaces, $d\gamma_{LA}(\cos\theta_0 + 1) = 0.9$, $\gamma_{LA}(\cos\theta_0 + \cos\alpha)/\sin\alpha = -0.2$, $\gamma_{LA}(\cos\theta_0 + (kL_0/2\pi)\cos\alpha)/\sin\alpha = -0.15$, $E_0 = 0.015$, $A = 0.1$ [241]

touch the horizontal part of the interface, and only the fraction $(\pi - \arccos(z/A))/\pi$ of the interface is liquid–air. In this case the energy change is given by

$$U(z) = \frac{z\gamma_{LA}}{\sin\alpha}\left[\cos\theta_0 + \frac{\pi - \arccos(z/A)}{\pi}\left(\frac{kL_{LA}}{2\pi}\right)\cos\alpha\right]$$

$$+ d\gamma_{LA}\left(\cos\theta_0 + \frac{kL_{LA}}{2\pi}\right)\frac{\pi - \arccos(z/A)}{\pi}H(z) + E_0, \quad A > z, \quad (7.9)$$

where L_{LA} is the length of the liquid–air part of the interface (the wave not touching the solid horizontal part of the interface) given by [241]

$$L_{LA} = \int_{-\arccos(z/A)/k}^{\arccos(z/A)/k}\sqrt{1 + (Ak)^2\sin^2(xk)}\,dx. \quad (7.10)$$

The energy change U as a function of liquid–air interface position z is presented in Fig. 7.3 for the smooth interface, (7.8), and for the wavy liquid–air interface, (7.7) and (7.9). It is noted that in the case when the waves are introduced, $U(z)$ has a local minimum at $z = 0$, which corresponds to the homogeneous solid–liquid interface, and in the case of $\cos\theta_0 + \cos\alpha kL_0/(2\pi) < 0$ it has another minimum at $z = a/(2\tan\alpha)$, which corresponds to the composite solid–liquid–air interface (liquid staying at the tops of the asperities). The interface position z is normalized in such a manner that the first equilibrium position ($z = 0$) corresponds to zero, and the second $z = a/(2\tan\alpha)$ corresponds to the unity in Fig. 7.3. In this case, the system has two equilibriums and may be, with a certain probability, in either one or another position. The interface consists of many asperities and valleys. Some of the valleys have homogeneous interface, whereas others have the composite interface (Fig 7.2b). It is assumed that probability p for the interface to be composite depends on geometrical parameters of the interface and values of energy which correspond to the metastable states [241].

7.1.2 Probabilistic Model

In this section, we consider the mechanism of destabilization of the composite interface due to liquid–air interface waves (Fig. 7.2(c)–(d)), and we discuss a probabilistic
model for the interface destabilization [241]. The previous section showed that the
interface may have two stable states. The first stable state corresponds to the homogeneous interface with energy level

$$U(0) = 0. \tag{7.11}$$

The second metastable state corresponds to the composite interface ($z = a/2 \tan \alpha$)
with energy level obtained from (7.7)

$$U(a/2 \tan \alpha) = \frac{a\gamma_{LA}}{\tan \alpha \sin \alpha}\left(\cos \theta_0 + \cos \alpha \frac{kL_0}{2\pi}\right)$$
$$+ d\gamma_{LA}\left(\cos \theta_0 + \frac{kL_0}{2\pi}\right) + E_0. \tag{7.12}$$

A certain probability p may be associated with each of the two stable states of energy. Assuming that waves with the energy E_0 have a similar effect on the system, as
the thermal fluctuation of an ideal gas with the energy kT, the Maxwell–Boltzmann
statistical distribution may be applied [110]. This implies that energetic barriers are
small enough so that the transition can be activated thermally. This assumption is justified at the nanoscale, as discussed in the first chapter of this book. For a large-scale
system, the behavior can still be qualitatively similar, since transition takes place with
a certain probability. Based on the Maxwell–Boltzmann distribution, probability is
exponentially dependent upon the energy level

$$p = B \exp\left(-\frac{U}{E_0}\right), \tag{7.13}$$

where B is a normalization constant [241]. Substituting (7.12) into (7.13) yields the
probability of the composite interface

$$p(\phi) = C \exp(-\phi/\phi_0), \tag{7.14}$$

where

$$\phi = d/a, \tag{7.15}$$

$$\phi_0 = \frac{E_0}{a\gamma_{LA}(\cos \theta_0 + \frac{kL_0}{2\pi})}, \tag{7.16}$$

$$C = B \exp\left[-\frac{a\gamma_{LA}}{E_0 \tan \alpha \sin \alpha}\left(\cos \theta_0 + \cos \alpha \frac{kL_0}{2\pi}\right) - 1\right]. \tag{7.17}$$

7.1.3 Analysis of Rough Profiles

In this section, we analyze a patterned rough surface. Consider a periodic sawtooth profile with a distance between asperities d and width a, as shown in Fig. 7.2. Let us assume that the probability of the interface to remain composite, p, decreases exponentially with the distance between asperities according to (7.14) [241]. The roughness factor for the homogeneous interface on the basis of (6.7) is given by

$$R_{\mathrm{f}} = \frac{d + a/\cos\alpha}{d + a} = \frac{\phi + 1/\cos\alpha}{\phi + 1}. \tag{7.18}$$

The total fraction of valleys which are covered with liquid (homogeneous interface) is given by $1 - p$, whereas the fraction of the valleys which have air pockets (composite interface) is given by p, obtained from (7.14). Based on this, the fractional areas are given by

$$f_{\mathrm{SL}} = \frac{(1-p)(d + a/\cos\alpha)}{(1-p)(d + a/\cos\alpha) + p(d + a)}$$
$$= \frac{(1-p)(\phi + 1/\cos\alpha)}{(1-p)(\phi + 1/\cos\alpha) + p(\phi + 1)}, \tag{7.19}$$
$$f_{\mathrm{LA}} = \frac{p(d + a)}{(1-p)(d + a/\cos\alpha) + p(d + a)}$$
$$= \frac{p(\phi + 1)}{(1-p)(\phi + 1/\cos\alpha) + p(\phi + 1)}. \tag{7.20}$$

Substituting (7.18)–(7.20) into the Cassie–Wenzel equation yields the expression for the contact angle

$$\cos\theta = \frac{(1-p)(\phi + 1/\cos\alpha)^2 \cos\theta_0 - p(\phi + 1)^2}{(1-p)(\phi + 1/\cos\alpha)(\phi + 1) + p(\phi + 1)^2}. \tag{7.21}$$

The results for the contact angle as a function of ϕ are presented in Fig. 7.4(a). It is observed that higher roughness (lower ϕ) corresponds to higher contact angles [241].

Comparison of the models based on (6.6) (homogeneous interface), (6.9) (solid–liquid–air composite interface), and (7.21) (stochastic interface) is presented in Fig. 7.4(b). It is observed that for high roughness (small ϕ), $R_{\mathrm{f}} \cos\theta_0 > 1$ and all the three models predict $\theta = 180°$. However, for higher distance between the asperities (higher ϕ), the composite interface model, which does not account for the possibility of destabilization, still predicts $\theta = 180°$, if $\theta_0 + \alpha > 180°$, whereas the homogeneous interface model predicts a rapid decrease of θ down to the value of θ_0, due to the decreasing roughness factor. The stochastic model yields values of the contact angle close to the composite interface model for short distances between the asperities (small ϕ); however, with increasing ϕ, the probability of destabilization of the composite interface grows, and eventually the values of the contact angle approach those predicted by the homogeneous interface model.

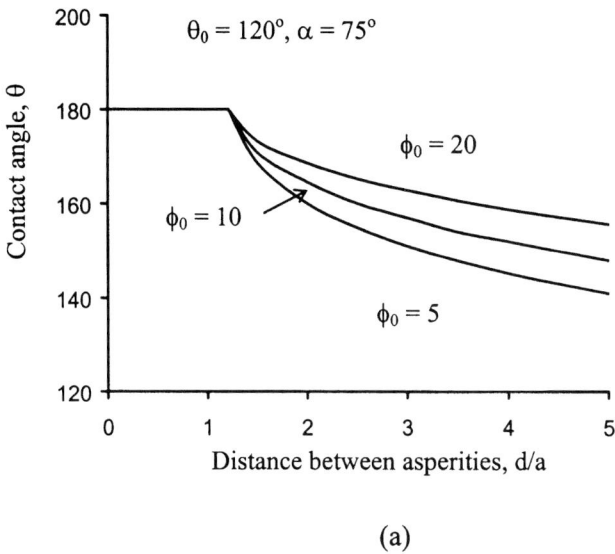

(a)

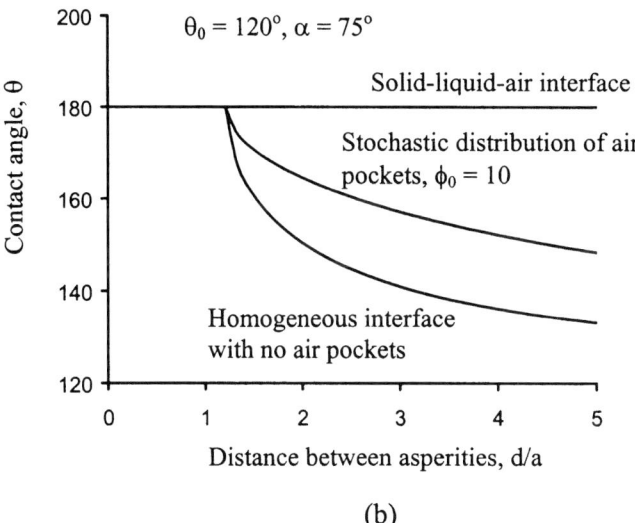

(b)

Fig. 7.4. Contact angle as a function of distance between asperities **a** for the stochastic model and **b** comparison of the interface with no air pockets, composite liquid–air interface and stochastic distribution of air pockets [241]

7.1.4 Effect of Droplet Weight

In the preceding analysis we ignored effect of the droplet weight by assuming that the gravity force is small compared to the surface tension forces. However, for big droplets, this assumption may not be correct. If the weight of a droplet exceeds the vertical component of the total surface tension force at the triple line, a droplet suspended at the tops of the asperities will collapse [108]. Thus a maximum critical size of the droplet R_{max} exists, above which the droplet cannot remain suspended on the tops of the asperities. Let us investigate how this maximum size depends on the period of asperities l. Consider a rough surface with roughness period l and amplitude $\hat{z}$, which corresponds to maximum droplet size R_{max}. The weight of the droplet is proportional to its volume and to the third power of R_{max}

$$W \propto R_{max}.$$ (7.22)

For the maximum value of the droplet radius, the weight is equal to the total vertical component of the surface tension, and thus it is proportional to the surface tension times the cosine of the contact angle times the total perimeter of the triple line

$$W \propto \gamma_{SL} t N \cos\theta,$$ (7.23)

where N is the number of asperities under the droplet [242].

Consider another rough surface with period αl and amplitude $\alpha \hat{z}$, which has the same roughness factor and the same contact angle. The length of the triple line for each asperity will be αt. The number of asperities under the droplet is proportional to the second power of R_{max}, divided by the second power of α

$$N \propto R_{max}^2 / \alpha^2.$$ (7.24)

Combining (7.22)–(7.24) yields

$$R_{max} \propto 1/\alpha.$$ (7.25)

This result suggests that with increasing size of asperities, the ability of a rough surface to form the composite interface decreases and greater droplets collapse. Therefore, smaller asperities make a composite interface more likely due to gravity. Increasing the droplet size has the same effect as increasing the period of roughness [242].

Let us now consider a superhydrophobic surface with asperities or pillars with pitch P, diameter D, and height H and density $\eta = 1/P$ asperities per area (Fig. 7.5). A droplet of radius R has a weight of $\rho g 4/(3\pi R^3)$ which acts upon the composite interface area of $\pi (R \sin\theta)^2$, assuming that $\sin\theta$ is small, which is justified for a superhydrophobic surface. Thus the pressure is

$$P_0 = \frac{4\rho g R}{3 \sin^2\theta}.$$ (7.26)

It is convenient to introduce nondimensional parameters, the spacing factor

$$S_f = \frac{D}{P}$$ (7.27)

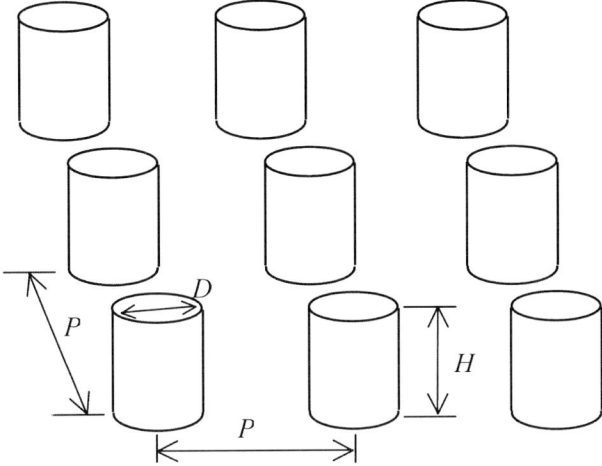

Fig. 7.5. Schematic of a patterned superhydrophobic surface with pillars with diameter D, height H, and pitch P

and the normalized droplet radius

$$R_n = \frac{R}{P}. \tag{7.28}$$

Since the fractional solid–liquid area is given by $\eta \pi D^2/4$ or

$$f_{SL} = \frac{\pi S_f^2}{4}, \tag{7.29}$$

and since for the flat-top pillars $R_f = 1$, the contact angle and contact angle hysteresis can easily be presented as functions of S_f using (6.9) and (6.59).

In the case of a composite interface, such a surface can support pressure given by the perimeter $\eta \pi D$, multiplied by $\sigma \cos \theta_0$, where σ is the surface tension, or

$$P_0 = \frac{\pi D \sigma \cos \theta_0}{P^2} = \frac{\pi \sigma S_f \cos \theta_0}{P}. \tag{7.30}$$

Comparing (7.26) and (7.30) yields the condition for transition from the composite interface to homogeneous interface due to the weight of the droplet

$$\frac{R_n}{S_f} = \frac{3\pi \sigma \sin^2 \theta_0 \cos \theta_0}{4\pi g}. \tag{7.31}$$

Thus the relation of the gravity and surface forces is controlled by the ratio R_n/S_f. Note that, as stated earlier, the ratio $R_n H/P$ controls the transition due to the droplet's curvature. Thus we have identified three parameters which control various modes of transition to the homogeneous interface: S_f is responsible for composite interface destabilization, R_n/S_f is responsible for the transition due to the gravity, and $R_n H/P$ is responsible for the transition due to the droplet's curvature.

7.2 Contact Angle with Three-Dimensional Solid Harmonic Surface

In this section, a simple model for the contact angle with a three-dimensional rough surface will be considered [242]. The model allows partial wetting, i.e., it is based on the probabilistic criterion. It also involves both asperities (pillars) and holes, so the deference of an effect of a rough structure and its negative replica on wetting will be discussed.

7.2.1 Three-Dimensional Harmonic Rough Surface

Consider a three-dimensional rough surface in coordinates $\hat{x}$, $\hat{y}$, $\hat{z}$

$$\hat{z} = z_0 \cos\left(\frac{2\pi\hat{x}}{l}\right) \cos\left(\frac{2\pi x\hat{y}}{l}\right). \tag{7.32}$$

The surface consists of periodical sets of asperities and valleys in $\hat{x}$- and $\hat{y}$-directions (Fig. 7.6(a)). For convenience, let us introduce the nondimensional coordinates

$$x = \frac{2\pi\hat{x}}{l}, \qquad y = \frac{2\pi x\hat{y}}{l}, \qquad z = \frac{\hat{z}}{z_0} \tag{7.33}$$

so that (7.32) in the nondimensional coordinates is given by

$$z = \cos x \cos y. \tag{7.34}$$

The profile (Fig. 7.6(a)) has peaks at $x = \pm 2\pi n$, $y = \pm 2\pi m$ and $x = \pi \pm 2\pi n$, $y = \pi \pm 2\pi m$ and valleys at $x = \pi \pm 2\pi n$, $y = \pm 2\pi m$ and $x = \pm 2\pi n$, $y = \pi \pm 2\pi m$, where m and n are integer numbers. At $x = \pi/2 \pm \pi n$ and at $y = \pi/2 \pm \pi n$ the surface cross zero ($z = 0$). There are two peaks and two valleys per area of $4\pi^2$ [242].

For a composite interface, the liquid–air interface forms angle θ_0 with the SL interface at the triple line. The slope or gradient of the surface is equal to θ_0 (Fig. 7.6(b))

$$\nabla\hat{z} = \tan\theta_0. \tag{7.35}$$

Substituting (7.32) into (7.35) yields a local equilibrium condition for the triple line [242]

$$\nabla\hat{z} = \frac{2\pi z_0}{l}\left[\left(\frac{\partial z}{\partial x}\right)^2 + \left(\frac{\partial z}{\partial y}\right)^2\right]^{1/2} = \frac{2\pi z_0}{l}\left[(\sin x \cos y)^2 + (\cos x \sin y)^2\right]^{1/2}$$

$$= \frac{2\pi z_0}{l}\left[\sin^2 x \cos^2 y + \left(1 - \sin^2 x\right)\left(1 - \cos^2 y\right)\right]^{1/2} = \tan\theta_0. \tag{7.36}$$

Equation (7.36) provides a dependence of $\cos y$ upon $\sin x$ at the triple line

$$\cos^2 y = \frac{K^2 - 1 + \sin^2 x}{2\sin^2 x - 1}, \tag{7.37}$$

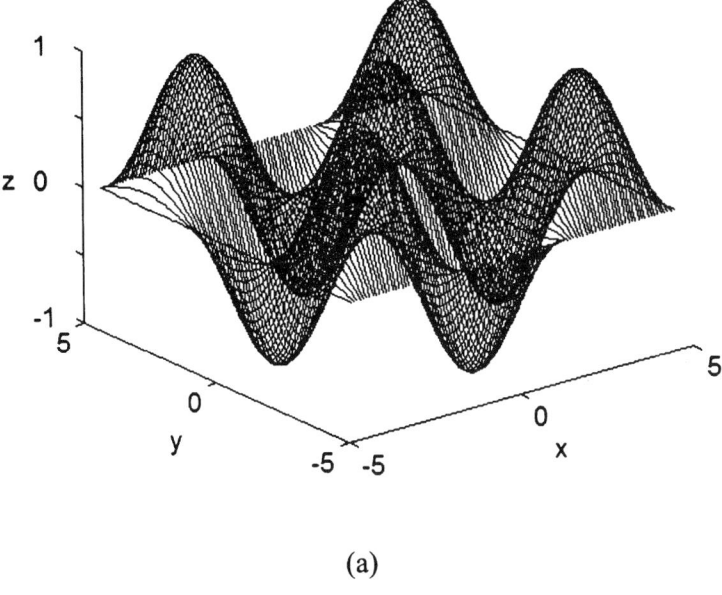

(a)

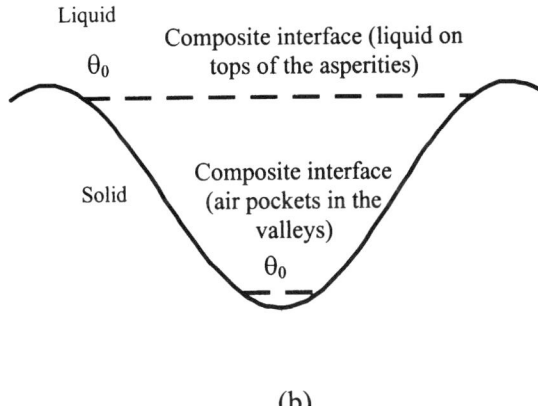

(b)

Fig. 7.6. a Three-dimensional periodic profile. b Composite interface with liquid staying at
the tops of the asperities and with air pockets in the valleys [242]

where

$$K = \frac{l \tan \theta_0}{2\pi z_0}.$$ (7.38)

It is observed that for small $|K|$ (corresponding to θ_0 close to π), two equilibrium
states of the composite interface are possible: one with the liquid staying close to
the tops of the asperities and one with liquid covering almost all the surface with air
pockets at the bottoms of valleys (Fig. 7.6(b)). In this study we ignore the effect of

air pressure in the pockets upon the position of the triple line, assuming that corresponding forces are smaller than the surface tension forces. It is assumed also that liquid–air interface is flat, which is reasonable in the case when the gravity effect is negligible [242].

7.2.2 Calculations of the Contact Areas

The length of the triple line t is given by the integral

$$t = \oint_{y(x)} \sqrt{1 + y'^2(x)}\, dx, \tag{7.39}$$

where $y(x)$ is determined from (7.37). The total area of the surface per surface area of 4π is given by a double integral, which is evaluated numerically

$$
\begin{aligned}
A_{tot} &= \int_0^{2\pi}\int_0^{2\pi} \sqrt{1 + \left(\frac{\partial z}{\partial x}\right)^2 + \left(\frac{\partial z}{\partial y}\right)^2}\, dx\, dy \\
&= \int_0^{2\pi}\int_0^{2\pi} \sqrt{1 + (\sin x \cos y)^2 + (\cos x \sin y)^2}\, dx\, dy \\
&\approx 1.2204 \times 4\pi^2.
\end{aligned}
\tag{7.40}
$$

The calculated value of A_{tot} corresponds to $R_f \approx 1.2204$. The flat and solid–liquid contact areas for a composite interface are given as [242]

$$A_{SL} = \oint\!\!\!\oint_{y(x)} \sqrt{1 + \left(\frac{\partial z}{\partial x}\right)^2 + \left(\frac{\partial z}{\partial y}\right)^2}\, dx\, dy, \tag{7.41}$$

$$A_F = \oint_{y(x)} y(x)\, dx. \tag{7.42}$$

The solid–air and liquid–air areas are given by subtracting A_{SL} and A_F from the corresponding total areas

$$A_{SA} = A_{tot} - A_{SL}, \tag{7.43}$$

$$A_{LA} = 4\pi^2 - A_F. \tag{7.44}$$

In order to avoid complicated integrals in (7.39) and (7.41)–(7.42), which should be calculated numerically, we will use approximate functions

$$t(K) = 2\pi \arcsin K + \left(4\pi - \pi^2\right)K^2, \tag{7.45}$$

$$A_F(K) = \pi K \arcsin K + \pi^2 K^2/2, \tag{7.46}$$

$$A_{SL}(K) = \pi K \arcsin K + \left(A_{tot}/4 - \pi^2/2\right)K^2. \tag{7.47}$$

Functions in (7.45)–(7.47) (Fig. 7.7) are selected in such a manner that they result in the same limiting values as (7.39) and (7.41)–(7.42), namely, for $K \to 0$, $t(K) =$

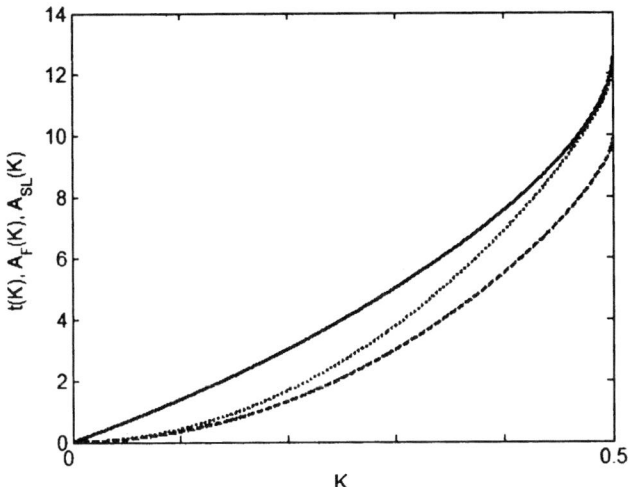

Fig. 7.7. Functions $t(K)$ (*solid line*), $A_F(K)$ (*dashed line*), and $A_{SL}(K)$ (*dotted line*) [242]

$2\pi K$, $A_F(K) = \pi K^2$, $A_{SL}(K) = \pi K^2$ and for $|K| \to 1$, $t(K) = 4\pi$, $A_F(K) = \pi^2$, $A_{SL}(K) = A_{tot}/4$ and derivatives $t'(K) \to \infty$, $A_F'(K) \to \infty$, $A_{SL}'(K) \to \infty$, and thus may serve as a reasonable approximation. Note that $|K| \to 1$ corresponds to the highest slope of the surface, whereas $K \to 0$ corresponds to zero slope.

The roughness factor is given by

$$R_f = \frac{A_{SL}}{A_F} = \frac{\arcsin K + (1 + K/2)\pi K}{\arcsin K + (A_{tot}/\pi^2 + K/2)\pi K}. \tag{7.48}$$

In (7.39)–(7.48) the contact parameters are calculated only for cases where liquid stays at the tops of the asperities. Corresponding values of these parameters for the case of air pockets at the valleys are determined easily based on the symmetry.

7.2.3 Metastable States

The total energy of the system is calculated as a sum of relevant contact areas times surface tensions

$$E_{tot} = A_{SL}\gamma_{SL} + A_{SA}\gamma_{SA} + A_{LA}\gamma_{LA}. \tag{7.49}$$

We assume here that the surface tensions are constant and scale-independent. This may not always be incorrect at nanoscale since some experimental data suggest that contact angle at nanoscale depends on the size of the droplet; however, we ignore this effect since it is far from being understood [75].

For the composite interface with liquid staying on top of the asperities, the total energy can be estimated as

$$E_1 = A_{SL}\gamma_{SL} + A_{SA}\gamma_{SA} + A_{LA}\gamma_{LA} = A_{SL}(\gamma_{SL} - \gamma_{SA}) + A_{tot}\gamma_{SA} + A_{LA}\gamma_{LA}$$

$$= (A_{LA} - A_{SL}\cos\theta_0)\gamma_{LA} + A_{tot}\gamma_{SA}$$

$$= (4\pi^2 - A_F - A_{SL}\cos\theta_0)\gamma_{LA} + A_{tot}\gamma_{SA}. \tag{7.50}$$

For the composite interface with air pockets A_{LA} has the same value as A_F for the previous case, due to the symmetry, whereas A_{SL} is given by the total area minus the value of A_{SL} in the previous case. The total area is

$$
\begin{aligned}
E_2 &= (A_{LA} - A_{SL} \cos \theta_0)\gamma_{LA} + A_{tot}\gamma_{SA} \\
&= \left(A_F - (A_{tot} - A_{SL}) \cos \theta_0 \right)\gamma_{LA} + A_{tot}\gamma_{SA}.
\end{aligned}
\tag{7.51}
$$

For the homogeneous interface the total energy is given

$$
E_3 = A_{tot}\gamma_{SL}.
\tag{7.52}
$$

The values of energy E_1, E_2, and E_3 correspond to three metastable states of equilibrium. It is noted that the second state of equilibrium involves isolated air pockets. Therefore, a composite interface is possible with some valleys filled with liquid and others filled with air pockets, if different probabilities are associated with different states and filling of the valleys with liquid occurs with probability p. Such a probability function should decrease with increasing energy of the state. Based on the Maxwell–Boltzmann statistics, it is assumed that probability p depends exponentially on the ratio of corresponding energies. This assumption is justified, on one hand, by the fact that change of the equilibrium state is random and, especially at smaller size scales, is similar in its nature to thermal fluctuations, which are usually governed by the Maxwell–Boltzmann statistics [110]. On the other hand, simplicity of the exponential function makes it convenient to use for calculations. If there are two metastable states with energies E_2 and E_3, the probability of finding the system in the state of E_3 is given by the exponent of the energy, divided by the total probability of the two states

$$
p = \frac{\exp(-E_3)}{\exp(-E_2) + \exp(-E_3)}.
\tag{7.53}
$$

Thus there are two different equilibrium states: the composite interface with liquid staying at the tops of the asperities (with energy E_1) and the composite interface with air pockets at some valleys and with other valleys filled with liquid (with energy $(1 - p)E_2 + pE_3$). These two states will be considered in the following sections.

7.2.4 Overall Contact Angle

The overall contact angle may be calculated for the two equilibrium states using (7.44) and (7.47)–(7.48). Note that (7.47)–(7.48) are formulated for individual asperities; however, there are two asperities and two valleys per rough surface area $2\pi \times 2\pi$, so the corresponding areas should be doubled. For the liquid staying on the tops of the asperities, the relevant contact areas are given by [242]

$$
A_{LA} = 2A_{LA}(K),
\tag{7.54}
$$

$$
A_{SL} = 2A_{SL}(K),
\tag{7.55}
$$

$$
A_F = A_F(K).
\tag{7.56}
$$

The contact angle θ_1 is given by [242]

$$\cos\theta_1 = \frac{A_{\rm SL}}{4\pi^2}\cos\theta_0 - \frac{4\pi^2 - A_{\rm F}}{4\pi^2} = \frac{2A_{\rm SL}(K)}{4\pi^2}\cos\theta_0 - \frac{4\pi^2 - 2A_{\rm F}(K)}{4\pi^2}. \quad (7.57)$$

For the air pockets at the valleys, relevant contact areas can be calculated based on the functions (7.46)–(7.47) due to the symmetry [242]. Namely,

$$A_{\rm LA} = 2A_{\rm F}(K), \quad (7.58)$$
$$A_{\rm SL} = A_{\rm tot} - 2A_{\rm SL}(K), \quad (7.59)$$
$$A_{\rm F} = 4\pi - 2A_{\rm F}(K). \quad (7.60)$$

The total contact angle θ_2 is given by

$$\cos\theta_2 = \frac{A_{\rm tot} - 2A_{\rm SL}(K)}{4\pi^2}\cos\theta_0 - \frac{2A_{\rm F}(K)}{4\pi^2}. \quad (7.61)$$

For the homogeneous interface, the contact angle θ_3 is given by

$$\cos\theta_3 = R_{\rm f}\cos\theta_0 = \frac{A_{\rm tot}}{4\pi^2}\cos\theta_0. \quad (7.62)$$

For the air pockets and liquid filling the valleys with probability p, the relevant areas are given by [242]

$$A_{\rm LA} = 2(1-p)A_{\rm F}(K), \quad (7.63)$$
$$A_{\rm SL} = A_{\rm tot} - 2(1-p)A_{\rm SL}(K), \quad (7.64)$$
$$A_{\rm F} = 4\pi - 2(1-p)A_{\rm F}(K). \quad (7.65)$$

The total contact angle θ_4 is given by

$$\cos\theta_4 = \frac{A_{\rm tot} - 2(1-p)A_{\rm SL}(K)}{4\pi^2}\cos\theta_0 - \frac{2(1-p)A_{\rm F}(K)}{4\pi^2}. \quad (7.66)$$

7.2.5 Discussion of Results

For wetting liquids ($\theta_0 < \pi/2$), no composite interface can be formed and the contact angle decreases with roughness, therefore the case of interest is nonwetting liquid ($\theta_0 > \pi/2$). Dependence of the contact angles θ_1, θ_2, θ_3, and θ_4 on θ_0 is presented in Fig. 7.8(a) for $l/(2\pi z_0) = 1$ and for a nonwetting liquid. It is observed that for $\theta_0 < 3\pi/4$, a solution for only θ_3 (homogeneous interface) exists. The contact angle for the homogeneous interface increases with increasing θ_0 in accordance with (7.61). It is noted that θ_3 approaches the value of $\theta_3 = \pi$ at $\theta_0 = \arccos(-4\pi/A_{\rm tot}) = 2.5312$. For $\theta_0 > 3\pi/4$, solutions with composite interface exist. For both the solution with liquid staying on the tops of the asperities (θ_1) and with air pockets at the valleys (θ_2), the contact angle approaches π with increasing θ_0. For the case of air pockets at the valleys (θ_4), the increase of the contact angle is much slower than for liquid

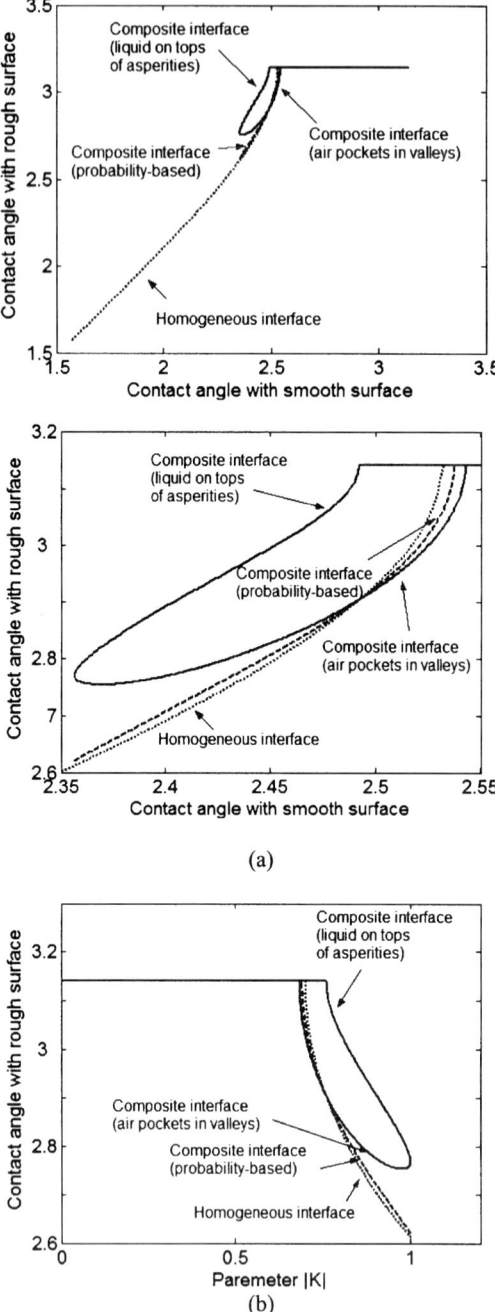

Fig. 7.8. a Dependence of the contact angle with rough surface on the contact angle with smooth surface for homogeneous and composite interfaces ($l/(2\pi z_0) = 1$, $\gamma_{SA} = 0$). **b** Dependence of the contact angle with rough surface on K for $\theta_d = 2\pi/3$ [242]

staying on the tops of the asperities. For $\theta_0 > 3\pi/4$, a solution exists (θ_4) with air pockets staying in some valleys and others filled with liquid with probability p. For this solution, the contact angle increases with increasing θ_0 and changes between the values of θ_3 and θ_4. Dependence of the contact angle on K is presented in Fig. 7.8(b) for $\theta_0 = 2\pi/3$ [242].

Experimental data on superhydrophobic states, obtained by Lafuma and Quèrè [197], show progressive sinking of the droplet, rather than instant transition between the homogeneous and composite contact, which supports the probability-based model.

7.2.6 The Similarity of Bubbles and Droplets

The interesting feature of the three-dimensional surface we studied is that it involves both asperities (pillars) and holes. These are two different types of roughness structures that constitute a negative replica of each other. It is interesting to compare their effect on hydrophobicity. Both types can lead to superhydrophobicity. They both can provide the same Wenzel roughness factor, R_f. However, there are a number of differences. First, the curvature is different, and convex and concave surfaces have different effects upon the stability of the composite interface. This phenomenon will be studied in detail in the following chapter. Second, the holes may include trapped air that has no way to leave, and therefore pressure in the air pockets may affect significantly the stability of the composite interface. Third, the triple line dynamics can be dependent upon micro- and nanoscale processes that govern the fine structure of the triple line (such as a thin liquid precursor next to a droplet). These processes may be very different for the case of the pillars and holes.

Another interesting similarity exists between the bubbles and the droplets. Consider a hydrophilic liquid with some dissolved gas (e.g., blood) in a vessel or channel with rough walls. Gas bubbles may be formed near the walls. The existence of nanobubbles has been demonstrated experimentally [206, 325]. Small gas bubbles with size comparable to the roughness size, analogous to air pockets considered in the preceding subsections, may form in the valleys with probability given by (7.53). The molecular dynamics simulation data show that with decreasing pressure, the formation of bubbles in the valleys is a probabilistic, rather than a deterministic, process.

Larger bubbles with a size greater than the roughness size may also form near the walls (Fig. 7.9). It is assumed that air pressure is constant, so the volume of the bubble does not change. The roughness-dependence of the contact angle for these bubbles is governed by the same equations, (7.31), (7.35)–(7.36), and (7.40), as the liquid droplets. Both the homogeneous and composite interfaces are possible. In the case of composite interface, liquid stays in the valleys under the droplet. For practical liquid transport applications it may be desirable for these bubbles to flow along the channel. Therefore the same approach, which is used for the liquid droplets in order to increase the contact angle and reduce contact angle hysteresis, may be applied to the bubbles in the liquid. The cavitation bubble effect on the capillary meniscus

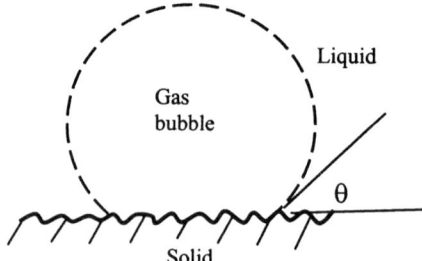

Fig. 7.9. Gas bubble in liquid near rough channel wall

force, in comparison with the condensed liquid bridge effect, will be discussed in Chap. 10.

In order to increase the bubbles' mobility, lower contact angles are desirable as well as lower contact angle hysteresis. The same recommendations can be made for optimization of roughness as for the case of droplets on a rough surface, namely, do not increase the maximum slope in order to maintain the homogeneous interface regime and decrease the contact angle hysteresis.

7.3 Capillary Adhesion Force Due to the Meniscus

When two solids come in contact, a meniscus can form due to the condensation of liquid or because the liquid film may be present initially, if the liquid is wetting. For nonwetting liquids, a meniscus will not be formed. The meniscus causes an increase of the friction force. This section considers the effect of surface roughness on the meniscus force for the case of a sphere in contact with a flat surface (single asperity contact) and for the case of multiple-asperity contact.

7.3.1 Sphere in Contact with a Smooth Surface

Consider a sphere with radius R in contact with a flat surface with a meniscus (Fig. 7.10(a)). The shape and size of the meniscus, as well as the total energy of the system, depend on the separation distance s between the flat surface and the center of the sphere. The normal meniscus force, F_m, which acts upon the sphere and the flat surface, can be calculated as the derivative of the energy E_{tot} by s

$$F_m = \frac{dE_{tot}}{ds}. \tag{7.67}$$

There are two solid–liquid interfaces, with the sphere and with the flat surface. The areas of these interfaces are approximately equal to πa^2, where a is the meniscus radius. Assuming that the ratio a/R is small and $A_{LA} \ll A_{SL}$, the total energy E_{tot} is equal to the sum of the surface energies at the SL interface. Based on these assumptions, E_{tot} can be calculated as

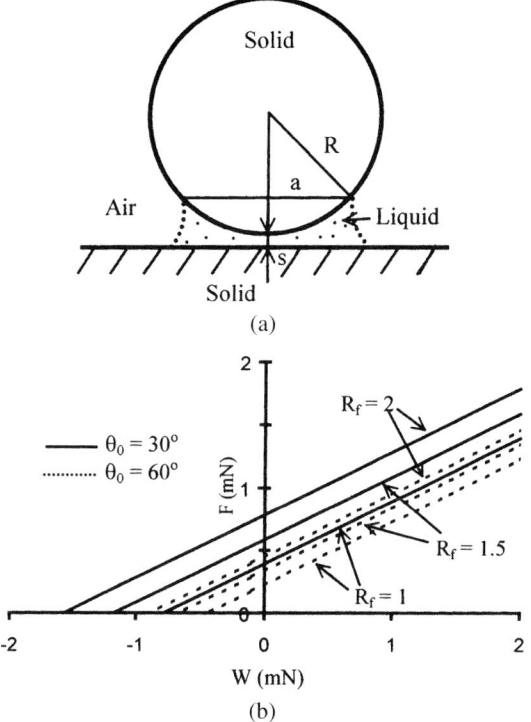

Fig. 7.10. a Meniscus formation during wet contact of a flat surface with a sphere. **b** Dependence of the friction force (F) on the normal load (W) for single-asperity contact, $\gamma_{LA} = 0.073$ J/m^2 (water), $R = 1$ mm, $\mu = 0.5$, for different values of $R_f = R_{f1} = R_{f2}$, $\theta_0 = \theta_1 = \theta_2$ [239, 240]

$$E_{tot} = \pi a^2 (\gamma_{SL1} - \gamma_{SA1} + \gamma_{SL2} - \gamma_{SA2}) = \pi a^2 \gamma_{LA}(\cos \theta_1 + \cos \theta_2), \quad (7.68)$$

where the indices 1 and 2 correspond to the sphere and the flat surface, and a is defined in Fig. 7.10(a) [240]. The volume of liquid V is a function of s and a and is given as the sum of the cylindrical volume with height $s + a^2/(2R)$ and cross-section area πa^2 minus a volume of the spherical segment of height $a^2/(2R)$

$$V = \pi a^2 s + \frac{\pi a^4}{4R}. \quad (7.69)$$

Assuming that the volume of the liquid remains constant during the contact (with is true if the evaporation and condensation processes are slow), so (7.69) may be viewed as a quadratic equation for a^2, which is solved as

$$a^2 = -2Rs \pm 2R\sqrt{s^2 + V/(\pi R)}. \quad (7.70)$$

The derivative of a^2 by s, $d(a^2)/ds$ for the sphere touching the plane $(s = 0)$ is given as

$$\frac{da^2}{ds} = -2R. \tag{7.71}$$

By using the derivative of (7.68) and the expression in (7.71) in (7.67), we get

$$F_m = 2\pi R \gamma_{LA}(\cos\theta_1 + \cos\theta_2). \tag{7.72}$$

Equation (7.72) (also known as equation for the Laplace pressure) provides the value of the normal force due to meniscus. If the sphere and surface are rough, with roughness factors of R_{f1} and R_{f2}, respectively, must be taken into account

$$F_m = 2\pi R \gamma_{LA1}(R_{f1}\cos\theta_1 + R_{f2}\cos\theta_2). \tag{7.73}$$

In the presence of meniscus, the friction force is given by [30, 32, 121]

$$F = \mu(W + F_m). \tag{7.74}$$

The coefficient of friction in the presence of the meniscus force, μ_{wet}, is calculated using only the applied normal load, as normally measured in the experiments

$$\mu_{wet} = \mu\left(1 + \frac{F_m}{W}\right). \tag{7.75}$$

Equation (7.75) shows that μ_{wet} is greater than μ, because F_m is not taken into account for calculation of the normal load in the wet contact.

The effect of meniscus on friction force for different surface roughness shown in Fig. 7.10(b). It is observed that a roughness factor of $R_f = 2$ may result in a significant change of the friction force due to meniscus. In applications, it is usually desirable to decrease the meniscus forces; therefore, a smooth surface is preferable in the case of single-asperity contact.

7.3.2 Multiple-Asperity Contact

In the case of multiple-asperity contact, a statistical approach is used to model the contact. For a random surface with a certain σ and β^*, the average peak radius R_p and number of contacts N depend on roughness due to the so-called scale effect [42]. Bhushan and Nosonovsky [43, 44] showed that the average peak radius $\overline{R_p}$ is related to σ and β^*, whereas the number of contacts for moderate loads is proportional to the load, divided by σ and β^*

$$\overline{R_p} \propto \frac{(\beta^*)^2}{\sigma}, \tag{7.76}$$

$$N \propto \frac{W}{\sigma\beta^*}. \tag{7.77}$$

For asperities of equal peak radius R_p, the meniscus force is given by [30, 32, 317]

$$F_m = 2\pi R_p \gamma_{LA}(\cos\theta_1 + \cos\theta_2)N \propto \frac{\beta^* W}{\sigma^2}(\cos\theta_1 + \cos\theta_2). \tag{7.78}$$

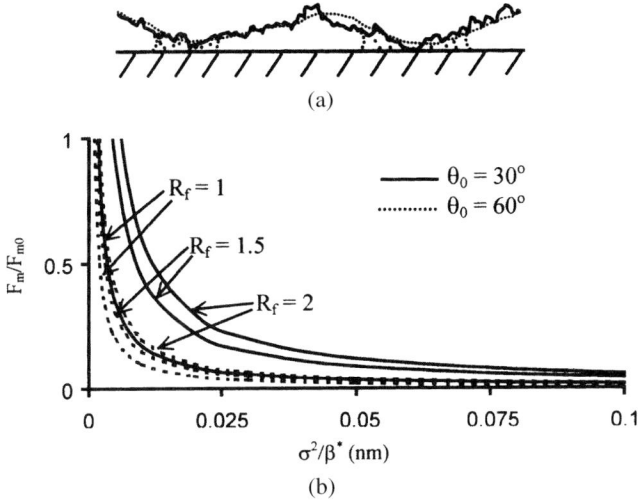

(a)

(b)

Fig. 7.11. a Menisci formation during wet contact of a smooth surface with a rough surface with a short wavelength roughness superimposed on long-wavelength roughness (*dotted line*). **b** Dependence of the meniscus force (normalized by F_{m0}, meniscus force value at $\sigma^2/\beta^* = 0.001$ nm) on roughness σ^2/β^*, for different values of $R_f = R_{f1} = R_{f2}$, $\theta_0 = \theta_1 = \theta_2$ [240]

The size of menisci is comparable with the size of individual contacts [240].

We consider a rough surface that consists of the short-wavelength roughness superimposed on the long-wavelength roughness with typical size of roughness smaller than the meniscus size (Fig. 7.11(a)). The nanoscale roughness of the two bodies is characterized by the roughness factors R_{f1} and R_{f2}. We further assume that the asperities have an average peak radius $\overline{R_p}$. Substituting the Wenzel equation into (7.78) results in

$$F_m = 2\pi \overline{R_p} \gamma_{LA} (\cos\theta_1 + \cos\theta_2) N \propto \frac{\beta^* W}{\sigma^2} (R_{fn1} \cos\theta_1 + R_{fn2} \cos\theta_2). \quad (7.79)$$

The meniscus force as a function of σ^2/β^*, which is a measure of roughness, is presented in Fig. 7.11(b) [240]. It is observed that with increasing roughness σ^2/β^*, the meniscus force decreases. A high nanoscale roughness factor may slightly increase the meniscus force. Since it is usually desirable to reduce the meniscus force, a rough surface with high σ^2/β^* is preferable in the case of multiple-asperity contact [240].

7.4 Roughness Optimization

In this section we are looking for optimization conditions of optimized biomimetic water-repellent surfaces. Optimization is a combination of conflicting requirements, so we will formulate these requirements first. Assuming that the adhesion hysteresis dominates in the contact angle hysteresis and using the same approach as in the derivation of (6.9), the difference of the cosines of the advancing and receding angles for composite interface, θ_{adv} and θ_{rec}, is related to the difference of those for the smooth surface, θ_{adv0} and θ_{rec0}, as

$$\cos\theta_{\text{adv}} - \cos\theta_{\text{rec}} = R_{\text{f}} f_{\text{SL}}(\cos\theta_{\text{adv0}} - \cos\theta_{\text{rec0}}). \tag{7.80}$$

It is observed from (7.80) that decreasing $f_{\text{SL}} \rightarrow 0$ results in increasing CA ($\cos\theta \rightarrow -1, \theta \rightarrow \pi$) and decreasing contact angle hysteresis ($\cos\theta_{\text{adv}} - \cos\theta_{\text{rec}} \rightarrow 0$). In the limiting case of a large contact angle close to π ($\cos\theta \approx -1 + (\pi - \theta)^2/2$, $\sin\theta \approx \theta - \pi$) and a small contact angle hysteresis ($\theta_{\text{adv}} \approx \theta \approx \theta_{\text{rec}}$), using simple trigonometric relations, the Cassie–Baxter equation (6.9) for the contact angle and (7.80) for the contact angle hysteresis are reduced to

$$\pi - \theta = \sqrt{2 f_{\text{SL}}(R_{\text{f}}\cos\theta_0 + 1)}, \tag{7.81}$$

$$\theta_{\text{adv}} - \theta_{\text{rec}} = f_{\text{SL}}R_{\text{f}}\frac{\cos\theta_{\text{a0}} - \cos\theta_{\text{r0}}}{-\sin\theta} = \sqrt{f_{\text{SL}}}R_{\text{f}}\frac{\cos\theta_{\text{r0}} - \cos\theta_{\text{a0}}}{\sqrt{2(R_{\text{f}}\cos\theta_0 + 1)}}. \tag{7.82}$$

For the homogeneous interface, $f_{\text{SL}} = 1$, whereas for the composite interface, f_{SL} is a small number. It is observed from (7.80)–(7.82) that for homogeneous interface, increasing roughness (high R_{f}) leads to increased contact angle hysteresis (high values of $\theta_{\text{adv}} - \theta_{\text{rec}}$), while for composite interface, small values of f_{SL} provide both high contact angle and small contact angle hysteresis. Therefore, a composite interface is essential for superhydrophobicity.

The superhydrophobic state, corresponding to composite interface, is likely to exist only for hydrophobic materials, since otherwise it will be energetically profitable for the liquid to spread and fill the valleys. However, it has been reported that a metastable composite interface was observed experimentally even for hydrophilic materials ($\theta_0 < 90°$), while the physical mechanism of this effect has not been adequately explained [245].

Water-repellent surfaces (Fig. 7.12) should have

- High contact angle. To achieve high contact angle, based on (7.81)–(7.82), either high R_{f} or low S_{f} are required.
- Low contact angle hysteresis. To achieve low contact angle hysteresis, based on (7.81)–(7.82), low R_{f} and S_{f}, and stable composite interface are required. Thus, based on this and the preceding requirement, composite interface is necessary.
- Ability to support high liquid pressure requires high spacing factor S_{f} and large perimeter (or diameter) of pillars, based on (7.31).
- Ability to support wide range of droplet sizes requires hierarchical roughness, which will be discussed later [244].

Stability of the composite interface implies ability to resist destabilizing factors, i.e., interface capillary waves, condensation of nanodroplets and surface inhomogeneity as well as liquid pressure. To resist capillary waves, high asperities are required. To resist nanoscale droplets, hierarchical roughness is required, and to resist

Optimized surfaces

Hemispherically topped cylindrical asperities

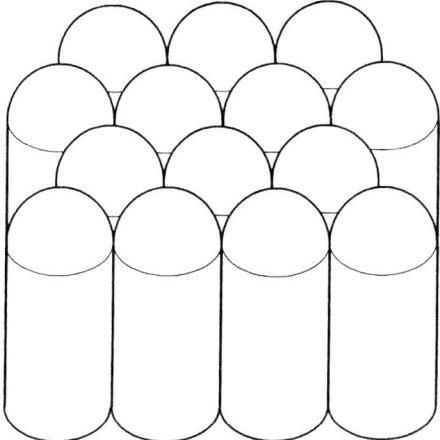

Hemispherically topped pyramidal asperities

Fig. 7.12. Optimized spaced roughness distribution—hemispherically-topped cylindrical asperities and pyramidal asperities with square foundation and rounded tops. Square base gives higher packing density but introduces undesirable sharp edges [240]

surface inhomogeneity nanobumps should be convex for the stability of the liquid–air interface, as discussed earlier. In addition, the surface should be initially hydrophobic, that is $\theta_0 > \pi/2$. Height of asperities is limited by their structural strength, and it is well known from mechanical consideration that the maximum height-to-width ratio required to support strength increases with decreasing scale. The requirement of low f_{SL} and high η and p are conflicting, in order to combine them, the pillars should be at least dense and wide enough to provide the spacing factor given by (7.31).

Based on the above analysis we propose the following set of requirements for optimized roughness distribution [247].

1. Roughness should be hierarchical, with several scale sizes, from microbumps to nanobumps. The largest asperities should be small compared to maximum droplet size, given by the capillary length.
2. Asperities should be high with their height limited by the requirement of their structural strength. It is known that the strength of geometrically similar structures increases with decreasing scale, since forces (e.g., weight) are usually proportional to volume and therefore the third power of length, while strength is proportional to the cross-section area and thus to the second power of length.
3. Asperities at each scale level should have a small width and a large distance between them (small S_f), however, this requirement is limited by some critical value of the spacing factor, providing the ability to support required pressure. The critical value found experimentally by Bhushan et al. [53] was in the range $0.083 < S_f < 0.111$. For liquid flow, a compromise between the ability to support pressure and the slip length is required. In addition, large pitch between asperities may lead to destabilization due to the capillary waves.
4. Nanoasperities should be convex (bumps rather than grooves) to stabilize the liquid–air interface, as it will be discussed in the next chapter.
5. For an initially hydrophilic surface a hydrophobic coating is required.

These requirements for a rough surface are summarized in Table 7.1. Remarkably, all these requirements are satisfied by biological water-repellent surfaces, such as leaves, which have hierarchical roughness with tightly packed convex papillae (asperities) and nanobumps/nano-"hairs" above them as well as hydrophobic wax coating.

Table 7.1. Roughness optimization for superhydrophobic surfaces [244]

Purpose	Method to achieve	Problems	Solution
High contact angle	Composite interface	Condensation of nanodroplets	Hierarchical roughness
Low contact angle hysteresis	(low S_f)	Surface inhomogeneity	Convex asperities
High slip length		Liquid–air interface waves	High asperities, but not exceeding structural strength limit
	High asperities	Exceeding maximum structural strength	
Support high pressure	High density and perimeter of asperities	Contradicts to low S_f	Not to exceed required spacing factor
Support both micro- and nanodroplets	Hierarchical roughness		

7.5 Effect of the Hierarchical Roughness

Many natural superhydrophobic surfaces have hierarchical or multiscale roughness, built usually of submicron-sized details superimposed on larger scale (usually, microns or dozens of microns) details. In the preceding sections, we considered the stability of the composite interface. In this section, we will discuss the hierarchical (multiscale) roughness and, in particular, its role in stabilizing the composite interface.

7.5.1 Hierarchical Roughness

Although there is significant literature about superhydrophobicity and numerous attempts have been made to produce artificial biomimetic roughness-induced hydrophobic surfaces [78, 105, 112, 255, 261, 262, 293, 308, 332, 350], many details of the mechanism of roughness-induced nonwetting are still not well understood. In particular, it is not clear why lotus leaves and other natural hydrophobic surfaces have a multiscale (or hierarchical) roughness structure, that is, nanoscale bumps superimposed on microscale asperities. Gao and McCarthy [124] recently suggested that multiscale roughness affects the kinetics of droplet motion and the Laplace pressure at which water intrudes between the bumps. In the present study we investigate the effect of the multiscale roughness on the stability of the roughness-induced hydrophobic interface.

Wetting of a solid by a liquid is characterized by the contact angle, which is the angle between the solid–air and the liquid–air interfaces. The greater the contact angle, the more hydrophobic is the material. The value of the contact angle is usually greater when the liquid is added (so-called advancing contact angle) than when it is removed (receding contact angle). The difference between the advancing and receding contact angle constitutes the contact angle hysteresis. The contact angle hysteresis is related to energy barriers, which a liquid droplet should overcome during its flow along a solid surface, and thus characterizes resistance to the flow. The lower the adhesion of a liquid droplet to the solid, the smaller the energy barriers, the lower the value of contact angle hysteresis, and the easier it is for the droplet to flow along the surface.

Several mechanisms are responsible for superhydrophobicity of natural surfaces, such as lotus leaves. First, these surfaces are coated with wax, which is hydrophobic itself (with contact angle of about 103° [175]), Second, they have a complicated geometrical structure with bumps or asperities (in the case of plant leaves called *papillae*) on the microscale (for the lotus leaf, the typical size of papillae is on the order of $10\,\mu m$) covered with much smaller nanoscale bumps or nanometer-scale structures [77, 274, 275]. In a similar manner, water strider legs are covered with a large number of oriented tiny hairs (microsetae) with fine nanogrooves [123]. Neinhuis and Barthlott [227] suggested that hierarchical surfaces are less vulnerable against mechanical damage of nanostructures and therefore maintain functionality even after damage. Wagner et al. [330] showed that hierarchically structured surfaces are more readily able to repel water even if the surface tension is drastically reduced as compared to surfaces with only one length scale of roughening. This might be of importance in wetlands or other aquatic habitats where water is often polluted due to decaying plant material and other contaminants that reduce surface tension [330]. Herminghaus [151] pointed out that certain self-affine profiles with multiscale roughness may result in superhydrophobic surfaces even for hydrophilic materials. However, the theoretical explanation of the predominance of hierarchically structured surfaces in nature remains an important task.

As discussed in the preceding chapters, in order to be superhydrophobic, a rough surface should be able to maintain a composite interface with air pockets or bubbles trapped in the valleys between the asperities [171, 197, 216, 260], as opposed to a homogeneous solid–liquid interface. In many cases both the composite interface and the homogeneous interface may exist for the same surface, however, only the composite interface provides the required superhydrophobic properties. Furthermore, the composite interface is much less stable than the homogeneous interface, and it may be destroyed by liquid filling the valleys between asperities and form an homogeneous interface, whereas the opposite transition has never been observed [274, 275]. The mechanisms of this transition have been the subject of intensive investigation [77, 197, 216, 232, 241]. Among the suggested factors which affect the transition are the effects of the droplet's weight and its curvature. For small droplets, surface effects dominate over the gravity, and the latter is hardly responsible for the transition, while the droplet's curvature may be responsible. The earlier discussion suggests that stability of a composite interface is a key issue for the design of roughness-induced superhydrophobic surfaces. In this chapter, we formulate a geometrical stability criterion and then investigate typical two-dimensional and three-dimensional surfaces with roughness at several scale levels. We show that a multiscale (hierarchical) roughness may enhance the stability of a composite interface.

7.5.2 Stability of a Composite Interface and Hierarchical Roughness

The spreading of liquid through porous media with a periodic geometry was studied by several authors [291, 322]; however, the stability of the composite interface has not been studied in detail in the literature. In this section, we will formulate a geometrical stability condition for a composite interface. The condition will be based on the free energy minimization using the Lagrange method of finding a minimum of a function of several variables with constraints. First, we will formulate the extremum criterion and show that it leads to the well-known Young equation, and then we will derive a stability criterion mathematically, and its physical meaning will be discussed.

The liquid–air interface is at equilibrium if the free energy of the solid–liquid–air system reaches its minimum. In order to find local conditional minima of the free surface energy $W = A_{SL}\gamma_{SL} + A_{SA}\gamma_{SA} + A_{LA}\gamma_{LA}$ with the constant volume constraint, $V = V_0$ (this requirement corresponds to the quasi-thermodynamic, i.e., slow evaporation/condensation limit), the Lagrange function is constructed

$$L(A_{SL}, A_{SA}, A_{LA}, V, \lambda) = A_{SL}\gamma_{SL} + A_{SA}\gamma_{SA} + A_{LA}\gamma_{LA} + p(V - V_0), \quad (7.83)$$

where A_{SL}, A_{SA}, and A_{LA} are areas of the solid–liquid, solid–air, and liquid–air interfaces and γ_{SL}, γ_{SA}, γ_{LA} are corresponding free energies, V_0 is the volume, and p is the Lagrange multiplier [135], having the dimension of pressure. The corresponding change of L is given by

$$\delta L = \delta A_{SL}\gamma_{SL} + \delta A_{SA}\gamma_{SA} + \delta A_{LA}\gamma_{LA} + \lambda \delta V + \delta p(V - V_0). \qquad (7.84)$$

Note that the arguments of L are interdependent with $\delta A_{SL} = -\delta A_{SA}$ whereas δA_{LA} consists of two terms, $\delta A_{LA} = \delta A_{LAT} + \delta A_{LAV}$. The first term, δA_{LAT}, is due to a change in position of the triple line (line of contact between solid, liquid and air), and the second, δA_{LAV}, is due to a change of the shape of the liquid–air interface. Furthermore, $\delta A_{LAT} = \delta A_{SL} \cos \theta$ from geometrical considerations [232].

Suppose the shape of the liquid–air interface is given parametrically by vector $\vec{r}(u, v)$, where u and v are parameters that uniquely characterize any point at a surface, and the shape changes slightly

$$\breve{r}(u, v) = \vec{r}(u, v) + \delta \vec{r}(u, v). \qquad (7.85)$$

The change due to the shape of the liquid–air interface is given by the area of an element of the liquid–air interface $A(u, v)\, du\, dv$ times the normal displacement multiplied by the sum of principal radii of curvature $\vec{n}\delta\vec{r}(1/R_1 + 1/R_2)$, where $\vec{n}$ is the normal vector and R_1, R_2 are the principal radii of curvature

$$\delta A_{LAV} = \iint_{A_{LA}} \vec{n}\delta\vec{r}(1/R_1 + 1/R_2)A\, du\, dv, \qquad (7.86)$$

where [135]

$$A(u, v) = \left[\left(\frac{\partial \vec{r}}{\partial u}\right)^2 \left(\frac{\partial \vec{r}}{\partial v}\right)^2 - \left(\frac{\partial \vec{r}}{\partial u}\frac{\partial \vec{r}}{\partial v}\right)^2\right]^{1/2}. \qquad (7.87)$$

The change of volume is given by

$$\delta V = \iint_{A_{LA}} \vec{n}\delta\vec{r} A\, du\, dv. \qquad (7.88)$$

Combining (7.86)–(7.88) and setting $\delta L(\delta A_{SL}, \delta\vec{r}, \delta V) = 0$ yields

$$\delta L = \delta A_{SL}\left[\cos \theta_0 - \frac{\gamma_{SA} - \gamma_{SL}}{\gamma_{LA}}\right]\gamma_{LA}$$
$$+ \iint_{A_{LA}} \left[\gamma_{LA}(1/R_1 + 1/R_2) + p\right]\vec{n}\delta\vec{r} A\, du\, dv + \lambda\delta V, \qquad (7.89)$$

which results in three equations that should be satisfied simultaneously. The first is the Young equation for the contact angle θ_0, which should be satisfied at the points of the triple line

$$\cos \theta_0 = \frac{\gamma_{SA} - \gamma_{SL}}{\gamma_{LA}}. \qquad (7.90)$$

The second equation for the Lagrange multipliers $p = -\gamma_{LA}(1/R_1 + 1/R_2)$ is satisfied only if the curvature $1/R_1 + 1/R_2$ is a constant independent of u and v throughout the entire liquid–air interface [232]. Physically, of course, this condition reflects Laplace pressure drop through a curved interface The third equation is just the condition of constant volume $V = V_0$ [232].

In order for the extremum to be a local minimum (rather than maximum) of W, the equilibrium should also satisfy the stability condition $d^2 W > 0$. Differentiating twice $W = A_{SL}\gamma_{SL} + A_{SA}\gamma_{SA} + A_{LA}\gamma_{LA}$ and using $\delta A_{LA} = \delta A_{SL} \cos\theta$ yields

$$d^2 W = d^2 A_{SL}\left[\cos\theta_0 - \frac{\gamma_{SA} - \gamma_{SL}}{\gamma_{LA}}\right]\gamma_{SL} + dA_{LA}d(\cos\theta) > 0. \qquad (7.91)$$

We ignored the effect of changing shape of the liquid–air interface (the term corresponding to δA_{LAV}), since it is known that $1/R_1 + 1/R_2 = \text{const}$ provides the minimum (rather than maximum) liquid–air interface area condition, and only the effect of moving triple line is of interest for us. Using (7.90), which is satisfied at the equilibrium, and the fact that $\cos\theta$ decreases monotonically with θ at the domain of our interest, $0 < \theta < 180°$, yields [232]

$$dA_{SL}\, d\theta < 0. \qquad (7.92)$$

In other words, for the interface to be stable, for advancing liquid (increasing A_{SL}) the value of the contact angle should decrease, whereas for receding liquid the contact angle should increase. Note also that for a liquid–air interface coming to the solid surface under the angle θ, an advance of the interface results in the change of energy

$$dW = dA_{SL}(\gamma_{SL} - \gamma_{SA}) + dA_{LA}\gamma_{LA} = dA_{SL}(\gamma_{SL} - \gamma_{SA}) + dA_{SL}\gamma_{LA}\cos\theta$$
$$= dA_{SL}\gamma_{LA}\left(-\frac{\gamma_{SA} - \gamma_{SL}}{\gamma_{LA}} + \cos\theta\right) = dA_{SL}\gamma_{LA}(\cos\theta - \cos\theta_0). \qquad (7.93)$$

Thus, if $\theta > \theta_0$, the energy decreases, and it is energetically profitable for the liquid to advance, whereas if $\theta < \theta_0$, the liquid would retreat. So, the physical meaning of (7.92) is that for a small advance/retreat of the liquid it should be more energetically profitable to return to the original position rather than to continue advancing/retreating [232].

For a two-dimensional surface, since a change of angle $d\theta$ is equal to the change of the slope of the surface, whether the configuration is stable or not depends on the sign of the curvature of the surface. The convex (bumpy) surface leads to a stable interface, whereas a concaved (groovy) surface leads to an unstable interface. The liquid keeps spreading until both (7.90) and (7.92) are satisfied at the triple line and $1/R_1 + 1/R_2 = \text{const}$ at the liquid–air interface, provided the volume of the liquid is conserved.

In the next section, we apply the stability criterion (7.91) to typical two-dimensional and three-dimensional surfaces with multiscale roughness.

7.5.3 Hierarchical Roughness

In this section, we consider several surfaces with nanoscale roughness superimposed on larger microscale pillars and investigate the effect of concaved and convex nanoroughness upon the stability of a composite interface. We study the case of an infinitely large reservoir of liquid on top of the pillars. In most applications, liquid droplets of finite size are in contact with a rough surface; however, the size of roughness details is small compared to the size of the droplets and for practical purposes droplet size can be considered infinite.

7.5.3.1 Two-Dimensional Roughness

Consider a two-dimensional structure with rectangular pillars of height h and width a separated by distance b, covered with small semicircular ridges and grooves of radius r (Fig. 7.13(a)). Since the distance between the pillars is small in comparison with the capillary length, and therefore the effect of gravity is negligible, we can assume that the liquid–air interface is a horizontal plane, and its position is characterized by the vertical coordinate z. The free energy is given by [232]

$$W = A_{SL}\gamma_{SL} + A_{SA}\gamma_{SA} + A_{LA}\gamma_{LA} = rL\gamma_{LA}(\sin\alpha - \alpha\cos\theta_0), \quad 0 < z < h,$$
(7.94)

where $\alpha = \arccos((r - z)/r) + 2\pi N$ is the angle corresponding to vertical position of the interface z, N is the number of a ridge or groove, and L is the length of the grooves in the y-direction, which is required based on the dimensional considerations. The dependence is presented in Fig. 7.13(b) for the cases of hydrophobic ($\theta_0 = 150°$) and hydrophilic ($\theta_0 = 30°$) materials for both the bumpy and the groovy surface. It is seen that, for the bumpy surface, there are many states of stable equilibrium (shown in Fig. 7.13(a) with dotted lines), separated by energy barriers, which correspond to every ridge, whereas for the groovy surface equilibrium states are unstable. Therefore, the ridges can pin the triple line and thus lead to a composite interface. In the case of a hydrophilic surface, each lower position of the equilibrium state corresponds to a lower value of W, therefore, when the liquid advances from one equilibrium state to the next, the total energy decreases, and thus the liquid's advance is energetically profitable. When the liquid reaches the bottom of the valley and completely fills the space between the pillars forming a homogeneous interface, the total energy decreases dramatically by the value of

$$\Delta W = bL(\gamma_{SA} + \gamma_{LA} - \gamma_{SL}) = bL\gamma_{LA}(1 + \cos\theta_0).$$
(7.95)

The opposite transition from a homogeneous interface to a composite interface requires high activation energy ΔW and is thus unlikely, making the transition from composite interface to homogeneous interface irreversible. Since the distance between the pillars b is much greater than r, the energy barriers which separate the equilibrium states $2\pi rL\gamma_{LA}\cos\theta_0$ are relatively small compared with ΔW, and low activation energy is required for the liquid to spread and propagate from one equilibrium state to the other [243, 244].

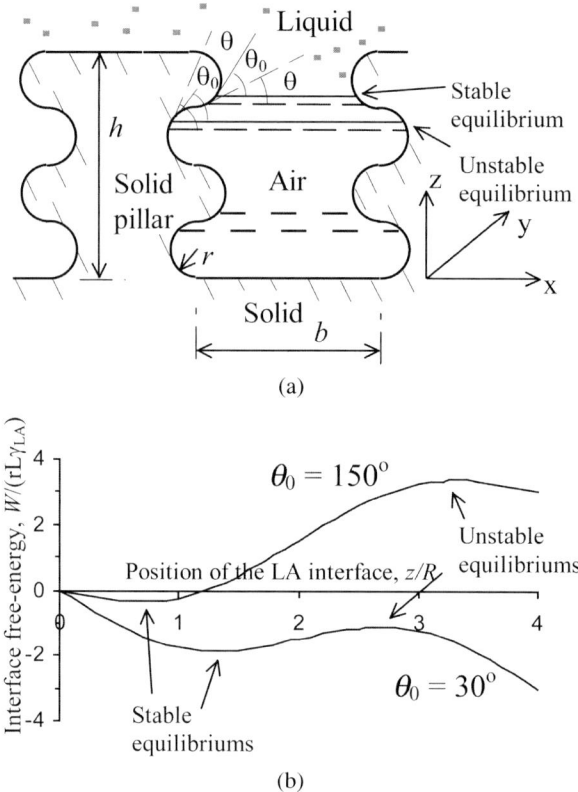

(a)

(b)

Fig. 7.13. Two-dimensional pillars with semi-circular bumps/grooves. **a** Schematics of the structure. The bumps may pin the triple line, because an advance of the liquid–air interface results in decrease of the contact angle ($\theta < \theta_0$), making equilibrium stable. Grooves provide equilibrium positions, which satisfy the Young equation, however, the equilibrium is unstable, because an advance of the liquid–air interface results in increase of the contact angle ($\theta > \theta_0$). **b** Energy profiles for configurations in Fig. 7.13(a) with bumps and grooves for hydrophilic ($\theta_0 = 30°$) and hydrophobic ($\theta_0 = 150°$) materials. Energy (normalized by $Lr\gamma_{LA}$) is shown as a function of vertical position of the interface z (normalized by the radius of bumps/grooves r). Bumps result in stable equilibriums (energy minima), whereas grooves result in unstable equilibriums (energy maxima) [232]

Since the change of angle $d\theta$ for a two-dimensional surface is equal to the change of surface slope, based on (7.92), it depends upon the sign of curvature of the surface whether the configuration is stable or not. The convex (bumpy) surface leads to a stable interface, whereas the concaved (groovy) surface leads to an unstable interface. The liquid keeps spreading until both (7.90) and (7.92) are satisfied at the triple line and $1/R_1 + 1/R_2 = \text{const}$ at the liquid–air interface, provided the volume of the liquid is conserved, which is the case for a slow thermodynamic process.

7.5.3.2 Three-Dimensional Pillars with Ridges and Grooves

Consider now a three-dimensional structure with circular pillars of height h and radius R separated by distance b and distributed hexagonally with the density of $\eta = 2/[\sqrt{3}(2R + b)^2]$ pillars per unit area, covered with small ridges and grooves of radius r (Fig. 7.13(a)). Similar to the preceding section, the free energy per area S is given by the circumference of a pillar $2\pi R$ times the number of pillars ηS times $r\gamma_{LA}(\sin\alpha - \alpha\cos\theta_0)$

$$W = 2\pi R\eta S r\gamma_{LA}(\sin\alpha - \alpha\cos\theta_0), \quad 0 < z < h. \qquad (7.96)$$

The similarity between (7.94) and (7.96) is noted: both energy profiles are different only in their normalization constant, so the dependence of the free energy upon the position of the interface presented in Fig. 7.13(b) for the case of two-dimensional pillars has qualitatively the same profile as for the case of three-dimensional pillars. In a manner similar to the case of two-dimensional pillars, the ridges can pin the triple line [232].

7.5.3.3 Three-Dimensional Surface

In the previous sections, we considered two-dimensional nanoscale ridges and grooves superimposed on two- and three-dimensional pillars. Real superhydrophobic surfaces, such as plant leaves, are three-dimensional with three-dimensional nanobumps. For three-dimensional surfaces, the shape of the liquid–air interface may be quite complex and thus the stability of the composite interface is difficult to analyze. In order to consider a three-dimensional configuration that allows for a plane horizontal liquid–air interface, we will investigate the surface composed of circular pillars of height h and radius R separated by distance b with the density of $\eta = 2/[\sqrt{3}(2R + b)^2]$ pillars per unit area (following the hexagonal distribution pattern shown in Fig. 7.14), which are formed of layers of small spheres of radius r, packed according to the hexagonal pattern (Fig. 7.15(a)). The packing density of the spheres is equal to $1/(2\sqrt{3}r^2)$ spheres per unit area in every horizontal layer. The liquid–air interface area is now given by the total flat area of the surface, A_0, minus the cross-sectional area of spheres under water. The latter is given by A_0 times the pillar density η times the pillar area πR^2 times the packing density of the spheres $1/(2\sqrt{3}r^2)$ times the cross-sectional area of individual sphere under water, $\pi(r\sin\alpha)^2$, which yields [232]

$$A_{LA} = A_0\left(1 - \frac{\eta\pi^2 R^2 \sin^2\alpha}{2\sqrt{3}}\right). \qquad (7.97)$$

The solid–liquid interface area is equal to the total surface area of the spheres under water, which is given by the number of spheres $\eta A_0\pi^2 R^2/(2\sqrt{3}r^2)$ times the spheres' surface area multiplied by the number of layers $4\pi r^2 N$ plus the area of the spheres in the layer, which is only partially under water $\pi(z^2 + 2z(2r - z))$

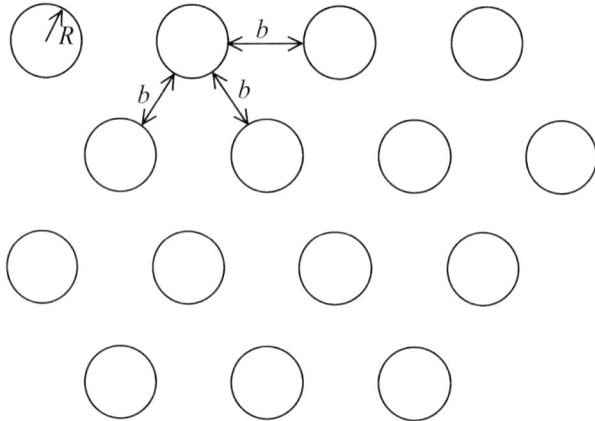

Fig. 7.14. Schematics of spatial distribution of three-dimensional pillars with semi-circular bumps/grooves upon a surface [232]

$$A_{SL} = \frac{\eta A_0 \pi^2 R^2}{2\sqrt{3}r^2}\left[4r^2 N + \left(z^2 + 2z(2r - z)\right)\right]. \tag{7.98}$$

Using $\sin^2 \alpha = 1 - \cos^2 \alpha = 1 - ((r - z)/r)^2 = 2z/r - (z/r)^2$, the free energy is now given by

$$W = A_{LA}\gamma_{LA} + A_{SL}(\gamma_{LA} - \gamma_{SA}) = \gamma_{LA}(A_{LA} + A_{SL}\cos\theta_0)$$
$$= A_0\gamma_{LA}\left(1 - \frac{\eta\pi^2 R^2}{2\sqrt{3}}\{2z/r - (z/r)^2\right.$$
$$\left. - \left[4\pi N + (z/r)^2 + 2(z/r)(2 - z/r)\right]\cos\theta_0\}\right). \tag{7.99}$$

The dependence of the free energy, normalized by $A_0\gamma_{LA}$, upon the vertical position z is presented in Fig. 7.15(b) for the cases of hydrophobic ($\theta_0 = 105°$) and hydrophilic ($\theta_0 = 75°$) materials [232].

7.5.4 Results and Discussion

We studied three different surface profiles with large-scale pillars and small-scale roughness superimposed on the pillars. Figures 7.13(b) and 7.15(b) show that for both the hydrophobic and hydrophilic materials, a convex surface leads to stable equilibriums, whereas a concaved surface leads to unstable equilibriums. Therefore, a convex small-scale roughness can pin the liquid–air interface even in the case of a hydrophilic material. This may be important for producing reliable superhydrophobic surfaces, since the factors destabilizing the liquid–air interface, such as nanodroplet condensation [78, 256, 339], chemical surface heterogeneity [75], and capillary waves [241] are scale-dependent and therefore multiscale roughness is required to control the stability.

Side view

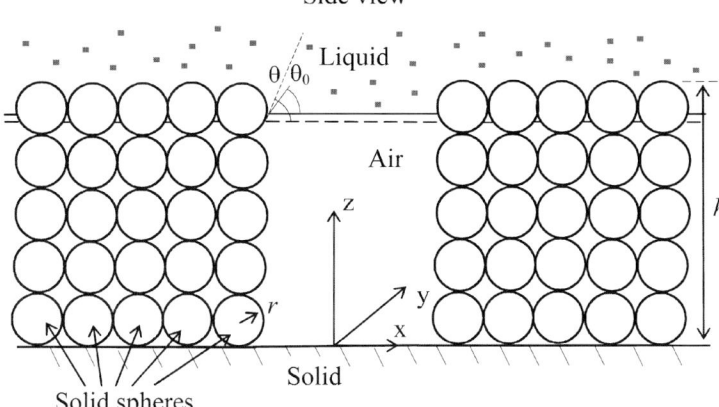

Top view

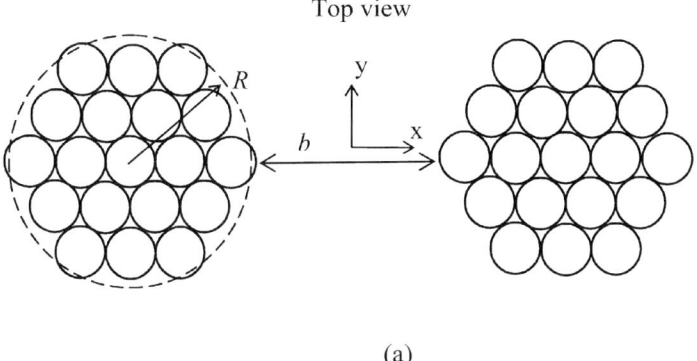

(a)

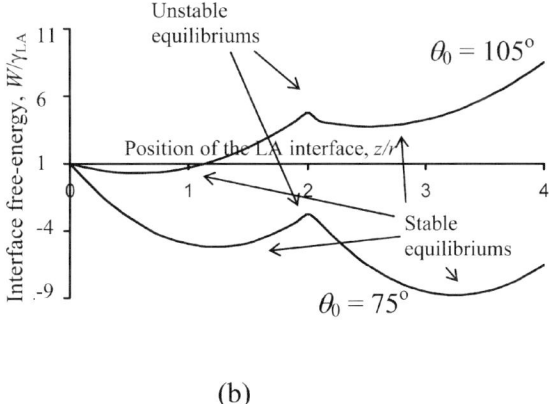

(b)

Fig. 7.15. Three-dimensional pillars consisting of small solid spheres. **a** Schematics of the structure. **b** Energy (normalized by $A_0\gamma_{LA}$) as a function of vertical position of the interface z (normalized by the radius of bumps/grooves r) for $\pi^2 R^2/(2\sqrt{3}r^2) = 1$ [232]

An experiment suggesting that the sign of curvature is indeed important for hydrophobicity was conducted by Sun et al. [308]. They produced both a positive and a negative replica of a lotus leaf surface by nanocasting, using poly(dimethylsiloxane), which has the contact angle with water of about 105°. This value is close to the contact angle of wax, which covers lotus leaves (about 103° [175]). The positive and negative replicas have the same roughness factor and thus should produce the same contact angle in the case of a homogeneous interface; however, the values of the surface curvature are opposite. The value of the contact angle for the positive replica was found to be 160° (same as for the lotus leaf), while for the negative replica it was only 110°. This result suggests that the high contact angle of the lotus leaf is due to the composite, rather than homogeneous interface and that the sign of surface curvature indeed plays a critical role for formation of the composite interface.

Natural and successful artificial superhydrophobic surfaces exhibit hierarchical multiscale roughness. Thus, the lotus leaf has microscale bumps (papillae) with a typical height and radius of $10\,\mu$m to $20\,\mu$m, which are covered with hydrophobic paraffin wax. Upon these bumps much smaller nanobumps are found of typical submicron sizes. Artificial biomimetic superhydrophobic surfaces should also have multiscale roughness.

To summarize, biomimetic superhydrophobic surfaces should satisfy the following requirements: they should have hydrophobic coating, high roughness factors, provide high contact angle, and the ability to form a composite interface. Achieving a stable composite interface requires a hierarchical roughness structure with nanoscale bumps upon microscale asperities and valleys.

The mechanism of roughness-induced hydrophobicity is complicated and involves effects at various scale ranges. For most superhydrophobic surfaces, it is important that composite solid–liquid–air interface is formed. A composite interface dramatically decreases the area of contact between liquid and solid and, therefore, decreases adhesion of a liquid droplet to the solid surface and contact angle hysteresis. Formation of a composite interface is also a multiscale phenomenon that depends upon the relative sizes of the liquid droplet and roughness details. The composite interface is fragile, since transition to a homogeneous interface is irreversible. Therefore, stability of a composite interface is crucial for superhydrophobicity and should be addressed for the successful development of superhydrophobic surfaces. We have demonstrated that a multiscale roughness can help resist the destabilization, with convex surfaces pinning the interface and thus leading to stable equilibrium, and prevents filling the gaps between the pillars even in the case of a hydrophilic material. The effect of roughness on wetting is scale dependent and mechanisms that lead to the destabilization of a composite interface are also scale dependent. To effectively resist these scale-dependent mechanisms, a multiscale roughness is required. Such multiscale roughness was found in natural and successful artificial superhydrophobic surfaces.

7.6 Summary

In this chapter, we studied several effects related to roughness-induced hydropho-bicity. It is not enough that the Young and Laplace equations are satisfied at the liquid–vapor interface. The interface should also be stable, so a stability criterion was formulated. In addition, we explored the possibility that the valleys between the asperities are only partially filled with water, that is, the filling of the valleys is a probabilistic process. In the proposed model, filling of the valleys is thermally ac-tivated (the probability is proportional to $\exp(kT)$). This may be correct for small (nanoscale) energy barriers between metastable states. For larger energy barriers, the probabilistic law may be different; however, the qualitative picture is expected to be similar. A more detailed discussion of the transition will be given in Chap. 8. We also studied a simple three-dimensional surface, which involved both peaks (asper-ities or pillars) and holes (valley). The three-dimensional surface is different from two-dimensional profiles, for example, in that the holes are isolated (and air pressure may be an issue, although ignored in our simplified analysis). Comparing the effects of the peaks and valleys is interesting, since they constitute a negative replica of each other. In a similar manner, the effects of bubbles and droplets may be compared (it will be discussed in more detail in Chap. 9). We also discussed dependence of the meniscus force, which is often the principal component of the friction force, upon roughness. In addition, we formulated design optimization considerations. The inter-actions involved in the studied processes act at various scale levels, making wetting essentially a multiscale process. We also considered the multiscale roughness and its effect upon the stability of the composite interface and found that the introduc-tion of small-scale roughness may pin the triple line and thus make the liquid–vapor interface stable even in the case of an initially hydrophilic surface ($\theta_0 < 90°$).

8

Cassie–Wenzel Wetting Regime Transition

Abstract Experimental results and theoretical considerations of the Cassie–Wenzel wetting regime transition are discussed. The transition can be considered as a multiscale phenomenon involving interaction at the molecular, micro (size of a solid surface pattern) and macroscale. The transition can also be viewed as a phase transition. The reversible superhydrophobicity (stimulated by light and UV irradiation, electric potential, or changing surface chemistry) is also discussed.

In this chapter we will discuss the practically important case of the destabilization of the composite interface: the transition from the Cassie (composite) to the Wenzel (homogeneous) wetting regime for micropatterned surfaces.

8.1 The Cassie–Wenzel Transition and the Contact Angle Hysteresis

It is known from experimental observations that the transition from the composite to homogeneous interface (called also Cassie and Wenzel states) is an irreversible event [22, 40, 197]. Whereas such a transition can be induced, for example, by applying pressure or force to the droplet [197], electric voltage [17, 195], light for a photocatalytic texture [112], and vibration [60], the opposite transition has never been observed, although there is no apparent reason for that. Several approaches have been proposed to investigate the Cassie–Wenzel transition. Lafuma and Quèrè [197] suggested that the transition takes place when the net surface energy of the Wenzel state becomes equal to that of the Cassie state or, in other words, when the contact angle predicted by the Cassie equation is equal to that predicted by the Wenzel equation. They noticed that in certain cases the transition does not occur even when it is energetically profitable and they considered such a Cassie state metastable. Extrand [109] suggested that the weight of the droplet is responsible for the transition and proposed the contact line density model, according to which the transition takes place when the weight exceeds the surface tension force at the triple line. Patankar [261, 262] suggested that which of the two states is realized might depend upon how the droplet

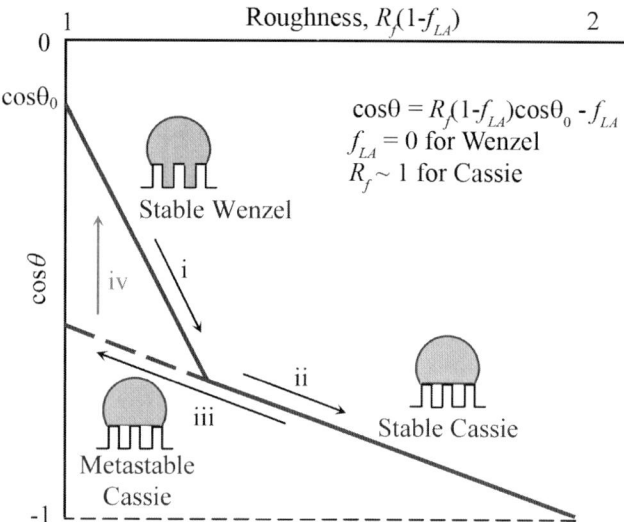

Fig. 8.1. Wetting hysteresis for a superhydrophobic surface. Contact angle as a function of roughness. The transition (i)–(ii) corresponds to equal Wenzel and Cassie states free energies, whereas the transition (iv) corresponds to a significant energy dissipation, and thus it is irreversible. The metastable Cassie state (iii) can abruptly transform (iv) into the stable Wenzel state [250]

was formed, that is upon the history of the system. Quéré [275] also suggested that the droplet curvature (which depends upon the pressure difference between inside and outside of the droplet) governs the transition. Nosonovsky and Bhushan [241] suggested that the transition is a dynamic process of destabilization and identified possible destabilizing factors. It has also been suggested that the curvature of multiscale roughness defines the stability of the Cassie state [232, 243] and that the transition is a stochastic gradual process [60, 160, 240]. Numerous experimental results support many of these approaches, but it is not clear which particular mechanism prevails.

There is an asymmetry between the wetting and dewetting processes, since droplet nucleation requires less energy than vapor bubbles nucleation (cavitation). During wetting, which involves the creation of the solid–liquid interface, less energy is released than the amount required for dewetting or destroying the solid–liquid interface due to the adhesion hysteresis. Adhesion hysteresis is one of the factors leading to contact angle hysteresis, and it also results in the hysteresis of the Wenzel–Cassie state transition. Figure 8.1 shows the contact angle of a rough surface as a function of surface roughness parameters, given by (6.12). Here it is assumed that $R_f \sim 1$ for the Cassie–Baxter regime if the liquid droplet with a stable composite interface sits flat over the surface. It is noted that at a certain point, the contact angles given by the Wenzel and Cassie–Baxter equations are the same, and $R_f = (1 - f_{LA}) - f_{LA}/\cos\theta_0$. At this point, the lines corresponding to the Wenzel and Cassie regimes intersect. This point corresponds to an equal net energy of the Cassie and Wenzel states. For a lower roughness (e.g., larger pitch between the pillars) the Wenzel state is more energetically profitable, whereas for a higher roughness the Cassie state is more energetically profitable.

Figure 8.1 shows that an increase of roughness may lead to the transition between the Wenzel and Cassie regimes at the intersection point. With decreasing roughness, the system is expected to transit to the Wenzel state. However, experiments show [22, 53] that—despite the energy of the Wenzel state being lower than that of the Cassie state—the transition does not necessarily occur, and the droplet may remain in the metastable Cassie state. This is because there are energy barriers associated with the transition, which occurs due to destabilization by dynamic effects (such as waves and vibration).

In order to understand contact angle hysteresis and the transition between the Cassie and Wenzel states, the shape of the free surface energy profile can be analyzed. The free surface energy of a droplet upon a smooth surface as a function of the contact angle has a distinct minimum, which corresponds to the most stable contact angle. As shown in Fig. 8.2(a), the macroscale profile of the net surface energy allows us to find the contact angle (corresponding to energy minimums); however, it fails to predict the contact angle hysteresis and Cassie–Wenzel transition, which are governed by micro- and nanoscale effects. As soon as the microscale substrate roughness is introduced, the droplet shape can no longer be considered as an ideal truncated sphere, and energy profiles have multiple energy minimums, corresponding to location of the pillars (Fig. 8.2(b)). The microscale energy profile (solid line) has numerous energy maxima and minima due to the surface micropattern. While the exact calculation of the energy profile for a 3D droplet is complicated, a qualitative shape may be obtained by assuming a periodic sinusoidal dependence [171] superimposed upon the macroscale profile, as shown in Fig. 8.2(b). Thus the advancing and receding contact angles can be identified as the maximum and the minimum possible contact angles corresponding to energy minimum points. However, the transition between the Wenzel and Cassie branches still cannot be explained. Note also that Fig. 8.2(b) explains qualitatively the hysteresis due to the kinetic effect of the pillars, but not the inherited adhesion hysteresis, which is characterized by the molecular scale length and cannot be captured by the microscale model.

The energy profile as a function of the contact angle does not provide any information on how the transition between the Cassie and Wenzel states occur, because these two states correspond to completely isolated branches of the energy profile in Fig. 8.2(a)–(b). However, the energy may depend not only upon the contact angle, but also upon micro/nanoscale parameters, such as the vertical position of the liquid–vapor interface under the droplet, h (assuming that the interface is a horizontal plane) or similar geometrical parameters (assuming a more complicated shape of the interface). In order to investigate the Wenzel–Cassie transition, the dependence of the energy upon these parameters should be studied. We assume that the liquid–vapor interface under the droplet is a flat horizontal plane. When such vapor layer thickness or the vertical position of the liquid–vapor interface, h, is introduced, the energy can be studied as a function of the droplet's shape, the contact angle, and h (Fig. 8.2(c)). For an ideal situation the energy profile has an abrupt

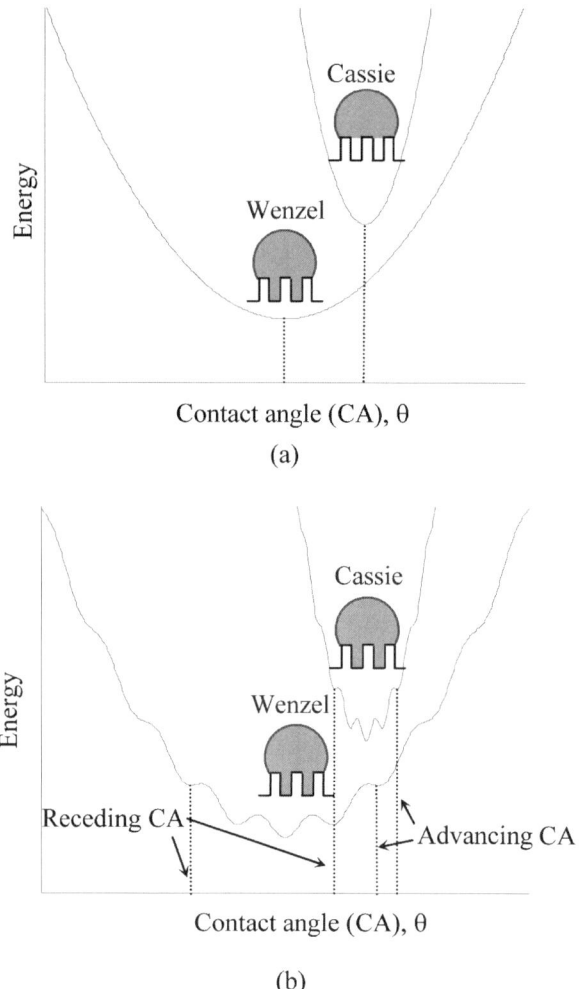

Fig. 8.2. Schematics of net free energy profiles. **a** Macroscale description; energy minimums correspond to the Wenzel and Cassie states. **b** Microscale description with multiple energy minimums due to surface texture. Largest and smallest values of the energy minimum correspond to the advancing and receding contact angles. **c** Origin of the two branches (Wenzel and Cassie) is found when a dependence of energy upon h is considered for the microscale description (*solid line*) and nanoscale imperfectness (*dashed line*) based on [241]. When the nanoscale imperfectness is introduced, it is observed that the Wenzel state corresponds to an energy minimum and the energy barrier for the Wenzel–Cassie transition is much smaller than for the opposite transition [250]

minimum at the point corresponding to the Wenzel state, which corresponds to the sudden net energy change due to destroying solid–vapor and liquid–vapor interfaces ($\gamma_{SL} - \gamma_{SV} - \gamma_{LV} = -\gamma_{LV}(\cos\theta + 1)$) times the interface area (Fig. 8.2c). In a

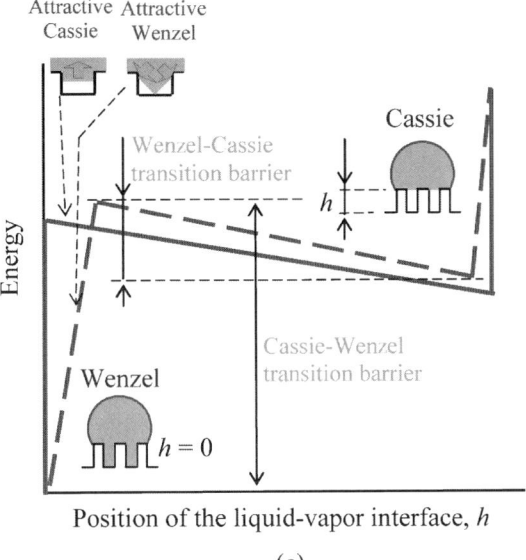

Position of the liquid-vapor interface, h

(c)

Fig. 8.2. (*Continued*)

more realistic case, the liquid–vapor interface cannot be considered horizontal due to nanoscale imperfectness or dynamic effects such as the capillary waves [241]. A typical size of the imperfectness is much smaller than the size of the details of the surface texture and thus belongs to the molecular scale level. The height of the interface, h, can now be treated as an average height. The energy dependence upon h is now not as abrupt as in the idealized case. For example, for the "triangular" shape of the interface, as shown in Fig. 8.2(c), the Wenzel state may become the second attractor for the system. We see that there are two equilibriums that correspond to the Wenzel and Cassie states, with the Wenzel state corresponding to a much lower energy level. The energy dependence upon h governs the transition between the two states, and it is observed that a much larger energy barrier exists for the transition from Wenzel to Cassie than for the opposite transition. This is why the first transition has never been observed experimentally.

To summarize, we showed that the contact angle hysteresis and Cassie–Wenzel transition cannot be determined from the macroscale equations and are governed by micro- and nanoscale phenomena. Our theoretical arguments are supported by our experimental data on micropatterned surfaces.

8.2 Experimental Study of the Cassie–Wenzel Transition

Jung and Bhushan [173, 174] studied two series of patterned Si surfaces covered with a monolayer of hydrophobic tetrahydroperfluorodecyltrichlorosilane (contact angle with a nominally flat surface, $\theta_0 = 109°$, advancing and receding contact angle

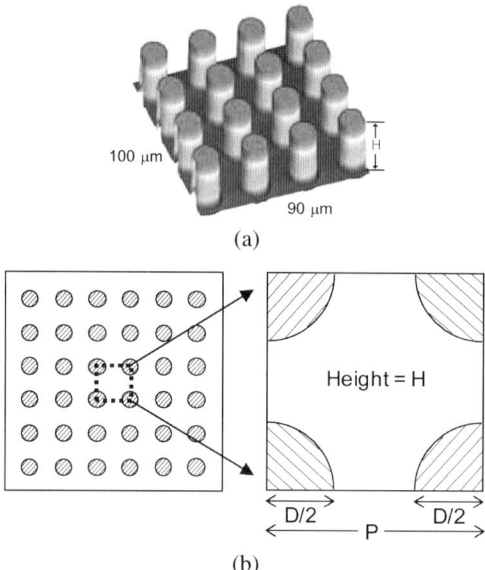

(a)

(b)

Fig. 8.3. a Optical profiler image and **b** schematic of the patterned surface [174]

are $\theta_{\text{adv0}} = 116°$ and $\theta_{\text{rec0}} = 82°$), formed by flat-top cylindrical pillars. Series
1 had pillars with the diameter $D = 5$ μm, height $H = 10$ μm, and pitch values
$P = (7, 7.5, 10, 12.5, 25, 37.5, 45, 60,$ and $75)$ μm, while series 2 had $D = 14$ μm,
$H = 30$ μm, $P = (21, 23, 26, 35, 70, 105, 126, 168,$ and $210)$ μm (Fig. 8.3). It is
convenient to introduce the spacing factor $S_{\text{f}} = D/P$ [243, 244]. The contact angle
and contact angle hysteresis of millimeter-sized water droplets upon the samples
were measured. In addition, we studied the contact angle and the Wenzel–Cassie
transition during evaporation of microscale droplets (Fig. 8.4). We found that the
contact angle hysteresis involves two terms: the term $S_{\text{f}}^2(\pi/4)(\cos\theta_{\text{adv0}} - \cos\theta_{\text{rec0}})$,
corresponding to the adhesion hysteresis (which is found even at a nominally flat
surface and is a result of molecular-scale imperfectness) and the term $H_{\text{r}} \propto D/P^2$
corresponding to microscale roughness and proportional to the edge line density.
Thus the contact angle hysteresis is given, based on (6.58) [53], by

$$\cos\theta_{\text{adv}} - \cos\theta_{\text{rec}} = \frac{\pi}{4}S_{\text{f}}^2(\cos\theta_{\text{adv0}} - \cos\theta_{\text{rec0}}) + H_{\text{r}}. \tag{8.1}$$

The data for contact angle hysteresis is shown in Fig. 8.5(a). It can be concluded
from the data that the contributions of both terms in the right-hand part of (8.1) are
of the same order of magnitude and, therefore, the interactions near the triple line
and at the solid–liquid contact area are of equal importance.

The droplet radius, R, at the Cassie–Wenzel transition was found to be proportional to P/D (or P/H) (Fig. 8.5(b)), which suggests that the transition is a linear
"one-dimensional" phenomenon and that neither droplet droop (that would involve
P^2/H) nor droplet weight (that would involve R^3) are responsible for the transition,

14-µm diameter, 30-µm height,and 105-µm pitch pillars

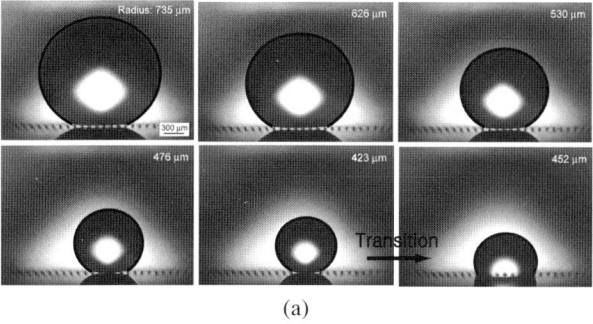

(a)

Droplet on the surface before and after transition

5-µm diameter, 10-µm height, and 37.5-µm pitch pillars

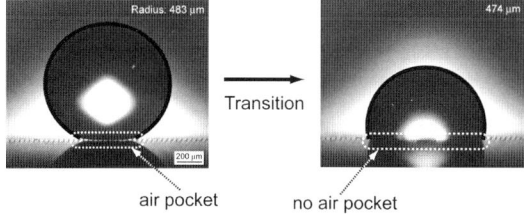

14-µm diameter, 30-µm height,and 105-µm pitch pillars

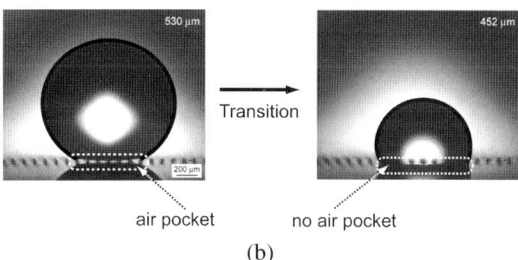

(b)

Fig. 8.4. Evaporation of a droplet on a patterned surface. **a** The initial radius of the droplet is about 700 µm, and the time interval between successive photos is 30 s. As the radius of droplet reaches 420 µm on the surface with 14 µm diameter, 30 µm height, and 105 µm pitch pillars, the transition from Cassie and Baxter regime to Wenzel regime occurs. **b** Before the transition air pocket is clearly visible at the bottom area of the droplet, but after transition, air pocket is not found at the bottom area of the droplet [174]

but rather linear geometric relations are involved. Note that the experimental values approximately correspond to the values of the ratio $RD/P = 50$ µm, or the total area of the pillar tops under the droplet $(\pi D^2/4)\pi R^2/P^2 = 6200$ µm^2.

Besides the contact angle hysteresis, the asymmetry of the Wenzel and Cassie states is the result of the wetting/dewetting asymmetry. While the fragile metastable Cassie state is often observed, as well as its transition to the Wenzel state, the op-

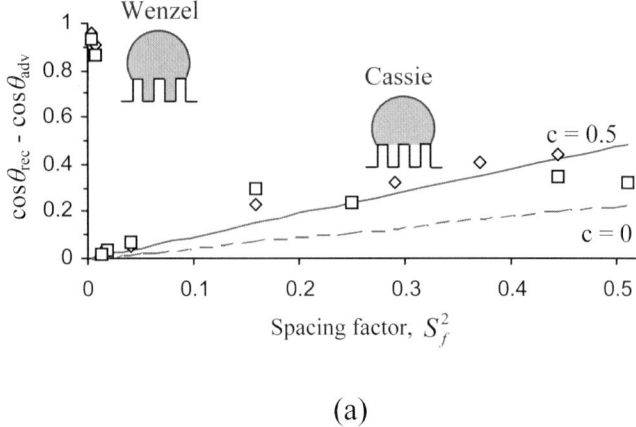

(a)

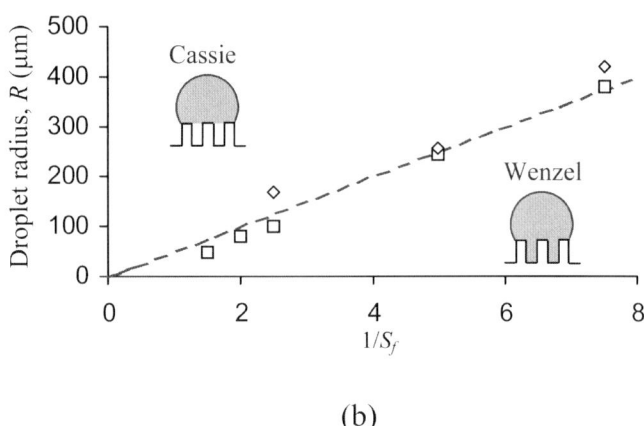

(b)

Fig. 8.5. a Contact angle hysteresis as a function of S_f for the 1st (*squares*) and 2nd (*diamonds*) series of the experiments compared with the theoretically predicted values of $\cos\theta_{adv} - \cos\theta_{rec}(D/P)^2(\eth/4)(\cos\theta_{adv0} - \cos\theta_{rec0}) + c(D/P)^2$, where c is a proportionality constant. It is observed that when only the adhesion hysteresis/interface energy term is considered ($c = 0$), the theoretical values are underestimated by about half, whereas $c = 0.5$ provides a good fit. Therefore, the contribution of the adhesion hysteresis is of the same order of magnitude as the contribution kinetic effects. **b** Droplet radius, R, for the Cassie–Wenzel transition as a function of $P/D = 1/S_f$. It is observed that the transition takes place at a constant value of $RD/P \sim 50$ μm (*dashed line*). This shows that the transition is a linear phenomenon [250]

posite transition never happens. Using (6.6) and (6.9), the contact angle with the patterned surfaces is given by [173]

$$\cos\theta = \left(1 + 2\pi S_f^2\right)\cos\theta_0 \qquad \text{(Wenzel state)}, \qquad (8.2)$$

$$\cos\theta = \frac{\pi}{4}S_f^2(\cos\theta_0 + 1) - 1 \quad \text{(Cassie state)}. \tag{8.3}$$

For a perfect macroscale system, the transition between the Wenzel and Cassie states should occur only at the intersection of the two regimes (the point at which the contact angle and net energies of the two regimes are equal, corresponding to $S_f = 0.51$). It is observed, however, that the transition from the metastable Cassie to stable Wenzel occurs at much lower values of the spacing factor $0.083 < S_f < 0.111$. As shown in Fig. 8.6(a), the stable Wenzel state (i) can transform into the stable Cassie state with increasing S_f (ii). The metastable Cassie state (iii) can abruptly transform (iv) into the stable Wenzel state. The transition (i–ii) corresponds to equal Wenzel and Cassie states' free energies, whereas the transition (iv) corresponds to a Wenzel energy much lower than the Cassie energy and thus involves significant energy dissipation and is irreversible. The solid and dashed straight lines correspond to the values of the contact angle, calculated from (8.2)–(8.3) using the contact angle for a nominally flat surface, $\theta_0 = 109°$. The two series of the experimental data are shown with squares and diamonds.

Figure 8.6(b) shows the values of the advancing contact angle plotted against the spacing factor. The solid and dashed straight lines correspond to the values of the contact angle for the Wenzel and Cassie states, calculated from (8.2)–(8.3) using the advancing contact angle for a nominally flat surface, $\theta_{adv0} = 116°$. It is observed that the calculated values underestimate the advancing contact angle, especially for big S_f (small distance between the pillars or pitch P). This is understandable, because the calculation takes into account only the effect of the contact area and ignores the effect of roughness and edge line density (it corresponds to $H_r = 0$ in (8.1)), while this effect is more pronounced for high pillar density (big S_f). In a similar manner, the contact angle is underestimated for the Wenzel state, since the pillars constitute a barrier for the advancing droplet.

Figure 8.6(c) shows the values of the contact angle after the transition took place (dimmed blue squares and diamonds) as it was observed during evaporation. For both series, the values almost coincided. For comparison, the values of the receding contact angle measured for millimeter-sized water droplets are also shown (squares and diamonds), since evaporation constitutes removing liquid, and thus the contact angle during evaporation should be compared with the receding contact angle. The solid and dashed straight lines correspond to the values of the contact angle, calculated from (8.2)–(8.3) using the receding contact angle for a nominally flat surface, $\theta_{rec0} = 82°$. Figure 8.6(c) demonstrates a good agreement between the experimental data and (8.2)–(8.3).

In this section, we showed that an abrupt transition from the metastable Cassie to Wenzel wetting regime is found for micropatterned surfaces. The transition can be observed during microdroplet evaporation. The droplet radius at the transition is linearly proportional to the pitch between pillars divided by their diameter. This suggests that interactions at the perimeter of the droplet (rather than at the bulk area beneath the droplet) dominate in the transition. We also showed that the transition could not be predicted from the macroscale equations for the contact angle and the contact angle hysteresis, such as (8.2)–(8.3), since it involves micro- and nanoscale

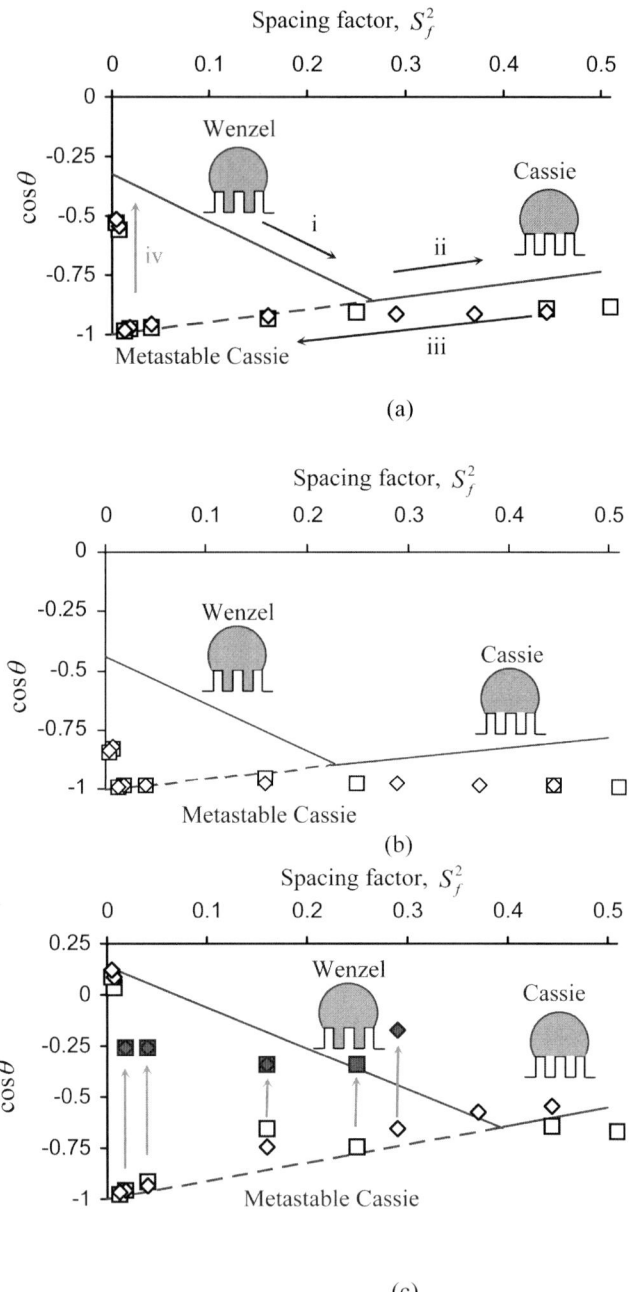

Fig. 8.6. Theoretical (*solid* and *dashed*) and experimental (*squares* for the 1st series, *diamonds* for the 2nd series) **a** contact angle as a function of the spacing factor, **b** advancing contact angle, and **c** receding contact angle and values of the contact angle observed after the transition during evaporation (*shaded*) [250]

interactions. We found, too, that the contact angle hysteresis can be explained as a result of two factors that act simultaneously. First, the changing contact area affects the hysteresis, since a certain value of the contact angle hysteresis is inherent for even a nominally flat surface. Decreasing the contact area by increasing the pitch between the pillars leads to a proportional decrease of the hysteresis. This effect is clearly proportional to the contact area between the solid surface and the liquid droplet. Second, the edges of the pillar tops prevent the motion of the triple line. This roughness effect is proportional to the contact line density, and its contribution was, in our experiment, comparable with the contact area effect. Interestingly, the effect of the edges is much more significant for the advancing than for the receding contact angle.

8.3 Wetting as a Multiscale Phenomenon

Nosonovsky and Bhushan [246] suggested that wetting be considered as a multi-scale process that involves effects at different scale levels. They pointed out that unlike most classical systems studied by thermodynamics and statistical physics, a droplet on a superhydrophobic surface is a multiscale system in a sense that its macroscale properties, such as the contact angle and contact angle hysteresis, cannot be determined from only macroscale equations. Thus it cannot be treated as a closed macroscale system. While macroscale thermodynamic analysis allows predicting the contact angle for both Wenzel and Cassie states, the transition between these two states and the contact angle hysteresis involve processes and instabilities at the micron scale (such as pinning of the triple line) and at the molecular scale (such as the inherent adhesion hysteresis). A truly multiscale approach would be to isolate the effects at each length scale and then incorporate the information at the other scales in some systematic manner.

Additional evidence that the superhydrophobic interface is a multiscale system is the possibility of self-organized behavior. Nosonovsky and Bhushan [251] suggested that contact angle hysteresis and the Cassie–Wenzel transition are natural candidates for SOC. Adding liquid to a droplet is similar to adding grains to a sandpile. The triple line tends to be pinned at the "critical locations," such as the edges of the microstructures, which serve as attractors (Fig. 8.7). After that the triple line suddenly advances to its new location in an "avalanche-like" behavior. The Cassie–Wenzel transition occurs in a similar manner with water either gradually filling some valleys, or in an "avalanche" manner filling all values at once. If contact angle hysteresis and the Cassie–Wenzel transition indeed involve SOC, relevant parameters (such as the number of valleys filled in a unit of time) should be related by a "power law" and the "one-over-frequency" noise law should be present. There is evidence that the Cassie–Wenzel transition occurs gradually through consequent filling of valleys, rather than abruptly [61] and there is evidence of power-law behavior in the contact angle hysteresis for regular surfaces [96]. However, it still remains to be verified experimentally by investigating the kinetics of the droplet flow and the Cassie–Wenzel transition, in order to determine whether SOC is involved.

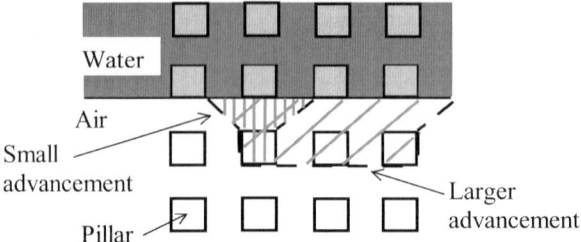

Fig. 8.7. Schematic of liquid advancement upon a micropatterned surface build of square pillars (top view). The wetted region is *gray*. The triple line tends to sit at the edges of the pillars (a "critical state"). When the advancement is triggered, the liquid can advance for a small distance or for a large distance, causing an "avalanche" [251]

Since fractals have been introduced into the surface mechanics [212], the argument continues on whether fractal geometry provides an adequate description of physical phenomena and scaling issues [138]. With advances in nanotechnology, different views on scale effect and scaling laws have emerged, as well as on whether the scaling laws are explained by the physics of the involved processes or by pure geometry [72], including fractal geometry. Unlike a fractal surface, which is a pure geometrical object and has no characteristic scale length, a hierarchical surface has a set of imminent scale lengths l_N ($l_1 \ll l_2 \ll \cdots \ll l_N$, where N is the number of hierarchy levels) which are related to the physics of relevant processes, rather than to the geometry. In biomimetic superhydrophobic surfaces various levels of hierarchy are related to various physical mechanisms that act simultaneously and have different characteristic length scales.

Roughness is the central property for bio-inspired superhydrophobic surfaces. It is very important to correctly assess the roughness effect on wetting in order to design successful superhydrophobic nonadhesive surfaces. This requires a roughness description that takes into account complicated wetting mechanisms, but at the same time is universal enough to include diverse natural and artificial forms. However, traditional roughness parameters, both statistical and fractal, do not fit well with the mechanisms of superhydrophobicity, such as pinning of the triple line, adhesion hysteresis, composite interface destabilization, and wetting regime transition. Therefore, new parameters are needed. Two simple roughness parameters that have been suggested are the roughness factor that characterizes the density change of surface energy due to the roughness, and the spacing factor for patterned surfaces that characterizes the change of the contact area for the composite interface. These simple parameters are useful; however, a deeper understanding of the effect of roughness on wetting requires an investigation of the multiscale nature of this process. The contact angle hysteresis and the Cassie–Wenzel transition are governed by micro- and nanoscale phenomena that involve a number of length parameters. Roughness description that is based on the hierarchical organization of surfaces and a hierarchy of length scales should be used to design superhydrophobic surfaces [252].

8.4 Investigation of Wetting as a Phase Transition

Since currently there is no comprehensive model to explain wetting, contact angle hysteresis, and wetting regime transition, it may be useful to develop phenomenological theory. Vedantam and Panchagunula [327] suggested developing such a model using the Landau–Ginzburg functional approach, which was used successfully to study various wetting transitions. The order-parameter $\eta(x, y)$ is selected in such a manner that $\eta = 0$ for the nonwetted regions of the surface and $\eta = 1$ for wetted regions, whereas $0 < \eta < 1$ for partially wetted regions. After that the energy function $f(\eta)$ is constructed in such a manner that the free surface energy

$$E = \gamma_{\mathrm{SL}} A_{\mathrm{SL}} + \gamma_{\mathrm{LV}} A_{\mathrm{LV}} + \gamma_{\mathrm{SV}} A_{\mathrm{SV}} \tag{8.4}$$

can be expressed as

$$E = \int_{A} f(\eta)\, \mathrm{d}A = \int_{A_{\mathrm{SV}}} f(0)\, \mathrm{d}A + \int_{A_{\mathrm{SL}}} f(1)\, \mathrm{d}A + \int_{A(0<\eta<1)} f(\eta)\, \mathrm{d}A, \tag{8.5}$$

where integration is performed by the area. The function $f(\eta)$ can be built in a somewhat arbitrary manner; however, its minima should correspond to the equilibrium states of the system (e.g., the Wenzel and Cassie states). After that, the energy functional is written as

$$L = \int_{A} \left\{ f(\eta) + \frac{\lambda}{2} |\nabla f(\eta)|^2 \right\} \mathrm{d}A, \tag{8.6}$$

where λ is the gradient coefficient. The functional that should be minimized involves the free energy and the gradient of the free energy. The latter term is needed to account for the fact that creating an interface between two phases is energetically unprofitable. The kinetic equation is given in the form

$$\beta \dot{\eta} = -\frac{\mathrm{d}L}{\mathrm{d}\eta} = \lambda \nabla^2 \eta - \frac{\partial f}{\partial \eta}, \tag{8.7}$$

where $\beta > 0$ is the kinetic coefficient. In the simplest case it may be assumed that $\beta = \mathrm{const}$, whereas in the general case a more complicated functional dependence upon η and its derivatives may exist. Vedantam and Panchagunula [327] showed that in the case of $\beta = \mathrm{const}$ for an axisymmetric droplet flowing with the velocity V, (8.7) leads to

$$\cos \theta_{\mathrm{adc}} - \cos \theta_{\mathrm{rec}} = 2\alpha\beta V. \tag{8.8}$$

In other words, assuming that the kinetic coefficient is constant, the contact angle hysteresis is expected to be proportional to the flow velocity. A more complicated form of the kinetic coefficient may lead to a more realistic dependence of the contact angle hysteresis on the velocity.

The phase field method also accounts in a natural way for the contact line tension term and leads to the Boruvka and Neumann [62] equation

$$\cos \theta = \frac{\gamma_{\mathrm{SV}} - \gamma_{\mathrm{SL}}}{\gamma_{\mathrm{LV}}} - \frac{\tau K}{\gamma_{\mathrm{LV}}}, \tag{8.9}$$

where K is the curvature and τ is the contact line tension [327]. And understandably so, because the gradient term in (8.6) implies that the transition between the wetted and unwetted parts of the surface is associated with excess energy or, in other words, that bending of the phase boundary line at the phase field plane is energetically unprofitable. The contact line term in (8.9) corresponds to this additional line energy. The phase field method also seems promising for the analysis of the Cassie–Wenzel transition.

8.5 Reversible Superhydrophobicity

The important area of application of superhydrophobic surfaces is reversible superhydrophobicity, that is, the ability of a surface to switch between the hydrophobic and hydrophilic properties under the influence of the electric potential, ultraviolet or light irradiation, or temperature [112, 296, 334, 342]. This area has emerged since 2004, and a number of important findings have been made, including the ability to switch between the Cassie and Wenzel states. Krupenkin et al. [196] reported that droplet behavior can be reversibly switched between the superhydrophobic Cassie state and the hydrophilic Wenzel state by the application of electrical voltage and current (electrowetting). Wang et al. [334] created a surface that can switch between stable superhydrophilic, metastable superhydrophobic, and stable superhydrophobic states. Interestingly, their switchable surface was driven by DNA nanodevices. A number of other approaches to reversible superhydrophobicity have been suggested as well [207].

8.6 Summary

We investigated the Cassie–Wenzel wetting regime transition of micropatterned superhydrophobic surfaces by water droplets and found several effects specific for the multiscale character of this process. First, we discussed applicability of the Wenzel and Cassie equations for average surface roughness and heterogeneity. These equations relate the local contact angle with the apparent contact angle of a rough/heterogeneous surface. However, it is not obvious what the size of roughness/heterogeneity averaging should be, since the triple line at which the contact angle is defined has two very different scale lengths: its width is of molecular size scale while its length is of the order of the size of the droplet (i.e., microns or millimeters). We presented an argument that, in order for the averaging to be valid, the roughness details should be small compared to the size of the droplet (and not the molecular size). We showed that, while we can apply the Wenzel and Cassie equations for the uniform roughness/heterogeneity, generalized equations should be used for the more complicated case of nonuniform heterogeneity. The proposed generalized Cassie–Wenzel equations are consistent with a broad range of available experimental data. The generalized equations are valid both in the cases when the classical Wenzel and Cassie equations can be applied as well as in the cases when the latter fails.

The macroscale contact angle hysteresis and Cassie–Wenzel transition cannot be determined from the macroscale equations and are governed by micro- and nanoscale effects, so wetting is a multiscale phenomenon. The kinetic effects associated with contact angle hysteresis should be studied at the microscale, whereas the effects of adhesion hysteresis and the Cassie–Wenzel transition involve processes at the nanoscale. Our theoretical arguments are supported by our experimental data on micropatterned surfaces. The experimental study of contact angle hysteresis demonstrates that two different processes are involved: the changing solid–liquid area of contact and pinning of the triple line. The latter effect is more significant for the advancing than for the receding contact angle. The transition between wetting states was observed for evaporating microdroplets, and the droplet radius was proportional to the linear geometric parameters of the micropattern. These findings provide new insights to the fundamental mechanisms of wetting and can lead to the creation of successful nonadhesive surfaces.

We investigated the practically important case of destabilization of the composite interface—the Cassie–Wenzel transition for micropatterned surfaces. We found that the mechanisms that control this transition act at various scale levels and thus the transition is a multiscale process (see also Table 6.1 in Chap. 6). Although the state of the droplet is described by a macroscale parameter, the value of the contact angle depends on the wetting state and, therefore, is determined by macro-, micro-, and nanoscale processes.

9

Underwater Superhydrophobicity and Dynamic Effects

Abstract The issues related to the superhydrophobicity and lotus effect are reviewed. This involves "underwater" superhydrophobicity and liquid flow slip near a superhydrophobic surface, hydrophobic interactions and nanobubbles, bouncing droplets, the Leidenfrost effect, and droplets on inclined solid surfaces.

In the preceding chapters we discussed how surface roughness affects wetting properties. With increasing roughness, it is possible to achieve superhydrophobicity—that is, a very high contact angle and low contact angle hysteresis. In this section, we will study several dynamic effects related to superhydrophobicity. First, we will discuss underwater hydrophobicity, or the possibility of a surface to remain water-repellent when immersed in water. Such a surface may have a thin film of gas upon it which dramatically decreases viscous friction. Underwater hydrophobicity involves the effects of pressure and viscosity, which have been ignored in the preceding chapters. We will discuss a conceptually interesting topic: the similarity and difference between droplets and vapor bubbles. After that, we will consider bouncing droplets as well as droplets upon hot and inclined surfaces.

9.1 Superhydrophobicity for the Liquid Flow

Liquid flow near rough solid walls, particularly in microscopic channels, is an important application of superhydrophobic surfaces. Recent experimental observations and theoretical analyses suggest that there is a correlation between wetting properties of a liquid and surface friction [79, 84, 93]. As we have discussed in the preceding chapters, the wetting behavior of a solid surface with a liquid droplet upon it is characterized by the contact angle and contact angle hysteresis. The contact angle characterizes the adhesion of the liquid to solid, whereas the hysteresis characterizes dissipation during the flow of the droplet. For liquid flow applications, other parameters are required in order to characterize the surface friction of liquid at the solid wall of a channel. The standard boundary condition for fluid flow along a wall is a no-slip condition. According to this condition, the tangential fluid velocity at the

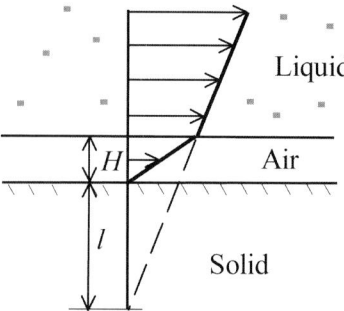

Fig. 9.1. Slip length l for a viscous liquid flow upon a thin gas film

wall is equal to zero. However, recent observations show that for some hydrophobic surfaces the slip may not be equal to zero.

For continuous macroscale liquid flow near a solid surface, the boundary condition of no-slip (or zero flow velocity at the wall) is usually assumed. However, it has been known on the basis of molecular dynamics simulations and experimental observations that liquid slip can exist over some materials, especially polymers [92] and surfaces due to a thin film of gas [84, 93, 359]. The slip is characterized by the so-called slip length or Navier length (Fig. 9.1), given by

$$l = H\left(\frac{\mu_L}{\mu_G} - 1\right), \tag{9.1}$$

where H is film thickness and μ_L and μ_G are viscosities of the liquid and gas. The concept of the slip length is based on the idea that flow velocity grows linearly with an increasing distance from the wall. For the nonslip boundary condition the Navier length is zero, $l = 0$. In the case when there is a thin layer of vapor, air, or another gas between the liquid and the solid wall, the gas viscosity is normally much lower than that of the liquid, which results in the effective nonzero slip condition.

It has been argued that a thin gas layer 5–80 nm can exist at the interface between a hydrophobic solid and water [356]. Formation of this layer was attributed to very small bubbles (nanobubbles) that can form at the solid surface and to other effects [202]. Rough superhydrophobic surfaces can also result in a gas layer trapped at the interface and thus lead to the nonzero slip [79]. The effect is very similar to the composite interface between a bubble and a rough surface, considered in detail in the preceding chapters (Fig. 9.2). The conclusions formulated regarding the stability of the composite interface in the case of a droplet upon a superhydrophobic surface are also valid with respect to the liquid flow [244]. Marmur [218] showed that for a hydrophobic surface made of convex pillars, the underwater superhydrophobicity is, in principle, feasible, and may be thermodynamically stable.

Equation (9.1) is valid for the case of a pure liquid–air interface ($f_{SL} = 0$). For a more realistic case of a composite interface we will assume that average slip length is given by averaging over the area of the slip length at the liquid–air interface (fractional area of $1 - f_{SL}$), given by (9.1) and zero slip velocity at the solid–liquid

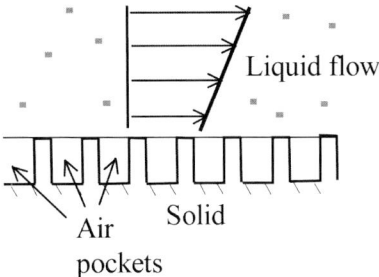

Fig. 9.2. Liquid flow upon a superhydrophobic surface

interface (fractional area of f_{SL})

$$l = H\left(\frac{\mu_L}{\mu_G} - 1\right)(1 - f_{SL}). \tag{9.2}$$

Whereas for a pure composite interface, f_{SL} can be calculated from (7.19)–(7.20), for a mixed interface f_{SL} can be calculated using probabilities of a liquid filling a valley or an air pocket staying there. The thickness of the layer H is assumed to be equal to the pillars' height. Based on (9.2), the higher the pillars, the larger the slip length and, therefore, the smaller the resistance to liquid flow. Note that maximum pressure, which can be supported by the surface, is given by (7.30).

Stability of the composite interface and the Cassie–Wenzel transition, discussed in the preceding chapter, is an important problem for producing successful superhydrophobic surfaces. Another important problem is the water pressure, since the pressure in the liquid flow is significantly higher than that created by the weight of a droplet. This liquid pressure should be balanced by the surface tension force and by the pressure in the gas. Choi and Kim [79] showed experimentally that a superhydrophobic "nanoturf" surface made by the black silicon method shows underwater superhydrophobicity. Voronov et al. [329] used molecular dynamics simulation and found a correlation between the slip length and the contact angle.

9.2 Nanobubbles and Hydrophobic Interaction

Nanobubbles (Fig. 9.3) remained a controversial topic until the early 2000s, because there are theoretical arguments against their existence. First, according to the Laplace equation, pressure in a gas bubble at the thermodynamic equilibrium is greater than that in the liquid outside. For bubbles with a very small radius, the pressure difference through the interface may be very big and thus the pressure inside is very high. Such high pressure would result in increasing gas solvability, so that the nanobubbles would rapidly dissolve. Second, the total free energy per unit area of a system in which water is in contact with a hydrophobic surface always increases when a gas layer or nanobubbles are formed, unless the surface is extremely rough, so the

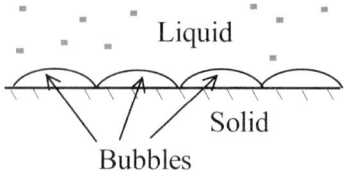

Fig. 9.3. Small gas bubbles form a layer on a solid wall, which can significantly reduce the friction

existence of the nanobubbles is not energetically favorable. Despite these thermo-dynamic objections, some experimental studies with the AFM and other equipment have demonstrated the existence of nanobubbles at the solid/water interface [356].

Nanobubbles can play a significant role in a number of processes, such as the long-range hydrophobic attractive force. Hydrophobic interaction is a well-known phenomenon. For example, when one adds small droplets of oil to water, the droplets combine to form a larger droplet. This comes about because water molecules are attracted to each other and are cohesive because they are polar molecules. Oil molecules are nonpolar and thus have no charged regions on them. This means that they are neither repelled nor attracted to each other. The attractiveness of the water molecules for each other then has the effect of squeezing the oil droplets together to form a larger droplet.

Singh et al. [298] showed that long-range hydrophobic interactions can exist underwater due to superhydrophobicity. They found that rough superhydrophobic surfaces experience attractive forces over the separation distance between the tip and sample of up to 3.5 μm, due to the spontaneous evaporation of the intervening, confined water. The effect was measured with an AFM and it is very similar to the capillary meniscus force. A capillary bridge is formed between a hydrophilic AFM tip and a sample due to the condensation of water vapor from air. Let us now assume that the tip and sample are hydrophobic and immersed in water. A gas bubble may form at the contact due to vapor bubble nucleation or due to gas dissolved in water. Such a bubble would result in an attractive force, the value of which is equal to the pressure difference inside and outside the bubble, ΔP, times the bubble foundation area, πR_1^2

$$F_{\text{cap}} = \pi R_1^2 \Delta P = \frac{\pi R_1^2 \gamma}{1/R_1 + 1/R_2}. \tag{9.3}$$

In a sense, such a bubble in water formed due to the cavitation between hydrophobic surfaces is a mirror image of a capillary bridge in air between hydrophilic surfaces formed due to the condensation (Fig. 9.4).

9.3 Bouncing Droplets

Another important phenomenon related to hydrophobicity is bouncing droplets (Fig. 9.5). When a droplet impacts a superhydrophobic surface with a certain velocity, in some cases it can be bounced off in an almost elastic manner [144, 275].

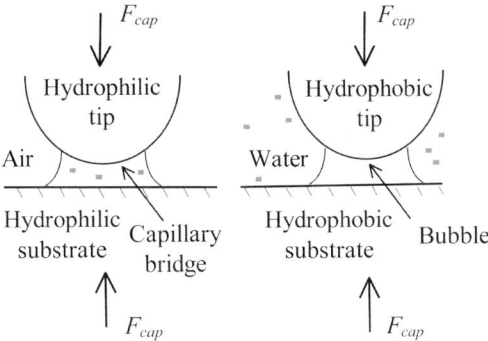

Fig. 9.4. Similarity of the attractive capillary force in the case of a condensed water bridge between hydrophilic surfaces in air and cavitation a bubble between hydrophobic surfaces in water

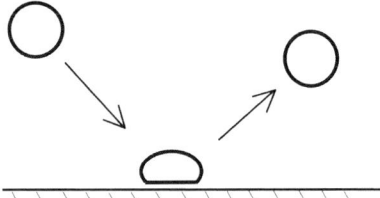

Fig. 9.5. A bouncing droplet impacts a solid hydrophobic surface

The kinetic energy of the droplet is stored in the surface deformation during the impact. A deformed droplet has a higher surface area and therefore higher free surface energy. Therefore, during the impact when the droplet is deformed, it can accommodate the kinetic energy. It was suggested that the effect might be of practical interest for agriculture, in particular, for treating leaves with pesticides [275]. Since the leaf is likely to repel the drop, it will not be treated; in addition, the scattering of the droplets contaminates the soil. If small amounts of a polymer soluble in water (such as polyoxyethylene) are added to water [26], the bulk viscosity of water remains the same, however, the so-called elongation viscosity (resistance of the liquid to large extensions) is high. Droplets of such a mixture do not bounce, and it has been used as an agricultural spray [275]. On the other hand, bouncing droplets are useful for waterproof fabrics or concrete that need to remain dry when exposed to rain.

The dissipation that prevents droplet bouncing is due to viscosity and due to the moving triple line. For the bouncing droplets, the triple lines are small or absent due to the presence of a thin film of compressed air during the impact. It was reported that full rebounding takes place for the surfaces with contact angles larger than 150°–160° with the ratio of the velocities after and before impact of 0.9 and higher [275]. To estimate the maximum velocities of impact, V, for which the effect takes place is obtained by comparison of the kinetic and free surface energy. The kinetic energy of a droplet of radius, R_0, and density ρ is given by

$$E_{\text{kin}} = \frac{(4/3)\pi\rho R_0^3 V^2}{2},$$
(9.4)

while the free surface energy is given by

$$E_{\text{surf}} = 4\pi R_0^2 \gamma_{\text{LV}}.$$
(9.5)

In order for the surface deformation to accommodate the kinetic energy, the latter should be of the same order of magnitude as the former, which yields

$$V^2 = \frac{6\gamma_{\text{LV}}}{\rho R_0}.$$
(9.6)

For a water ($\gamma_{\text{LV}} = 0.072\,\text{N/m}$, $\rho = 1000\,\text{kg/m}^2$) droplet of $R_0 = 0.001$, the corresponding impact velocity is $V = 0.66\,\text{m/s}$. Indeed, with the increasing velocity, the elastic behavior decreases [281]. The ratio of the kinetic and surface energy (or inertial to the capillary forces) is also be characterized by the nondimensional Weber number

$$We = \frac{\rho R_0 V^2}{\gamma_{\text{LV}}}.$$
(9.7)

Small We corresponds to law kinetic energy in comparison with the free surface energy. Two other important nondimensional numbers are the capillary number

$$Ca = \frac{\mu_{\text{L}} V}{\gamma_{\text{LV}}},$$
(9.8)

that characterizes the ratio of the viscous to capillary forces, and the Reynolds number

$$Re = \frac{\rho V R_0}{\mu_{\text{L}}},$$
(9.9)

that characterizes the ratio of the inertial to viscous forces (note that $Re = We/Ca$). Bouncing usually requires small We, Ca, and Re. Big velocity (high We) would result in the rupture of the droplet into parts, while big viscosity (high Ca and Re) would result in dissipation and eventual wetting of the surface (Fig. 9.6).

The energy barrier that corresponds to the division of a droplet with the radius R_0 into two droplets with the radius

$$R_1 = 2^{(-1/3)} R_0$$
(9.10)

which provides the same total volume, can be estimated by comparison of the net energy. The energy of the big droplet is given by the sum of the interface energy and the volumetric energy

$$E_0 = 4\pi R_0^2 \gamma_{\text{LV}} + \frac{4}{3}\pi R_0^3 \frac{2\gamma_{\text{LV}}}{R_0} = \frac{20\pi R_0^2 \gamma_{\text{LV}}}{3}.$$
(9.11)

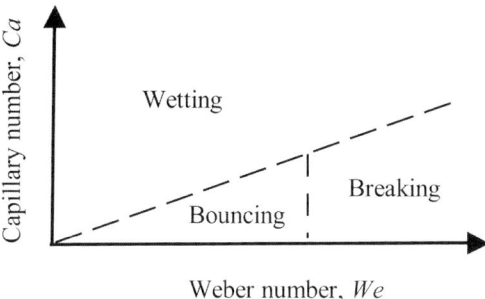

Fig. 9.6. Schematic of various regimes of a droplet impacting a hydrophobic surface. For high We, droplet will break into parts, whereas for high $Re = We/Ca$, the energy will be dissipated due to the viscosity and the droplet can wet the surface

The energy of a smaller droplet is given by

$$E_1 = \frac{20\pi R_1^2 \gamma_{LV}}{3} = 2^{(-2/3)} \frac{20\pi R_0^2 \gamma_{LV}}{3}. \tag{9.12}$$

The energy change as a result of the division of the big droplet into two is given by

$$\Delta E = 2E_1 - E_0 = \left(2^{(1/3)} - 1\right) \frac{20\pi R_0^2 \gamma_{LV}}{3} = 1.26 E_0. \tag{9.13}$$

Comparison with the kinetic energy given by (9.10) yields

$$V^2 = \frac{10(2^{(1/3)} - 1)\gamma_{LV}}{\rho R_0}. \tag{9.14}$$

For a water ($\gamma_{LV} = 0.072$ N/m, $\rho = 1000$ kg/m^2) droplet of $R_0 = 0.001$, the corresponding impact velocity is $V = 0.95$ m/s, which provides another estimate for a maximum velocity. Experimental observations show that small satellite droplets form in certain cases, when the Weber number is of the order of 10 or greater [275]. Formation of a small satellite droplet involves smaller energy barriers rather than the division of the large droplet into two equal parts.

We can conclude that water droplets can bounce off superhydrophobic surfaces in an elastic manner. The impact velocity should not be very high (small We requirement), so that the surface tension can accommodate the kinetic energy. On the other hand, the velocity should not be too low (small Re requirement), so that impact time is short, a thin film of compressed air can form, and the dissipation due to the triple line contact and viscosity is small.

9.4 A Droplet on a Hot Surface: the Leidenfrost Effect

An interesting effect of nonwetting that corresponds to an extreme nonequilibrium situation is the so-called Leidenfrost effect named after German physician J.G. Leidenfrost (1715–1794), who discovered it in his work published in 1756. When a

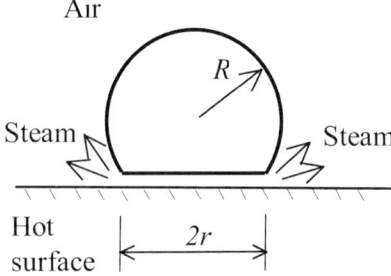

Air

R

Steam   Steam

Hot
surface

2r

Fig. 9.7. Schematic of a droplet upon a hot surface, supported by a thin layer of vapor (the Leidenfrost effect)

liquid droplet is placed on a hot surface with the temperature higher than $220\,°C$ (the so-called Leidenfrost point), a thin (0.1–0.2 mm) vapor film is formed under the droplet due to the extensive boiling [276]. The pressure in the film can be estimated as

$$P = P_0 + \frac{2\gamma_{LV}}{R},$$ (9.15)

where P_0 is the ambient atmospheric pressure and R is the droplet radius (Fig. 9.7). Equation (9.14) is based on the Laplace equation and the assumption that the interface under the droplet is flat (which is not exactly true, the interface is concave). The Laplace equation is valid for the thermodynamic equilibrium; however, we can assume that there is no pressure difference through the flat interface under the droplet. The additional vapor pressure creates the force $\pi r^2 (P - P_0)$, where r is the droplet foundation radius. This force should be balanced by the weight of the droplet, which is equal, assuming $r \ll R$, to $4/3\pi \rho g R^3$ and, therefore

$$3r^2\gamma_{LV} = 2\rho g R^4.$$ (9.16)

The droplet comes to the surface under the angle of θ, so that $\sin\theta = r/R$ or

$$\sin\theta = R\sqrt{\frac{2\rho g}{3\gamma_{LV}}}.$$ (9.17)

Thus the contact angle approaches 180° with decreasing droplet radius.

9.5 A Droplet on an Inclined Surface

For a droplet of radius R moving along the inclined plane with tilt angle α (Fig 9.8), the energy gain corresponding to the inclined distance l is given by $lmg \sin\alpha$, where mg is the weight of the droplet. The contact angle hysteresis is related to the tilt angle by

$$t\sigma(\cos\theta_{adv} - \cos\theta_{rec}) = gm\sin\alpha,$$ (9.18)

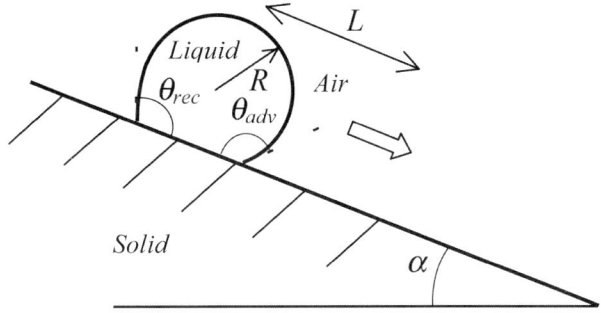

Fig. 9.8. Schematics of a droplet moving along an inclined surface with tilt angle α, advancing contact angle θ_{adv} and receding contact angle θ_{rec}

where σ is the surface tension and t is the length of the triple line [117]. Equation (9.17) was formulated by Macdougall and Ockrent [210] based on their experimental observations of droplets on inclined surfaces and independently by Frenkel [115] in his theoretical study of the contact angle hysteresis. Equation (9.17) is based on several assumptions, in particular, that the front edge of the droplet exhibits the advancing and the back end the receding contact angle and that the droplet foundation is well approximated by the cylindrical or rectangular shape. Marmur [216] and Krasovitski and Marmur [192] showed that the droplet on an inclined plane does not necessarily exhibit the receding or advancing contact angle, since the former is based on the geometry of the drop, whereas the latter is based on the nature of the solid surface.

We will use the energy balance method to determine the tilt angle, that is, we will assume that the energy loss due to adhesion hysteresis is equal to the energy gain due to gravity. The energy loss is given by ΔW times the contact area $lR \sin \theta$. Combining these equations with (7.80), the tilt angle is given by

$$\sin \alpha = \frac{\Delta W_0 f_{\mathrm{SL}} R_{\mathrm{f}} R \sin \theta}{mg}. \tag{9.19}$$

Using the trigonometric relationship

$$\cos \theta_{\mathrm{adv}} - \cos \theta_{\mathrm{rec}} = -2 \sin \frac{\theta_{\mathrm{adv}} + \theta_{\mathrm{rec}}}{2} \sin \frac{\theta_{\mathrm{adv}} - \theta_{\mathrm{rec}}}{2}, \tag{9.20}$$

and substituting into (6.55)

$$\sin \frac{\theta_{\mathrm{adv}} - \theta_{\mathrm{rec}}}{2} = -\frac{mg}{2R\gamma_{\mathrm{LA}} \sin \theta \sin \frac{\theta_{\mathrm{adv}} + \theta_{\mathrm{rec}}}{2}} \sin \alpha. \tag{9.21}$$

For θ and $(\theta_{\mathrm{adv}} + \theta_{\mathrm{rec}})/2$ close to π, it is found from (7.80) that the tilt angle is proportional to the contact angle hysteresis.

9.6 Summary

In this chapter we discussed several dynamic effects related to superhydrophobicity. Underwater hydrophobicity with a small resistance to the liquid flow is possible.

However, it requires a gas layer or bubbles to form at the solid wall of the vessel. The existence of nanobubbles has been recently confirmed experimentally. The cavitation that leads to nanobubble formation plays a role in various effects, such as long-range hydrophobic interactions. Other important considerations are droplets bouncing off a superhydrophobic surface, and droplets' behavior on inclined and hot surfaces (Leidenfrost effect).

Part III

Biological and Biomimetic Surfaces

"We live in an age of surfaces."
Oscar Wilde

Lotus-Effect and Water-Repellent Surfaces in Nature

Abstract Superhydrophobic surfaces of water-repellent plants, such as lotus leaf, are reviewed in this chapter. Experimental techniques and measurement results for wetting properties (contact angle, contact angle hysteresis), adhesion and surface topography are reviewed.

The preceding chapters presented the theory of wetting of rough surfaces. In the third part of the book, we will discuss experimental and practical issues related to water-repellent and nonadhesive surfaces. We will start in the present chapter with biological superhydrophobic surfaces. Many biological surfaces are known to be water-repellent and hydrophobic. The most common example is the leaf of the lotus plant, that gave us the term lotus-effect, which denotes roughness-induced superhydrophobicity and self-cleaning abilities of biological and artificial surfaces. In this chapter, we will discuss water-repellent plants, their roughness, and wax coatings in relation to their hydrophobic and self-cleaning properties.

10.1 Water-Repellent Plants

The hydrophobic and water-repellent properties of many plant leaves have been known for a long time. Scanning electron microscope (SEM) studies in the past 30 years revealed that the hydrophobicity of the leaf surface is related to its microstructure. All primary parts of plants are covered by a cuticle composed of soluble lipids embedded in a polyester matrix, which makes the cuticle hydrophobic in most cases [24]. The hydrophobicity of the leaves is related to another important effect, the ability of the hydrophobic leaves to remain clean after being immersed in dirty water, known as self-cleaning. This ability is best known for the lotus (*Nelumbo nucifera*) leaf that is considered by some oriental cultures as "sacred" due to its purity. The lotus flower is quoted extensively in sacred texts, for example: *"One who performs his duty without attachment, surrendering the results unto the Supreme Lord, is unaffected by sinful action, as the lotus leaf is untouched by water"* (Bhagavad Gita 5.10). Borrowing from Hinduism, in Buddhism, the lotus represents purity of

Fig. 10.1. Ancient Buddhist image of a hermitage with Lotus as a symbol of purity suspending on sides, Mathura (India), 2nd century BC

body, speech, and mind, floating above the muddy waters of attachment and desire (Fig. 10.1). Not surprisingly, the ability of lotus-like surfaces for self-cleaning and water repellency was dubbed the "lotus-effect."

The outer single-layered group of cells covering a plant, especially the leaf and young tissues, is called the epidermis. The protective waxy covering produced by the epidermal cells of leaves are called cuticles. The cuticle is composed of an insoluble cuticular membrane covered with epicuticular waxes, which are mixtures of hydrophobic aliphatic components, hydrocarbons with chain lengths typically in the range C16 to C36, such as paraffins [19]. The SEM study reveals that the lotus leaf surface is covered by "bumps," more exactly called *papillae* (papillose epidermal cells) which, in turn, are covered by an additional layer of epicuticular waxes [24]. The wax is hydrophobic with water contact angle of about 95°–110°, whereas the papillae provide the tool to magnify the contact angle in accordance with the Wenzel model, discussed in the preceding chapters. The experimental values of the static water contact angle with the lotus leaf were reported as about 160° [24]. Indeed, taking the papillae density of 3400 per square millimeter, the average radius of the hemispherical asperities $r = 10$ μm, and the aspect ratio $h/r = 1$ provides, based on (6.35), the value of the roughness factor $R_f \approx 4$ [240]. Taking the value of the contact angle for wax, $\theta_0 = 104°$, our naive calculation with the Wenzel equation yields $\theta = 165°$, which is not far from the experimentally observed values [240].

However, the simple Wenzel model may be not sufficient to explain the lotus-effect, since the lotus leaf also exhibits low contact angle hysteresis, apparently forming the composite interface. Moreover, its structure has hierarchical roughness. So, a number of more sophisticated models have been developed [243–246].

Neinhuis and Barthlot [227] systematically studied surfaces and wetting properties of about 200 water-repellent plants. They reported several types of epidermal relief features and epicuticular wax crystals. Among the epidermal relief features are the papillose epidermal cells. Either every epidermal cell forms a single papilla or cell being divided into papillae. The scale of the epidermal relief ranged from 5 μm in multipapillate cells to 100 μm in large epidermal cells. Some cells also were convex (rather than having real papillae) and/or had hairs (trichomes). They also found various types and shapes of wax crystals at the surface [330]. Interestingly, the hairy surfaces with a very thin film of wax exhibited water-repellency for short periods (minutes), after which water penetrated between the hairs, whereas waxy trichomes showed strong water-repellency. The wax crystal creates roughness, in addition to the roughness created by the papillae. The chemical structure of the epicuticular waxes has been studied extensively by plant scientists and lipid chemists in recent decades [165]. Apparently, roughness plays the dominant role in the lotus effect since superhydrophobicity can be achieved independently of type of wax or other hydrophobic coating.

Although it is intuitive that water-repellency and self-cleaning are related to each other, because the ability to repel water is related to the ability to repel contaminants, it is difficult to quantify self-cleaning. Therefore, a quantitative relation of the two properties remains to be established. A qualitative explanation of how was proposed by Barthlot and Neinhuis [24], who suggested that on a smooth surface contamination particles are mainly redistributed by a water droplet, whereas on a rough surface they adhere to the droplet and are removed from the leaves when the droplet rolls off. A detailed model of this process has not been developed but, obviously, whether the particle adheres to the droplet depends upon the interactions at the triple line and on whether the magnitude and direction of the surface force allows the particle to be detached from the surface and flown away (Fig. 10.2).

The role of surface hierarchy in the lotus effect is also not completely clear, although a number of explanations for why most natural surfaces are hierarchical has been suggested [124, 232, 243–245]. Nosonovsky and Bhushan [243–245] showed that the mechanisms involved with superhydrophobicity are scale-dependent and thus the roughness must be hierarchical in order to respond to these mechanisms. It may have to do also with the simple fact that the surface must be able to repel both macroscopic and microscopic droplets. Experiments with artificial fog (microdroplets) and artificial rain (large droplets) show that surfaces with only one scale of roughness repel rain droplets well, however, they cannot repel small fog droplets which are trapped in the valleys between the bumps [118].

As far as the biological implications of the lotus-effect, Barthlot and Neinhuis [24] suggested that self-cleaning plays an important role in the defense against pathogens, such as spores and conidia of pathogenic microorganisms.

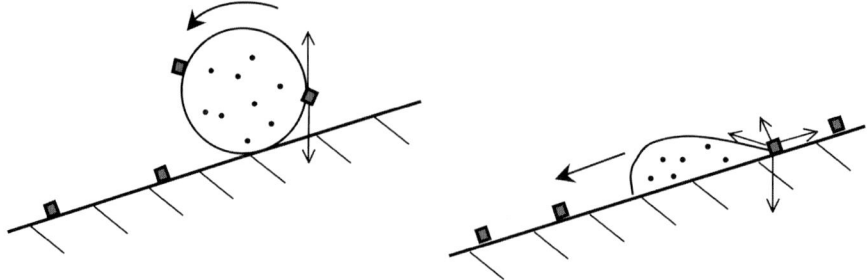

Fig. 10.2. Self-cleaning of a superhydrophobic surface and the two modes of droplet motion. A "rolling" droplet can wash away a contamination particle, whereas a "sliding" droplet can only drag a particle for a short distance. Thus, superhydrophobicity leads to self-cleaning. Forces are shown: surface tension, particle weight, reaction of the surface, and friction

10.2 Characterization of Hydrophobic and Hydrophilic Leaf Surfaces

In order to completely understand the nature of hydrophobic and hydrophilic leaves, a comprehensive characterization of the surface and its properties must be carried out. Using the various characterization techniques discussed previously, the surfaces of the leaves were measured by Bhushan and Jung [39] in order to understand the factors that are responsible for its hydrophobic/hydrophilic nature. In the following, we discuss the findings of that study. The idea of this study was a comparative investigation of hydrophobic and hydrophilic plant leaves, which would allow the researchers to identify the features that lead to superhydrophobicity.

10.2.1 Experimental Techniques

The static contact angles were measured using a Rame–Hart model 100 contact angle goniometer with droplets of deionized water [39, 68]. Droplets of about 5 μL in volume (with diameter of a spherical droplet about 2.1 mm) were gently deposited on the substrate using a microsyringe for the static contact angle. All measurements were made by five different points for each sample at $22 \pm 1\,°C$ and $50 \pm 5\%$ RH. The measurement results were reproducible within $\pm 3\%$.

An optical profiler (NT-3300, Wyko Corp., Tucson, AZ) was used to measure surface roughness for different surface structures [68, 39]. A greater Z-range of the optical profiler of 2 mm is a distinct advantage over the surface roughness measurements with an AFM, which has a Z-range of 7 μm, but it has a maximum lateral resolution of approximately 0.6 μm [30, 32]. A commercial AFM (D3100, Nanoscope IIIa controller, Digital Instruments, Santa Barbara, CA) was used for additional surface roughness measurements with a high lateral resolution and for adhesion and friction measurements [39, 67]. The measurements were performed with a square pyramidal Si(100) tip with a native oxide layer which had a nominal radius of 20 nm on a rectangular Si(100) cantilever with a spring constant of $3\,N\,m^{-1}$ in the tapping mode.

Adhesion and friction force at various relative humidity (RH) were measured using a 15 μm radius borosilicate ball. A large tip radius was used to measure contributions from several microbumps and a large number of nanobumps. Friction force was measured under a constant load using a 90° scan angle at the velocity of 100 μm/s in 50 μm and at a velocity of 4 μm/s in 2 μm scans. The adhesion force was measured using the single point measurement of a force calibration plot.

10.2.2 Hydrophobic and Hydrophilic Leaves

Figure 10.3 shows the SEM micrographs of two hydrophobic leaves—lotus (*Nelumbo nucifera*) and elephant ear or taro plant (*Colocasia esculenta*), referred to as lotus and colocasia, respectively. The figure also shows micrographs for two hydrophilic leaves—beech (*Fagus sylvatica*) and magnolia (*Magnolia grandiflora*), referred to as fagus and magnolia, respectively [39]. Lotus and colocasia are characterized by papillose epidermal cells responsible for the creation of papillae (or bumps) on the surfaces. In addition, they are covered by a layer of three-dimensional epicuticular waxes which are a mixture of very long-chain fatty acid molecules (compounds with chains >20 carbon atoms). Fagus and magnolia are characterized by tabular cells

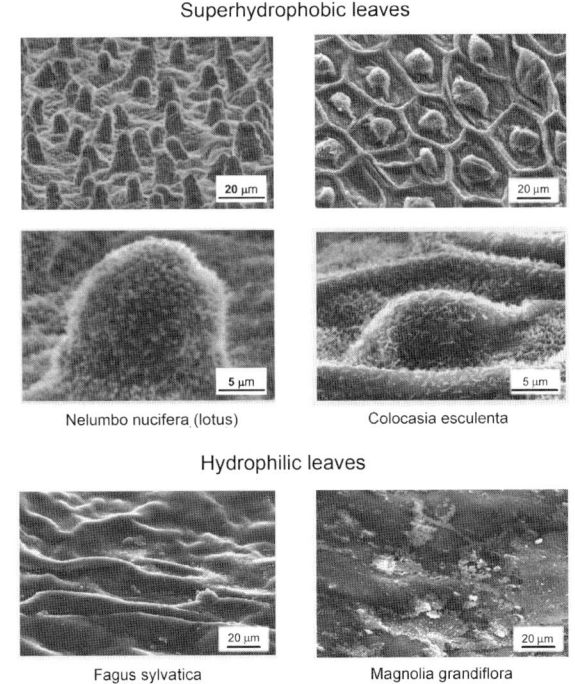

Fig. 10.3. Scanning electron micrographs of the relatively rough, water-repellent leaf surfaces of *Nelumbo nucifera* (*lotus*) and *Colocasia esculenta* and the relatively smooth, wettable leaf surfaces of *fagus sylvatica* and *magnolia grandiflora* [39]

with a thin wax film with a 2D structure [24]. The leaves are not self-cleaning and contaminant particles are accumulated which make them hydrophilic.

10.2.3 Contact Angle Measurements

In order to remove the epicuticular wax, acetone was applied to the leaves. Figure 10.4(a) shows the contact angles for the hydrophobic and hydrophilic leaves before and after using acetone. After using acetone, the contact angle dramatically reduced for the hydrophobic leaves; however, for the hydrophilic leaves, the contact angle remained almost unchanged. It is known that there is a 2D very thin wax layer on the hydrophilic leaves which introduces little roughness. As opposed to that, the hydrophobic leaves are known to have a thin 3D wax layer on their surface which consists of nanoscale roughness over microroughness created by the papillae, which

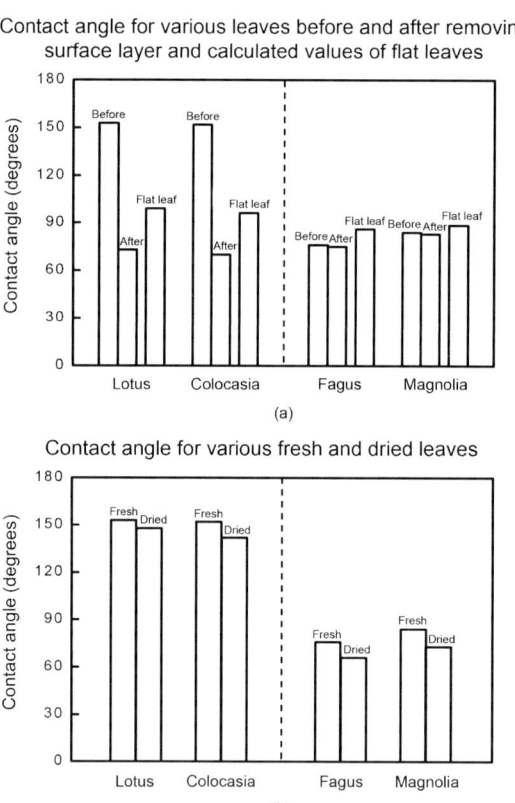

Fig. 10.4. Contact angle measurements and calculations for the leaf surfaces, **a** before and after removing surface layer as well as calculated values, and **b** fresh and dried leaves. The contact angle on a smooth surface for the four leaves was obtained using the roughness factor calculated [39]

results in a hierarchical roughness. The combination of the wax coating and surface roughness leads to the superhydrophobicity.

In order to test various hypotheses regarding the mechanisms of the superhydrophobicity, Bhushan and Jung [39] first assumed that the Wenzel regime occurs. They calculated the Wenzel roughness factors and contact angles for leaves with a smooth surface for the four leaves. These results are presented in Fig. 10.4(a). The approximate values of R_f for lotus and colocasia are 5.6 and 8.4, whereas for fagus and magnolia it was 3.4 and 3.8, respectively. The contact angles for the smooth surface for the four leaves were calculated using these values. The calculated contact angles with smooth surfaces were approximately 99° for lotus and 96° for colocasia. For fagus and magnolia, the contact angles with the smooth surfaces were found as approximately 86° and 88°.

The contact angles were measured for fresh and dried leaves. Figure 10.4(b) shows the contact angles for both fresh and dried states for the four leaves. There was a decrease in the contact angle for all four leaves after they had been dried. For lotus and colocasia, this decrease took place because a fresh leaf had taller bumps than a dried leaf, which gave a larger contact angle in accordance with the Wenzel equation. To understand the reason for the decrease of contact angle after drying of hydrophilic leaves, dried magnolia leaves were also measured using an AFM. It was found that the dried leaf (P-V height $= 7\,\mu m$, mid-width $= 15\,\mu m$, and peak radius $= 18\,\mu m$) had taller bumps than a fresh leaf (P-V height $= 3\,\mu m$, mid-width $= 12\,\mu m$, and peak radius $= 15\,\mu m$). This increased the roughness and decreased the contact angle leading to a more hydrophilic surface [39].

10.2.4 Surface Characterization Using an Optical Profiler

An optical profiler allowed the researchers to conduct measurements on fresh leaves with a large P-V distance. Three different surface height maps for hydrophobic and hydrophilic leaves are presented in Figs. 10.5 and 10.6 [39]. Each figure shows a 3D map and a flat map along with a 2D profile at a given location of the flat 3-D map. A scan size of $60\,\mu m \times 50\,\mu m$ was used, since it had a sufficient number of bumps to characterize the surface but also to maintain a significant resolution for accurate measurements [39].

The structures obtained with the optical profiler correlate well with the SEM images shown in Fig. 10.3. The bumps on the lotus leaf are distributed on the entire surface, and the colocasia leaf shows a very different structure to that of the lotus. The surface of colocasia not only has bumps, similar to the lotus, but also present are ridges around each bump that keep the bumps separated. With these ridges, the bumps have a hexagonal (honeycomb) packing arrangement that allows for the maximum density of bumps per area. The bumps of lotus, and both bumps and ridges of colocasia, contribute to the hydrophobic nature since they both increase the R_f factor and result in air pockets between the droplet of water and the surface. In fagus and magnolia height maps, short bumps on the surface can be seen [39].

For each leaf a second-order curve fit has been given to the profiles to show how closely the profile is followed, as shown in 2D profiles of hydrophobic and

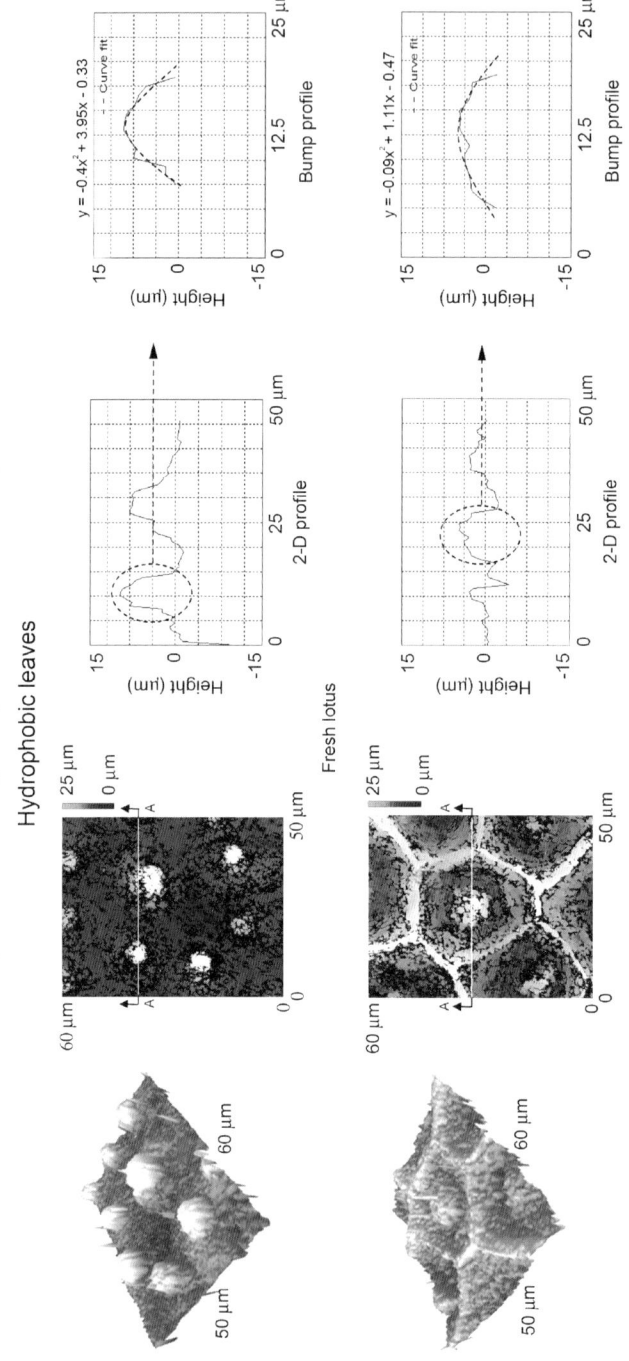

Fig. 10.5. Surface height maps and 2-D profiles of hydrophobic leaves using an optical profiler. For lotus leaf, a microbump is defined as a single, independent microstructure protruding from the surface. For colocasia leaf, a microbump is defined as the single, independent protrusion from the leaf surface, whereas a ridge is defined as the structure that surrounds each bump and is completely interconnected on the leaf. A curve has been fitted to each profile to show exactly how the bump shape behaves. The radius of curvature is calculated from the parabolic curve fit of the bump [39]

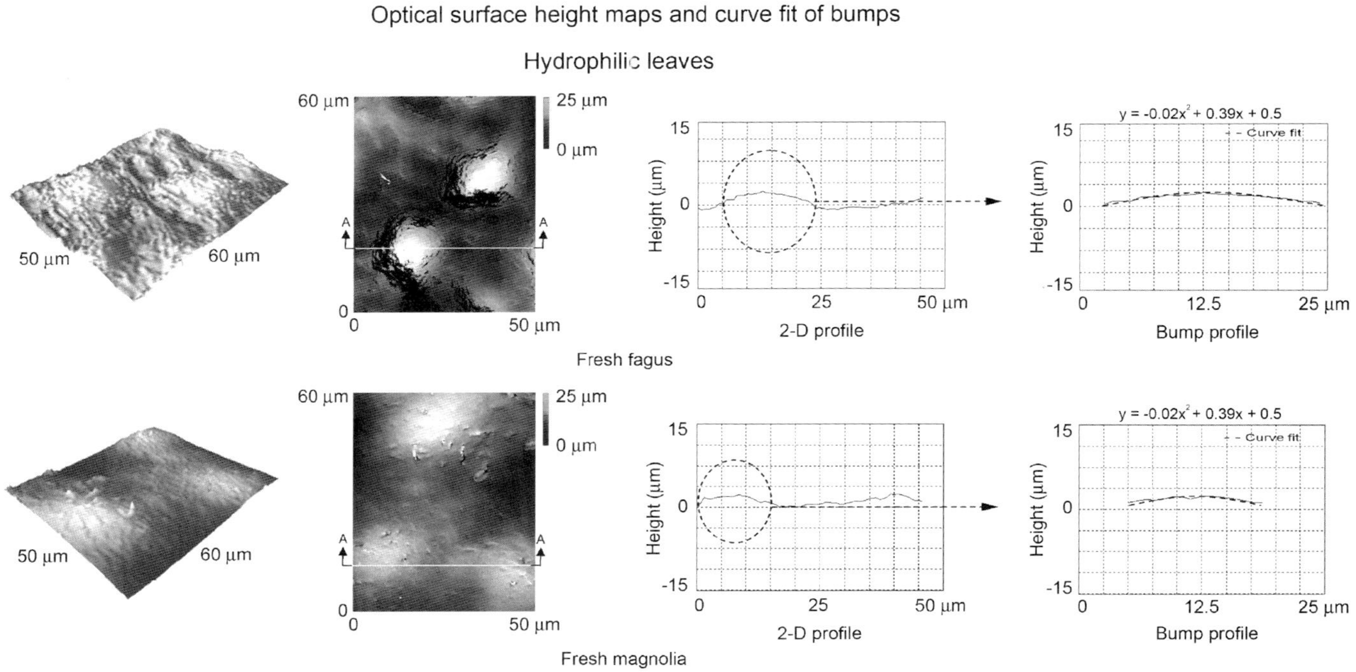

Optical surface height maps and curve fit of bumps

Hydrophilic leaves

Fig. 10.6. Surface height maps and 2-D profiles of hydrophilic leaves using an optical profiler. For fagus and magnolia leaves, a microbump is defined as a single, independent microstructure protruding from the surface. A curve has been fitted to each profile to show exactly how the bump shape behaves. The radius of curvature is calculated from the parabolic curve fit of the bump [39]

Table 10.1. Microbump and nanobump map statistics for hydrophobic and hydrophilic leaves, measured both fresh and dried leaves using an optical profiler and AFM [39]

Leaf		Microbump (μm) Scan size ($50 \times 50\ \mu$m)			Nanobump (μm) Scan size ($2 \times 2\ \mu$m)		
		P-V height	Mid-width	Peak radius	P-V height	Mid-width	Peak radius
Lotus							
Fresh		13*	10*	3*	0.78**	0.40**	0.15**
Dried		9**	10**	4**	0.67**	0.25**	0.10**
Colocasia							
Fresh	Bump	9*	15*	5*	0.53**	0.25**	0.07**
	Ridge	8*	7*	4*	0.68**	0.30**	0.12**
Dried	Bump	5**	15**	7**	0.48**	0.20**	0.06**
	Ridge	4**	8**	4**	0.57**	0.25**	0.11**
Fagus							
Fresh		5*	10*	15*	0.18**	0.04**	0.01**
		4**	5**	10**			
Magnolia							
Fresh		4*	13*	17*	0.07**	0.05**	0.04**
		3**	12**	15**			

*Data measured using optical profiler
**Data measured using AFM

hydrophilic leaves in Figs. 10.5 and 10.6. The radius of curvature for any function $y(x)$ is known to be

$$R(x) = \frac{(1 + y'(x)^2)^{3/2}}{y''(x)}, \tag{10.1}$$

where $R(x)$ is the radius of curvature. By using the second-order curve fit of the profiles, the radius of curvature can be found [39].

Using the optical surface height maps, different statistical parameters can be determined to characterize the surface, including the peak-to-valley (P-V) height, mid-width, and peak radius [30, 32]. The mid-width is defined here as the width of the bump at a height equal to half of peak-to-mean value. Table 10.1 shows these quantities found in the optical height maps for four leaves. Comparing the hydrophobic and hydrophilic leaves, it is found that the P-V height for the bumps of lotus and colocasia is much taller than that for the bumps of fagus and magnolia. The peak radius for the bumps of lotus and colocasia is also smaller than that for the bumps of fagus and magnolia. However, the values of mid-width for the bumps of four leaves are similar [39].

10.2.5 Leaf Characterization with an AFM

An AFM can operate in various modes, including contact, noncontact, dynamic, etc. To measure topographic imaging of the leaf surfaces, the contact and tapping modes

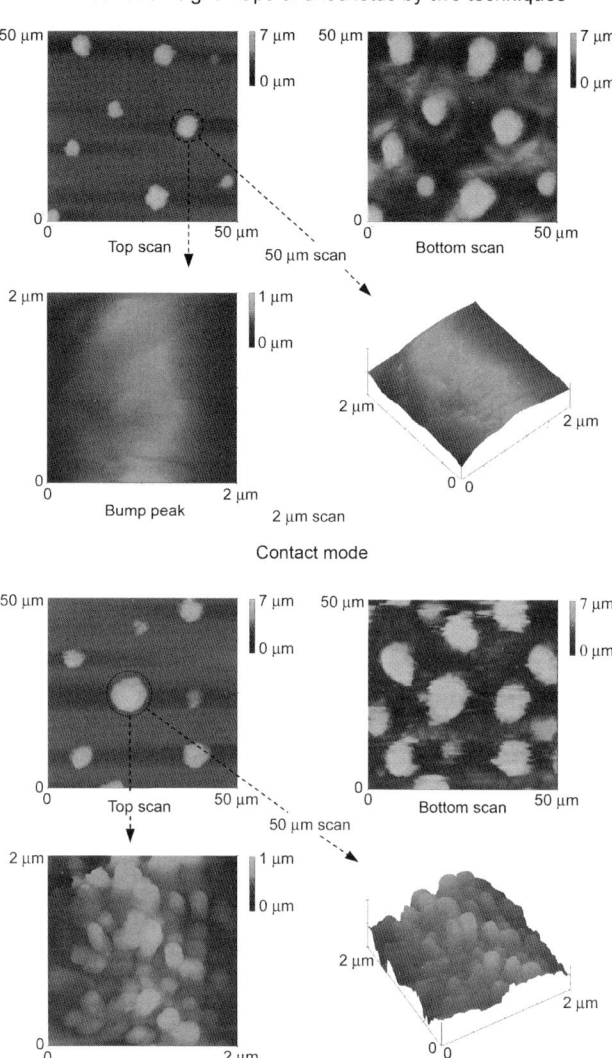

Fig. 10.7. Surface height maps showing the top scan and bottom scan in a 50 μm scan size and the bump peak scan selected in a 2 μm scan size for a lotus leaf in contact mode and tapping mode. Two methods were tested to get high resolution of nanotopography for a lotus leaf [39]

are the most appropriate. Both modes of the AFM were used to characterize the lotus leaf [39]. Figure 10.7 shows surface height maps of dried lotus obtained using the two techniques. In contact mode, local height variation for the lotus leaf was observed in 50 μm scan size. However, little height variation was obtained in a 2 μm scan even at loads as low as 2 nN. This could be due to the substantial frictional force generated as the probe scanned over the sample in the contact mode. The friction force can damage the sample. The tapping mode technique allows high-resolution topographic imaging of sample surfaces that could otherwise be easily damaged, are loosely held to their substrate, or are difficult to image by other AFM techniques [30, 32]. As shown in Fig. 10.7, with the tapping mode technique, the soft and fragile leaves can be imaged successfully. Therefore the tapping mode technique was selected to examine the surface roughness of the hydrophobic and hydrophilic leaves using an AFM [39].

The AFM has a Z-range of about 7 μm, and thus it cannot be used to make measurements in the conventional way because of the lotus leaf's large P-V distance. Burton and Bhushan [68] developed a new method to fully determine the bump profiles. In order to compensate for the large P-V distance, two scans were made for each height: one measurement that scanned the tops of the bumps and another measurement that scanned the bottoms or valleys of the bumps. The total height of the bumps was embedded within the two scans. Figure 10.8 shows the 50 μm surface height maps obtained using this method [39]. The 2D profiles in the right side column took the profiles from the top scan and the bottom scan for each scan size and spliced them together to get the total profile of the leaf. The 2 μm surface height maps for both fresh and dried lotus can also be seen in Fig. 10.8. This scan area was selected on the top of a microbump obtained in the 50 μm surface height map. It can be seen that nanobumps are randomly and densely distributed on the entire surface of lotus.

Bhushan and Jung [39] also obtained the surface height maps for the hydrophilic leaves in both 50 μm and 2 μm scan sizes as shown in Fig. 10.9. For fagus and magnolia, microbumps were found on the surface, and the P-V distance of these leaves was lower than that of lotus and colocasia. It can be seen in the 2 μm surface height maps that nanobumps selected on the peak of the microbump have an extremely low P-V distance.

Using the AFM surface height maps, different statistical parameters of bumps and ridges can be obtained, such as the P-V height, mid-width, and peak radius. These quantities for four leaves are listed in Table 10.1. It can be seen that these values correlate well with those obtained from optical profiler scans except for the bump heights, which decreased by more than half because of leaf shrinkage [39].

10.2.6 Adhesion Force and Friction

The adhesion force and coefficient of friction for the hydrophobic and hydrophilic leaves are presented in Fig. 10.10 on the basis of the AFM data. For each type of leaves, adhesive force measurements were done for both fresh and dried leaves using a 15 μm radius AFM tip. It was found that the dried leaves had lower values of the adhesion force than the fresh leaves. The adhesion force arises from several sources including a thin liquid film, such as adsorbed water layers that causes meniscus bridges to build up around the contacting and near-contacting bumps as a result of surface energy effects [30, 32]. When the leaves are fresh there is moisture within

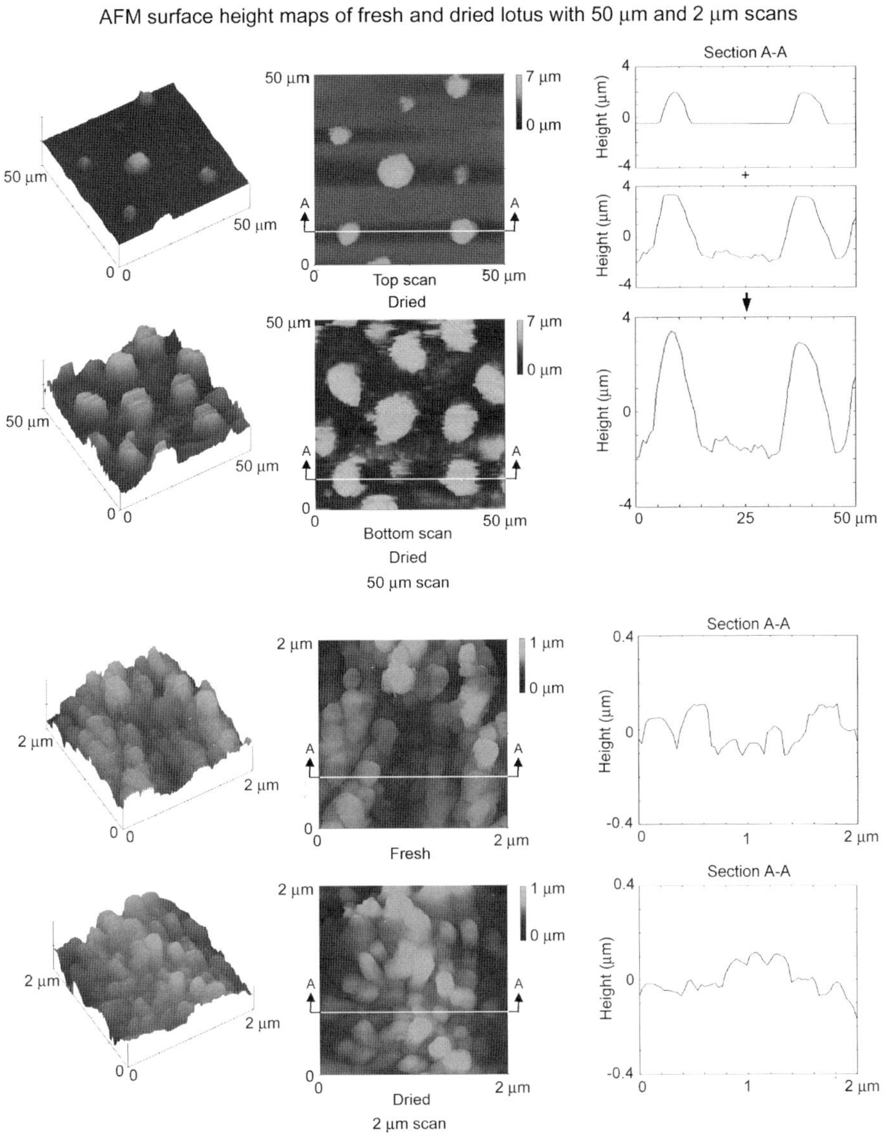

AFM surface height maps of fresh and dried lotus with 50 μm and 2 μm scans

Fig. 10.8. Surface height maps and 2-D profiles showing the top scan and bottom scan of a dried lotus leaf in 50 μm scan (because the P-V distance of a dried lotus leaf is greater than the Z-range of an AFM), and both fresh and dried lotus in a 2 μm scan [39]

the plant material that causes the leaf to be soft, so that when the tip comes into contact with the leaf sample, the sample deforms and a larger real area of contact between the tip and sample occurs, and thus the adhesive force increases. After the leaf has been dried, the moisture that was in the plant material is gone, and there is

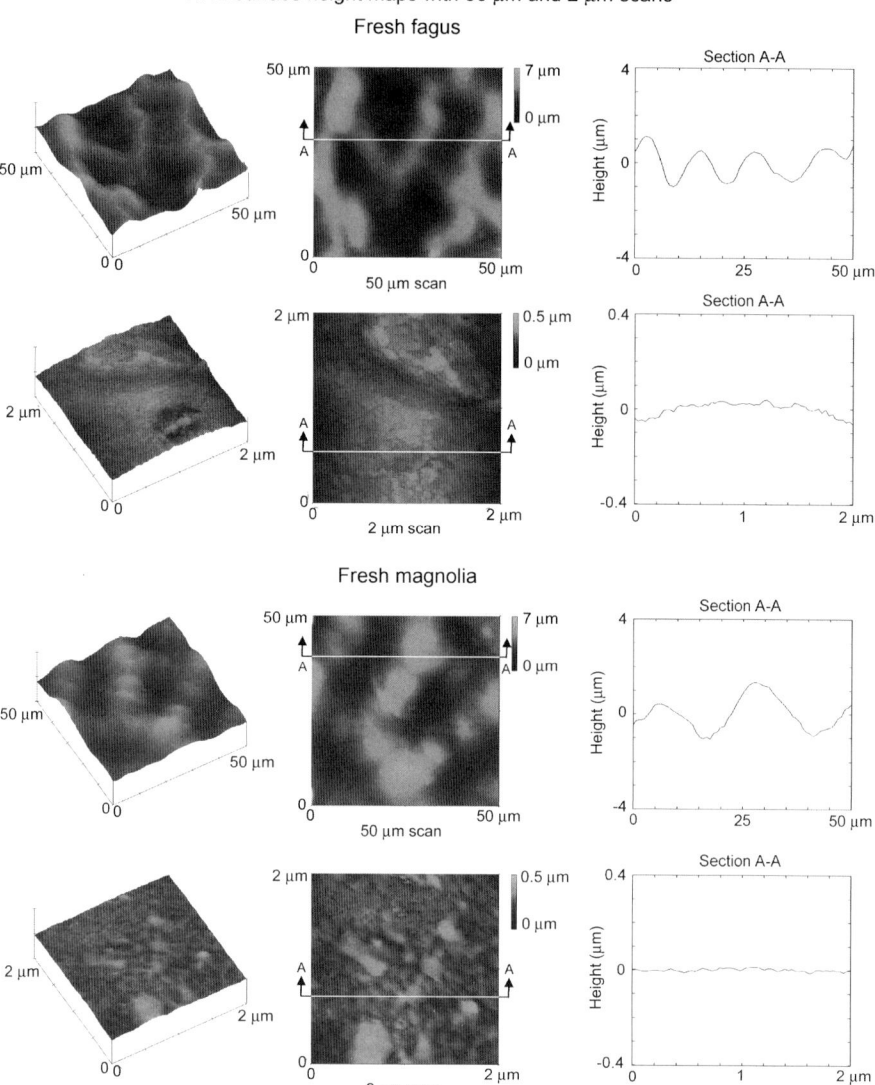

Fig. 10.9. Surface height maps and 2-D profiles of fagus and magnolia using an AFM in both 50 μm and 2 μm scans [39]

not as much deformation of the leaf when the tip comes into contact with the leaf sample. Hence, the adhesive force is decreased because the real area of contact has decreased [39].

The adhesion force of fagus and magnolia is higher than that of lotus and colocasia. The reason is that the real area of contact between the tip and leaf sample is

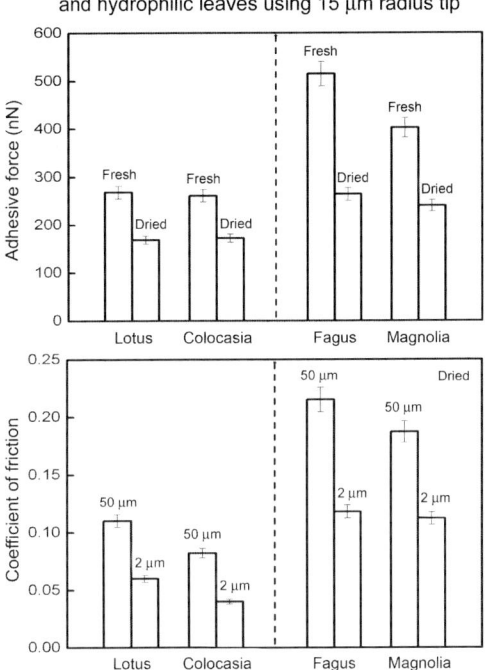

Fig. 10.10. Adhesive forces for fresh and dried leaves, and the coefficient of friction for dried leaves for 50 µm and 2 µm scan sizes for hydrophobic and hydrophilic leaves. All measurements were made using a 15 µm radius borosilicate tip. Reproducibility for both adhesive force and coefficient of friction is ±5% for all measurements [39]

expected to be higher in hydrophilic leaves than that in hydrophobic leaves. In addition, the fagus and magnolia are hydrophilic and have high affinity to water. The combination of high real area of contact and affinity to water are responsible for higher meniscus forces [31, 32].

The coefficient of friction was measured only on dried plant surfaces with the constant sliding velocity of the tip (10 µm/s) in different scan sizes. The fresh surfaces were excluded because the P-V was too large to scan back and forth with the AFM to obtain friction force. As expected, the coefficient of friction for hydrophobic leaves is lower than that for hydrophilic leaves due to the decrease in the real area of contact between the tip and leaf sample, similar to the adhesion force results. When the scan size decreases from microscale to nanoscale, the coefficient of friction also decreases in each leaf. The reason for such dependence is the scale-dependent nature of the roughness of the leaf surface. Figures 10.8 and 10.9 show AFM topography images and 2D profiles of the surfaces for different scan sizes. The scan-size dependence of the coefficient of friction has been reported previously [42, 190, 268, 312].

Table 10.2. Roughness factor and contact angle ($\Delta\theta = \theta - \theta_0$) calculated using R_f on the smooth surface for hydrophobic and hydrophilic leaves measured using an AFM, both microscale and nanoscale [39]

Leaf (contact angle)	Scan size	State	R_f	$\Delta\theta$
Lotus (153°)	50 μm	Dried	5.6	54*
	2 μm	Fresh	20	61**
		Dried	16	60**
Colocasia (152°)	50 μm	Dried	8.4	56*
	2 μm bump	Fresh	18	60**
		Dried	14	59**
	2 μm ridge	Fresh	18	60**
		Dried	15	59**
Fagus (76°)	50 μm	Fresh	3.4	−10*
	2 μm	Fresh	5.3	2**
Magnolia (84°)	50 μm	Fresh	3.8	−4*
	2 μm	Fresh	3.6	14**

*Calculations made using Wenzel equation
**Calculations made using Cassie–Baxter equation. We assume that the contact area between the droplet and air is a half of the whole area of the rough surface

10.2.7 Role of the Hierarchy

The approximation of the roughness factor for the leaves on the micro- and nanoscale was made using AFM scan data [39]. Roughness factors for various leaves are presented in Table 10.2. As mentioned earlier, the open space between asperities on a surface has the potential to collect air, and its probability appears to be higher in nanobumps as the distance between bumps in the nanoscale is smaller than those in microscale. Using roughness factor values along with the contact angles (θ) from both hydrophobic and hydrophilic surfaces—153° and 152° in lotus and colocasia, and 76° and 84° in fagus and magnolia, respectively—the contact angles (θ_0) for the smooth surfaces can be calculated using the Wenzel and Cassie–Baxter equations for nanobumps. Contact angle θ, calculated using R_f on the smooth surface, can be found in Table 10.2. It can be seen that the roughness factors and the differences ($\Delta\theta$) between θ and θ_0 on the nanoscale are higher than those in the microscale. This means that nanobumps on the top of a microbump increase contact angle more effectively than microbumps. In the case of hydrophilic leaves, the values of R_f and $\Delta\theta$ change very little on both scales.

Based on the data in Fig. 10.10, the coefficient of friction values in the nanoscale are much lower than those in the microscale. It is clearly observed that friction values are scale dependent. The height of a bump and the distance between bumps in the microscale is much larger than those in the nanoscale, which may be responsible for larger values of friction force on the microscale.

A difference between microbumps and nanobumps for surface enhancement of water repellency is the effect on contact angle hysteresis, in other words, the ease with which a droplet of water can roll on the surface. It was stated earlier that contact angle hysteresis decreases and contact angle increases due to the decreased contact with the solid surface caused by the air pockets beneath the droplet. The surface with nanobumps has a high roughness factor compared with that of microbumps. With large distances between microbumps, the probability of air pocket formation decreases and is responsible for high contact angle hysteresis. Therefore, on the surface with nanobumps, the contact angle is high and contact angle hysteresis is low, and drops rebound easily and can set into a rolling motion with a small tilt angle [39].

10.3 Other Biological Superhydrophobic Surfaces

While water-repellent plants are the most familiar examples of biological superhydrophobic surfaces, many other examples exist, including bird and butterfly wings, and water strider legs. Bormashenko et al. [61] studied the wetting of pigeon feathers, which are known to repel water. They found that the tissue constituting the pigeon pennae was hydrophilic; however, the water drop, supported by a network formed by barbs and barbules, sits partially on air pockets due to the two-fold structure of a feather that favors large contact angles and provides its water repellency. They also noted that the Cassie–Wenzel wetting regime transition has been observed under drop evaporation.

Gao and Jiang [123] studied the legs of water striders. They found that the leg has a hierarchical structure, with microhairs covered by nanogrooves, and suggested that the hierarchical structure is responsible for water-repellent properties.

10.4 Summary

In this chapter we reviewed experimental methods for the study of water-repellent plant surface topography and other properties, such as the contact angles, friction, adhesion, and wetting regime transition. Among these methods are the use of the SEM and ESEM, AFM, optical profiler and goniometers for the contact angle measurement. We also discussed experimental data for plant surfaces leaves, with the emphasis on the hierarchical structure, as well as other biological water repellent surfaces such as water strider legs and bird feathers.

11

Artificial (Biomimetic) Superhydrophobic Surfaces

Abstract Artificial (biomimetic) superhydrophobic surfaces utilizing the Lotus effect are reviewed in this chapter. First, modern ways of production of superhydrophobic surfaces are discussed, including lithography, deposition, stretching, itching, evaporation, sol-gel, and others. The variety of materials used to make superhydrophobic surfaces (metals, polymers, semiconductors, nanotubes, nanoparticles) is discussed. Then wetting and self-cleaning properties of micro- and nanopatterned silicon and polymer biomimetic surfaces are presented on the basis of experimental measurements. After that, commercially available superhydrophobic products (paints, textiles, glasses) are reviewed as well as future applications in industry, bio- and nanotechnology.

In the preceding chapter, we discussed biological superhydrophobic and self-cleaning surfaces and experimental methods of their study. In this chapter, we will review artificial (biomimetic) superhydrophobic surfaces. First we will discuss various ways to produce rough superhydrophobic surfaces (Table 11.1). Second, we will review properties of these surfaces and present some applications that are coming to the market.

In recent years, fabrication of superhydrophobic surfaces has become an area of active fundamental research. This chapter will discuss a number of new approaches, and there is no doubt that in the near future new technological concepts will emerge. In general, the same techniques that are used for micro- and nanostructure manufacturing, such as lithography, etching, and deposition, have been used for producing superhydrophobic surfaces. Advantages and shortcomings of these techniques are summarized in Table 11.2. One especially interesting development is the creation of reversible surfaces that can be turned from hydrophobic to hydrophilic by applying electric potential, heat, or ultraviolet (UV) irradiation [112, 198, 207, 308, 342]. Another important task is to create transparent superhydrophobic surfaces, which may have numerous potential applications for optics and self-cleaning glasses. In order for the surface to be transparent, roughness details should be smaller than the wavelength of the visible light (about 400–700 nm) [226]. While the fundamental research is very active, a number of attempts to produce commercial products using the lotus effect have been made [118, 308]. This includes glasses, textile, paints, aerosols, etc.

Table 11.1. Typical materials and corresponding techniques to produce micro/nanoroughness [54]

Material	Technique	Contact angle	Notes	Source
Teflon	Plasma	168		Zhang et al. [354]; Shiu et al. [294]
Fluorinated block polymer solution	Casting under humid environment	160	Transparent	Yabu and Shimomura [343]
PFOS	Electro- and chemical polymerization	152	Reversible (electric potential)	Xu et al. [342]
PDMS	Laser treatment	166		Khorasani et al. [179]
PS-PDMS Block copolymer	Electrospinning	>150		Ma et al. [208]
PS, PC, PMMA	Evaporation	>150		Bormashenko et al. [58]
PS nanofiber	Nanoimprint	156		Lee et al. [203]
PET	Oxygen plasma etching	>150		Teshima et al. [314]
Organo-triethoxysilanes	Sol-gel	155	Reversible (temperature)	Shirtcliffe et al. [296]
Al	Chemical etching	>150		Qian and Shen [273]
Copper	Electrodeposition	160	Hierarchical	Shirtcliffe et al. [294]
Si	Photolithography	170		Bhushan and Jung [40]
Si	E-beam lithography	164		Martines et al. [219]
Si	X-ray lithography	>166		Fürstner et al. [118]
Si	Casting	158	Plant leaf replica	Sun et al. [308]; Fürstner et al. [118]
Si (Black Si)	Plasma etching	>150	For liquid flow	Jansen et al. [164]
Silica	Sol-gel	150		Hikita et al. (2005); Shang et al. (2005)
Polyelectrolyte multilayer surface overcoated with silica nanoparticles	Self assembly	168		Zhai et al. [352]

Table 11.1. (*Continued*)

Material	Technique	Contact angle	Notes	Source
Nano-silica spheres	Dip coating	105		Klein et al. [187]
Silica colloidal particles in PDMS	Spin coated	165	Hierarchical	Ming et al. [223]
Au clusters	Electrochemical deposition	>150		Zhang et al. [355]
Carbon nanotubes	CVD	159		Huang et al. [157]
ZnO, TiO$_2$ Nanorods	Sol-gel	>150	Reversible (UV irradiation)	Feng et al. [112]

11.1 How to Make a Superhydrophobic Surface

There are two main requirements for a superhydrophobic surface: the surface should be rough and it should have a hydrophobic (low surface energy) coating. These two requirements lead to two methods of producing a superhydrophobic surface. The first method is to make a rough surface from an initially hydrophobic material, and the second method is to modify an initially rough surface by changing the surface chemistry or applying a hydrophobic material. Note that roughness is usually a more critical property than low surface energy, since both moderately hydrophobic and very hydrophobic materials exhibit similar wetting behavior when roughened [207]. And understandably so, based on the simple Wenzel model, the cosine of the contact angle is given by $R_f \cos \theta_0$, so even small (negative) $\cos \theta_0$ will result in a high contact angle when combined with a big roughness factor.

Fabrication of superhydrophobic surfaces has been an area of active research since the mid-1990s. In general, the same techniques that are used for micro- and nanostructure fabrication, such as lithography, etching, and deposition, have been used to produce superhydrophobic surfaces (see Fig. 11.1).

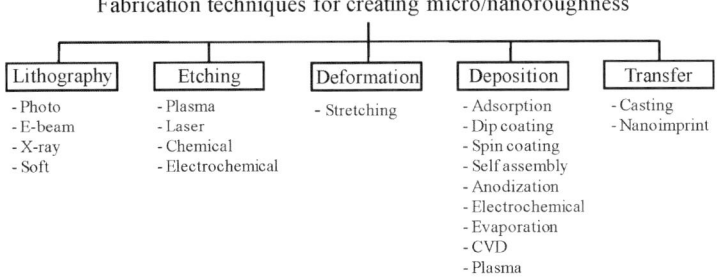

Fig. 11.1. Typical methods to fabricate micro/nanoroughening on a surface [54]

Table 11.2. Advantages and shortcomings of various fabrication techniques [54]

Techniques	Advantages	Shortcomings
Lithography	Accuracy, large area	Slow process, high cost
Etching	Fast	Chemical contamination, less control
Deposition	Flexibility, cheap	Can be high temperature, less control

11.1.1 Roughening to Create One-Level Structure

Lithography is a well-established technique for creating large areas of periodic micro/nanopatterns. It includes photo, E-beam, X-ray, and soft lithography. Bhushan and Jung [40] produced patterned Si using photolithography. To obtain a sample that is hydrophobic, a SAM of $1,1,-2,2,$-tetrahydroperfluorodecyltrichlorosilane (PF$_3$) was deposited on the sample surfaces using the vapor phase deposition technique. They obtained a superhydrophobic surface with a contact angle up to 170°. Martines et al. [219] fabricated ordered arrays of nanopits and nanopillars by using electron beam lithography. They obtained a superhydrophobic surface with a contact angle of 164° and hysteresis of 1° for a surface consisting of tall pillars with cusped tops after a hydrophobization with octadecyltrichlorosilane (OTS). Fürstner et al. [118] created silicon wafers with regular patterns of spikes by X-ray lithography. The wafer was hydrophobized by sputtering a layer of gold and subsequent immersion in a hexadecanethiol solution. Jung and Bhushan [172] created low aspect ratio asperities (LAR, 1 : 1 height-to-diameter ratio), high aspect ratio asperities (HAR, 3 : 1 height-to-diameter ratio), and lotus pattern (replica from the lotus leaf), all on a PMMA surface using soft lithography. A self-assembled monolayer (SAM) of perfluorodecyltriethoxysilane (PFDTES) was deposited on the patterned surfaces using vapor phase deposition technique.

One well-known and effective way to make rough surfaces is etching using either plasma, laser, chemical, or electrochemical techniques [207]. Jansen et al. [164] etched a silicon wafer using a fluorine-based plasma by using the black silicon method to obtain isotropic, positively and negatively tapered as well as vertical walls with smooth surfaces. Coulson et al. [85] described an approach in plasma chemical roughening of poly(tetrafluoroethylene) (PTFE) substrates followed by the deposition of low surface energy plasma polymer layers, which give rise to high repellency toward polar and nonpolar probe liquids. A different approach was taken by Shiu et al. [294], who treated a Teflon film with oxygen plasma and obtained a superhydrophobic surface with a contact angle of 168°. Fluorinated materials have a limited solubility, which makes it difficult to roughen them. However, they may be linked or blended with other materials, which are often easier to roughen, in order to make superhydrophobic surfaces. Teshima et al. [314] obtained a transparent superhydrophobic surface from a poly(ethylene terephthalate) (PET) substrate via selective oxygen plasma etching followed by plasma-enhanced chemical vapor deposition using tetramethylsilane (TMS) as the precursor. Khorasani et al. [179] produced porous PDMS surfaces with a contact angle of 175° using a CO_2-pulsed laser etching method as an excitation source for the surface. Qian and Shen [273] described a simple surface roughening method by dislocation selective chemical etching on polycrystalline metals such as aluminum. After treatment with fluoroalkylsilane, the etched metallic surfaces exhibited superhydrophobicity. Xu et al. [342] fabricated a reversible superhydrophobic surface with a double-roughened perfluorooctanesulfonate (PFOS) doped conducting polypyrrole (PPy) film by a combination of electropolymerization and chemical polymerization. The reversibility was achieved by switching between superhydrophobic doped or oxidized states and superhydrophilicity dedoped or neutral states with changing the applied electrochemical potential.

The stretching method can be used to produce a superhydrophobic surface. Zhang et al. [354] stretched a Teflon film and converted it into fibrous crystals with a large fraction of void space in the surface, leading to high roughness and superhydrophobicity.

Deposition methods also make a substrate rough from the bulk properties of the material and enlarge potential applications of superhydrophobic surfaces. There are several ways to make a rough surface including adsorption, dip coating, electrospinning, anodization, electrochemical, evaporation, chemical vapor deposition (CVD), and plasma. Solidification of wax can be used to produce a superhydrophobic surface. Shibuichi et al. [293] used alkylketene dimer (AKD) wax on a glass plate to spontaneously form a fractal structure in its surfaces. They obtained a surface with a contact angle larger than 170° without any fluorination treatments. Klein et al. [187] obtained superhydrophobic surfaces by simply dip-coating a substrate with slurry containing nano-silica spheres, which adhered to a substrate after a low temperature heat treatment. After reaction of the surface with a fluoroalkyltrichlorosilane, the hydrophobicity increased with decreasing area fraction of spheres. Ma et al. [208, 209] produced block copolymer poly(styrene-b-dimethylsiloxane) fibers with submicrometer diameters in the range 150–400 nm by electrospinning from solution in tetrahydrofuran and dimethylformamide. They obtained superhydrophobic nonwoven fibrous mats with a contact angle of 163°. Shiu et al. [294] showed that self-organized, close-packed superhydrophobic surfaces can be easily achieved by spin-coating the monodispersed polystyrene beads solution on substrate surfaces. The sizes of the beads were reduced by controlling the etching conditions. After plasma treatment, the surfaces were coated with a layer of gold and eventually a layer of octadecanethiol SAM to render hydrophobicity. Abdelsalam et al. [1] studied the wetting of structured gold surfaces formed by electrodeposition through a template of submicrometer spheres and discussed the role of the pore size and shape in controlling wetting. Bormashenko et al. [58] used evaporated polymer solutions of polystyrene (PS), polycarbonate (PC), and polymethylmethacrylate (PMMA) dissolved in chlorinated solvents, dichloromethane (CH_2Cl_2), and chloroform ($CHCl_3$), to obtain a self-assembled structure with hydrophobic properties. Chemical/physical vapor deposition (CVD/PVD) has been used to modify surface chemistry as well. Lau et al. [201] created superhydrophobic carbon nanotube forests by modifying the surface of vertically aligned nanotubes with plasma enhanced chemical vapor deposition (PECVD). Superhydrophobicity was achieved down to the microscopic level where essentially spherical, micrometer-sized water droplets can be suspended on top of the nanotube forest. Zhu et al. [360] and Huang et al. [157] prepared surfaces with two-scale roughness by controlled growth of carbon nanotube (CNT) arrays by CVD. Zhao et al. [358] also synthesized the vertically aligned multiwalled carbon nanotube (MWCNT) arrays by chemical-vapor deposition on Si substrates using thin film of iron (Fe) as catalyst layer and aluminum (Al) film.

Attempts to create a superhydrophobic surface by casting and nanoimprint methods have been successful. Yabu and Shimomura [343] prepared a porous superhydrophobic transparent membrane by casting a fluorinated block polymer solution in a humid environment. The transparency was achieved because the honeycomb-patterned films had subwavelength pore size. Sun et al. [308] reported a nanocasting method to make a superhydrophobic PDMS surface. They first made a negative PDMS template using the lotus leaf as an original template and then used the negative template to make a positive PDMS template—a replica of the original lotus leaf. Zhao et al. [358] prepared a superhydrophobic surface by casting a micellar solution of a copolymer poly(styrene-b-dimethylsiloxane) (PS-PDMS) in humid air based on the cooperation of vapor-induced phase separation and surface enrichment of PDMS block. Lee et al. [203] produced vertically aligned PS nanofibers by using nanoporous anodic aluminum oxide as a replication template in a heat- and pressure-driven nanoimprint pattern transfer process. As the aspect ratio of the polystyrene (PS) nanofibers increased, the nanofibers could not stand upright but formed twisted bundles resulting in a three-dimensionally rough surface with advancing and receding contact angles of 155.8° and 147.6°, respectively.

11.1.2 Coating to Create One-Level Hydrophobic Structures

Modifying the surface chemistry with a hydrophobic coating widens the potential applications of superhydrophobic surfaces. There are several ways to modify the chemistry of a surface, including sol-gel, dip coating, self-assembly, electrochemical, and chemical/physical vapor deposition. Shirtcliffe et al. [296] prepared porous sol-gel foams from organo-triethoxysilanes which exhibited switching between superhydrophobicity and superhydrophilicity when exposed to different temperatures. The critical switching temperature was between 275 °C and 550 °C for different materials, and when the foam was heated above the critical temperature, complete rejection of water by the cavities switched to complete filling of the pores. Hikita et al. [152] used colloidal silica particles and fluoroalkylsilane as the starting materials and prepared a sol-gel film with superliquid-repellency by hydrolysis and condensation of alkoxysilane compounds. Feng et al. [112] produced superhydrophobic surfaces using ZnO nanorods by sol-gel method. They showed that superhydrophobic surfaces can be switched into hydrophilic surfaces by alternation of ultraviolet (UV) irradiation. Shang et al. [290] did not blend low surface energy materials in the sols, but described a procedure to make transparent superhydrophobic surfaces by modifying silica-based gel films with a fluorinated silane. In a similar way, Wu et al. [341] made a microstructured ZnO-based surface via a wet chemical process and obtained superhydrophobicity after coating the surface with long-chain alkanoic acids.

Zhai et al. [352] used a layer-by-layer (LBL) self-assembly technique to create a poly(allylamine hydrochloride)/poly(acrylic acid) (PAH/PAA) multilayer which formed a honeycomb-like structure on the surface after an appropriate combination of acidic treatments. After cross-linking the structure, they deposited silica nanoparticles on the surface via alternating dipping of the substrates into an aqueous suspension of the negatively charged nanoparticles and an aqueous PAH solution, followed by a final dipping into the nanoparticle suspension. Superhydrophobicity was obtained after the surface was modified by a chemical vapor deposition of (tridecafluoro-1,1,2,2-tetrahydrooctyl)-1-trichlorosilane followed by a thermal annealing.

Zhang et al. [355] showed that the surface covered with dendritic gold clusters, which was formed by electrochemical deposition onto indium tin oxide (ITO) electrode modified with a polyelectrolyte multilayer, showed superhydrophobic properties after further deposition of an n-dodecanethiol monolayer. Han et al. [142] described the fabrication of lotus leaf-like superhydrophobic metal surfaces by using electrochemical reaction of Cu or Cu–Sn alloy plated on steel sheets with sulfur gas, and subsequent perfluorosilane treatment. The chemical bath deposition (CBD) has also been used to make nanostructured surfaces, thus, Hosono et al. [155] fabricated a nanopin film of brucite-type cobalt hydroxide (BCH) and achieved a contact angle of 178° after further modification with lauric acid (LA). Shi et al. [292] described the use of galvanic cell reaction as a facile method to chemically deposit Ag nanostructures on the p-silicon wafer on a large scale. When the Ag covered silicon wafer was further modified with a self-assembled monolayer of n-dodecanethiol, a superhydrophobic surface was obtained with a contact angle of about 154° and a tilt angle lower than 5°.

11.1.3 Methods to Create Two-Level (Hierarchical) Superhydrophobic Structures

Two-level (hierarchical) roughness structures are typical for superhydrophobic surfaces in nature, as we discussed earlier. Recently, much effort has been devoted to fabricating these hierarchical structures in various ways. Shirtcliffe et al. [294] prepared a hierarchical (double-roughened) copper surface by electrodeposition from acidic copper sulfate solution onto flat copper and patterning technique of coating with a fluorocarbon hydrophobic layer. Another way to obtain a rough surface for superhydrophobicity is assembly from colloidal systems. Ming et al. [223] prepared a hierarchical (double roughened) surface consisting of silica-based raspberry-like particles which were made by covalently grafting amine-functionalized silica particles of 70 nm to epoxy-functionalized silica particles of 700 nm via the reaction between epoxy and amine groups. The surface became superhydrophobic after being modified with PDMS. Northen and Turner [231] fabricated arrays of flexible silicon dioxide platforms supported by single high aspect ratio silicon pillars down to 1 μm in diameter and with heights up to ~50 μm. When these platforms were coated with polymeric organorods approximately $2\,\mu m$ tall and 50–200 nm in diameter, it showed that the surface is highly hydrophobic with a water contact angle of 145°. Chong et al. [81] used the combination of the porous anodic alumina (PAA) template with microsphere monolayers to fabricate hierarchically ordered nanowire arrays, which have periodic voids at the microscale and hexagonally packed nanowires at the nanoscale. They created the arrays by selective electrodeposition using nanoporous anodic alumina as a template and a porous gold film as a working electrode that is patterned by microsphere monolayers. Wang et al. [333] also developed a novel precursor hydrothermal redox method with $Ni(OH)_2$ as the precursor to fabricate a hierarchical structure of nickel hollow microspheres with nickel nanoparticles as the in situ formed building units. Wang et al.'s hierarchical hollow structure exhibited enhanced coercivity and remnant magnetization as compared with hollow nickel submicrometer spheres, hollow nickel nanospheres, bulk nickel, and free Ni nanoparticles. Kim et al. [185] fabricated a hierarchical structure that looks like the same structure as a lotus leaf. First, nanoscale porosity was generated by anodic aluminum oxidation, then the anodized porous alumina surface was replicated by polytetrafluoroethylene. The polymer sticking phenomenon during the replication created the submicrostructures on the negative polytetrafluoroethylene nanostructure replica. The contact angle of the created hierarchical structure was obtained about 160° and the tilting angle is less than 1°. Del Campo and Greiner [95] reported that SU-8 hierarchical patterns composed of features with lateral dimensions ranging from 5 mm to 2 mm and heights from 10 to $500\,\mu m$ were obtained by photolithography which comprises a step of layer-by-layer exposure in soft contact printed shadow masks which are embedded into the SU-8 multilayer.

11.2 Experimental Techniques

11.2.1 Contact Angle, Surface Roughness, and Adhesion

The static and dynamic (advancing and receding) contact angles were measured using a Rame–Hart model 100 contact angle goniometer and water droplets of deionized water [39, 68, 172]. For the measurement of static contact angle, the droplet size should be small but larger than the dimension of the structures present on the surfaces. Droplets of about $5\,\mu L$ in volume (with a diameter of a spherical droplet about 2.1 mm) were gently deposited on the substrate using a microsyringe for the static contact angle. The receding contact angle was measured by the removal of water from a DI water sessile droplet ($\sim 5\,\mu L$) using a microsyringe. The advancing contact angle was measured by adding additional water to the sessile droplet ($\sim 5\,\mu L$) using the microsyringe. The contact angle hysteresis was calculated by the difference between the measured advancing and receding contact angles. The tilt angle was measured by a simple stage-tilting experiment with the droplets of $5\,\mu L$ volume [40, 41]. All measurements were made using five different points for each sample at $22 \pm 1\,°C$ and $50 \pm 5\%$ RH. The measurements were reproducible to with $\pm 3\%$.

For surface roughness, an optical profiler (NT-3300, Wyko Corp., Tucson, AZ) was used for different surface structures [39–41, 53, 68, 174]. A greater Z-range of the optical profiler of 2 mm is a distinct advantage over the surface roughness measurements using an AFM which has a Z-range of 7 μm, but it has a maximum lateral resolution of approximately 0.6 μm [31, 32]. Experiments were performed using three different radii tips to study the effect of scale dependence. Large radii atomic force microscopy (AFM) tips were primarily used in this study. Borosilicate ball with 15 μm and a silica ball with 3.8 μm radius were mounted on a gold-coated triangular Si_3N_4 cantilever with a nominal spring constant of $0.58 \, N \, m^{-1}$. A square pyramidal Si_3N_4 tip with a nominal radius 30–50 nm on a triangular Si_3N_4 cantilever with a nominal spring constant of $0.58 \, N \, m^{-1}$ was used for smaller radius tip. Adhesive force was measured using the single point measurement of a force calibration plot [31, 32, 34].

11.2.2 Measurement of Droplet Evaporation

Droplet evaporation was observed and recorded by a digital camcorder (Sony, DCRSR100) with a 10 X optical and 120 X digital zoom for every run of the experiment. Then the decrease in the diameter of the droplets with time was determined [173, 174]. Time resolution of the camcorder was 0.03 s per frame. An objective lens placed in front of the camcorder during recording gave a total magnification of between 10 to 20 times. Droplet diameters as small as a few hundred microns could be measured with this method. Droplets were gently deposited on the substrate using a microsyringe, and the whole process of evaporation was recorded. The evaporation starts right after the deposition of the droplets. Images obtained were analyzed using Imagetool® software (University of Texas Health Science Center) for the contact angle. To find the dust trace remaining after droplet evaporation, an optical microscopy with a CCD camera (Nikon, Optihot-2) was used. All measurements were made in a controlled environment at $22 \pm 1 \,°C$ and $45 \pm 5\%$ RH [173, 174].

11.2.3 Measurement of Contact Angle Using ESEM

A Philips XL30 ESEM equipped with a Peltier cooling stage was used to study smaller droplets [174]. ESEM uses a gaseous secondary electron detector (GSED) for imaging. The SESM column is equipped with a multistage differential pressure-pumping unit. The pressure in the upper part is about 10^{-6} to 10^{-7} Torr, but the pressure of about 1 to 15 Torr can be maintained in the observation chamber. When the electron beam (primary electrons) ejects secondary electrons from the surface of the sample, the secondary electrons collide with gas molecules in the ESEM chamber, which in turn acts as a cascade amplifier, delivering the secondary electron signal to the positively biased GSED. The positively charged ions are attracted toward the specimen to neutralize the negative charge produced by the electron beam. Therefore, the ESEM can be used to examine electrically isolated specimens in their natural state. In ESEM, adjusting the pressure of the water vapor in the specimen chamber and the temperature of the cooling stage will allow the water to condense on the sample in the chamber. For the measurement of the static and dynamic contact angles on patterned surfaces, the video images were recorded. The voltage of the electron beam was 15 kV and the distance of the specimen from the final aperture was about 8 mm. If the angle of observation is not parallel to the surface, the electron beam is not parallel to the surface but inclined at an angle, this will produce a distortion in the projection of the droplet profile. A mathematical model to calculate the real contact angle from the ESEM images was used to correct the tilting of the surfaces during imaging [65, 174].

11.3 Wetting of Micro- and Nanopatterned Surfaces

In this section, we will discuss experimental observations of wetting properties of micro- and nanopatterned surfaces on the basis of the experimental data by Jung and Bhushan [173] and other groups.

11.3.1 Micro- and Nanopatterned Polymers

Jung and Bhushan [172] studied two types of polymers: poly(methyl methacrylate) (PMMA) and polystyrene (PS). PMMA and PS were chosen because they are widely used in MEMS/NEMS devices. Both hydrophilic and hydrophobic surfaces can be produced using these two polymers, as PMMA has polar groups with high surface energy (hydrophilic) while PS has electrically neutral and nonpolar groups (hydrophobic) with low surface energy. Furthermore, a PMMA structure can be made hydrophobic by treating it appropriately, for example, by coating with a hydrophobic self-assembled monolayer (SAM).

Four types of surface patterns were fabricated from PMMA: a flat film, low aspect ratio asperities (LAR, 1 : 1 height-to-diameter ratio), high aspect ratio asperities (HAR, 3 : 1 height-to-diameter ratio), and the lotus pattern (replica of the lotus leaf). Two types of surface patterns were fabricated from PS: a flat film and the lotus pattern. Figure 11.2 shows SEM images of the two types of nanopatterned structures, LAR and HAR, and the one type of micropatterned structure, lotus pattern, all on a PMMA surface [67, 172]. For nanopatterned structures, PMMA film was spin-coated on the silicon wafer. A UV cured mold (PUA mold) with nanopatterns of interest was made which enables one to create sub-100-nm patterns with high aspect ratio [80]. The mold was placed on the PMMA film and a slight pressure of $\sim$10 g/cm^2 ($\sim$1 kPa) was applied and annealed at 120 °C. Finally, the PUA mold was removed from PMMA film. For micropatterned structures, a polydimethylsiloxane (PDMS) mold was first made by casting PDMS against a lotus leaf followed by heating. As shown in Fig. 11.2, it can be seen that only microstructures exist on the surface of lotus pattern [172].

Since PMMA by itself is hydrophilic, in order to obtain a hydrophobic sample, a self-assembled monolayer (SAM) of perfluorodecyltriethoxysilane (PFDTES) was deposited on the sample surfaces using vapor phase deposition technique. PFDTES was chosen because of its hydrophobic nature. The deposition of PFDTES took place at a temperature of 100 °C, pressure 400 atm, with 20 min deposition time, and 20 min annealing time. The polymer surface was exposed to an oxygen plasma

Nanopatterns
PMMA low aspect ratio (LAR)

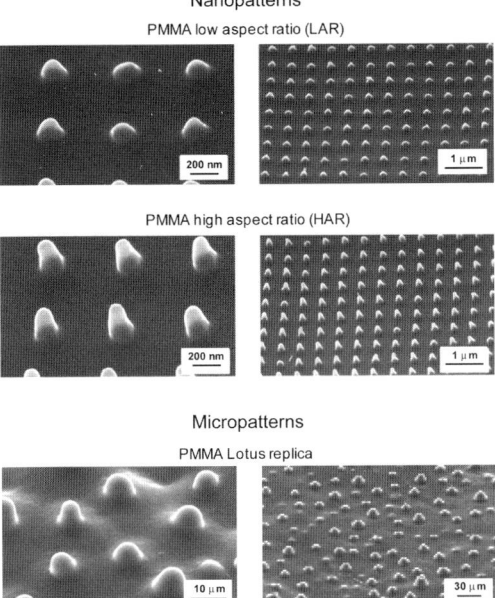

PMMA high aspect ratio (HAR)

Micropatterns
PMMA Lotus replica

Fig. 11.2. Scanning electron micrographs of the two nanopatterned polymer surfaces (shown using two magnifications to see both the asperity shape and the asperity pattern on the surface) and the micropatterned polymer surface (lotus pattern, which has only microstructures on the surface) [67, 172]

treatment (40 W, O_2 187 Torr, 10 s) prior to coating. The oxygen plasma treatment is necessary to oxidize any organic contaminants on the polymer surface and to also alter the surface chemistry to allow for enhanced bonding between the SAM and the polymer surface [172].

11.3.1.1 Contact Angle Measurements

Jung and Bhushan [172] measured the static contact angle of water with the patterned PMMA and PS structures; see Fig. 11.3. Since the Wenzel roughness factor is the parameter that often determines wetting behavior, we calculated the roughness factor and it is presented in Table 11.3 for various samples. The data show that the contact angle of the hydrophilic materials decreases with an increase in the roughness factor, as predicted by the Wenzel model. When the polymers were coated with PFDTES, the film surface became hydrophobic. Figure 11.3 also shows the contact angle for various PMMA samples coated with PFDTES. For a hydrophobic surface, the standard Wenzel model predicts an increase of contact angle with roughness factor, which is what happens in the case of patterned samples. We also present the calculated values of the contact angle for various patterned samples based on the contact angle of the smooth film and the Wenzel equa-

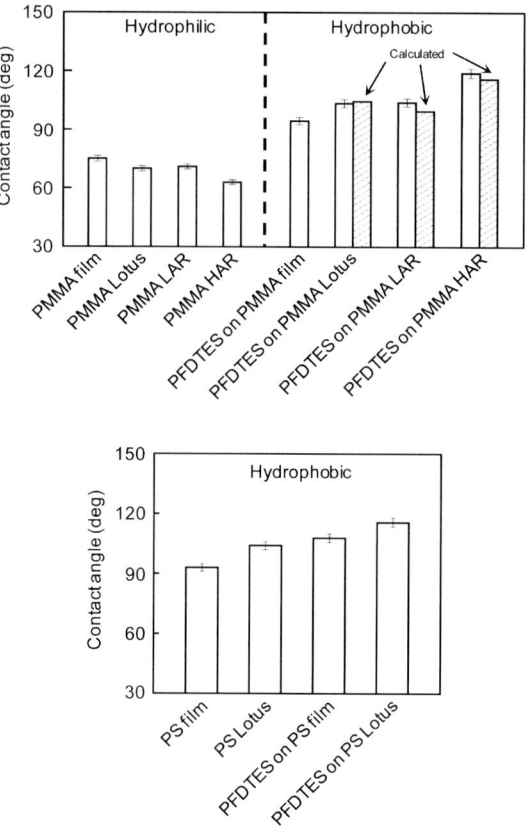

Fig. 11.3. Contact angles for various patterned surfaces on PMMA and PS polymers [172]

Table 11.3. Roughness factor for micro- and nanopatterned polymers [172]

	LAR	HAR	Lotus
R_f	2.1	5.6	3.2

tion. The measured contact angle values for the lotus pattern were comparable with the calculated values, whereas for the LAR and HAR patterns they are higher. It suggests that nanopatterns benefit from air pocket formation. For the PS material, the contact angle of the lotus pattern also increased with increased roughness factor.

11.3.1.2 Scale Dependence on Adhesive Force

Jung and Bhushan [172] found that scale-dependence of adhesion and friction are important for this study because the tip/surface interface area changes with size. The meniscus force will change due to either changing tip radius, the hydrophobicity of

PMMA polymers with different surface roughness patterns (22 °C, 50% RH)

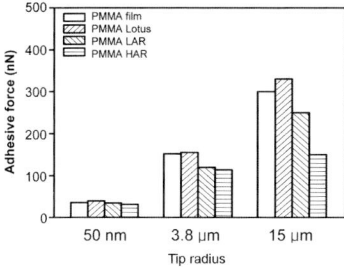 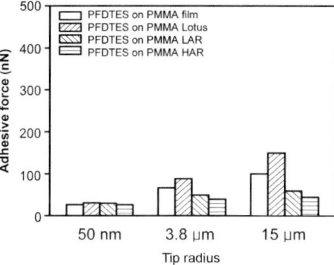

Fig. 11.4. Scale dependent adhesive force for various patterned surfaces measured using AFM tips of various radii [172]

the sample, or the number of contact and near-contacting points. Figure 11.4 shows the dependence of the tip radius and hydrophobicity on the adhesive force for PMMA and PFDTES coated on PMMA [172]. When the radius of the tip is changed, the contact angle of the sample is changed, and asperities are added to the sample surface, the adhesive force will change due to the change in the meniscus force and the real area of contact.

The two plots in Fig. 11.4 show the adhesive force on a linear scale for the different surfaces with varying tip radius. The left bar chart in Fig. 11.4 is for hydrophilic PMMA film, lotus pattern, LAR, and HAR, and shows the effect of tip radius and hydrophobicity on adhesive force. For increasing radius, the adhesive force increases for each material. With a larger radius, the real area of contact and the meniscus contribution increase, resulting in increased adhesion. The right bar chart in Fig. 11.4 shows the results for PFDTES coated on each material. These samples show the same trends as the film samples, but the increase in adhesion is not as dramatic. The hydrophobicity of PFDTES on material reduces meniscus forces, which in turn reduces adhesion from the surface. The dominant mechanism for the hydrophobic material is real area of contact and not meniscus force, whereas with hydrophilic material there is a combination of real area of contact and meniscus forces [172].

11.3.2 Micropatterned Si Surfaces

Micropatterned surfaces produced from single-crystal silicon (Si) by electrolithography and coated with a self-assembled monolayer (SAM) were used in the study by Jung and Bhushan [173, 174]. Silicon has traditionally been the most commonly used structural material for micro/nanocomponents. A Si surface can be made hydrophobic by coating with a SAM. One purpose of this investigation was to study the transition for Cassie–Baxter to Wenzel regimes by changing the distance between the pillars. To create patterned Si, two series of nine samples each were fabricated using photolithography [22]. Series 1 has 5-μm diameter and 10-μm height flat-top, cylindrical pillars with different pitch values (7, 7.5, 10, 12.5, 25, 37.5, 45, 60, and 75) μm, and Series 2 has 14-μm diameter and 30-μm height flat-top, cylindrical pillars with different pitch values (21, 23, 26, 35, 70, 105, 126, 168, and 210) μm. The pitch is the spacing between the centers of two adjacent pillars. The Si chosen were initially hydrophilic, so to obtain a sample that is hydrophobic, a self-assembled monolayer (SAM) of 1,1,−2,2,-tetrahydroperfluorodecyltrichlorosilane (PF$_3$) was deposited on the sample surfaces using the vapor phase deposition technique [22]. PF$_3$ was chosen because of the hydrophobic nature of the surface. The thickness and rms roughness of the SAM of PF$_3$ were 1.8 nm and 0.14 nm, respectively [176].

An optical profiler was used to measure the surface topography of the patterned surfaces [40, 53, 173]. One sample each from the two series was chosen to characterize the surfaces. Two different surface height maps can be seen for the patterned Si in Fig. 11.5. In each case, a 3D map and a flat map along with a 2D profile in a given location of the flat 3D map are shown. A scan size of 100 μm × 90 μm was used to obtain a sufficient number of pillars to characterize the surface but also to maintain enough resolution to get an accurate measurement.

The images found with the optical profiler indicate that the flat-top, cylindrical pillars on the Si surface are distributed on the entire surface. These pillars were distributed in a square grid with different pitch values. Each sample series has the same series of Wenzel roughness factors ($R_f = 1 + \pi DH/P^2$) and other relevant geometric parameters (e.g., the spacing factor $S_f = P/H$). Keeping these parameters constant means that Cassie and Baxter's and Wenzel's theoretical models predict exactly the same series of contact angle values for all two series of nine samples [173].

11.3.2.1 Contact Angle for Flat-Top, Cylindrical Pillars

Let us consider the geometry of flat-top, cylindrical pillars of diameter D, height H, and pitch P, distributed in a regular square array as shown in Fig. 11.5. For the special case of the droplet size much larger than P (of interest in this study), a droplet contacts the flat-top of the pillars forming the composite interface, and the cavities are filled with air. For this case,

$$f_{LA} = 1 - \frac{\pi D^2}{4P^2} = 1 - f_{SL}$$

and the contact angles for the Wenzel and Cassie–Baxter regimes are given by corresponding equations.

Geometrical values of the flat-top, cylindrical pillars in series 1 and 2 are used for calculating the contact angle for the above-mentioned two cases. Figure 11.6 shows the plot of the predicted values of the contact angle as a function of pitch between the pillars for the two cases. Wenzel's and Cassie and Baxter's equations present two possible equilibrium states for a water droplet on the surface. This indicates that there is a critical pitch below which the composite interface dominates and above which the homogeneous interface dominates the wetting behavior. The process to design superhydrophobic surfaces is important in determining the equilibrium water droplet. Therefore, one needs to find the critical point that can be used to design superhydrophobic surfaces. It should also be noted that even in cases where the liquid droplet does not contact the bottom of the cavities, the water droplet in a metastable

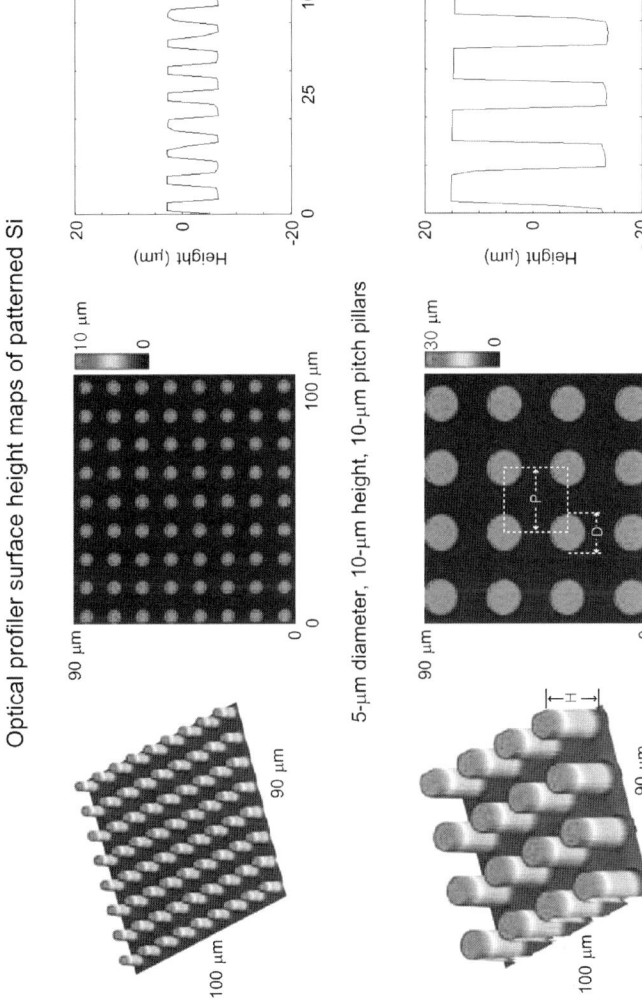

Fig. 11.5. Surface height maps and 2-D profiles of the patterned surfaces using an optical profiler [40]

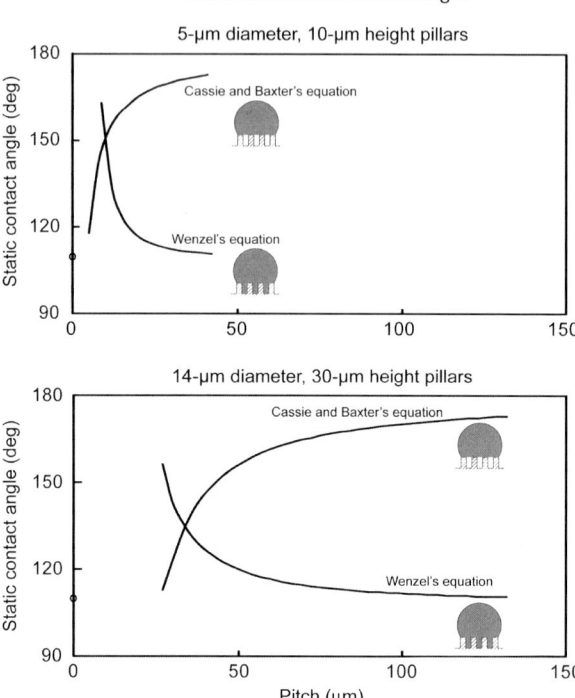

Fig. 11.6. Calculated static contact angle as a function of geometric parameters for a given value of θ_0 using Wenzel and Cassie and Baxter equations for two series of the patterned surfaces with different pitch values [40]

state becomes unstable and transition from the Cassie–Baxter regime to the Wenzel regime occurs if the pitch is large.

11.3.2.2 Curvature-Based Cassie–Wenzel Transition Criteria

A stable composite interface is essential for the successful design of superhydrophobic surfaces. However, the composite interface is fragile, and it may transform into the homogeneous interface. What triggers the transition between the regimes remains a subject of debate, although a number of explanations have been suggested. Nosonovsky and Bhushan [243] studied destabilizing factors for the composite interface and found that a convex surface (with bumps) leads to a stable interface and high contact angle. Also, they suggested that a droplet's weight and curvature are among the factors which affect the transition.

Bhushan and Jung [40, 41] and Jung and Bhushan [173, 174] investigated the effect of droplet curvature on the Cassie–Wenzel regime transition. First, they considered a small water droplet suspended on a superhydrophobic surface consisting of a regular array of circular pillars with diameter D, height H, and pitch P as shown

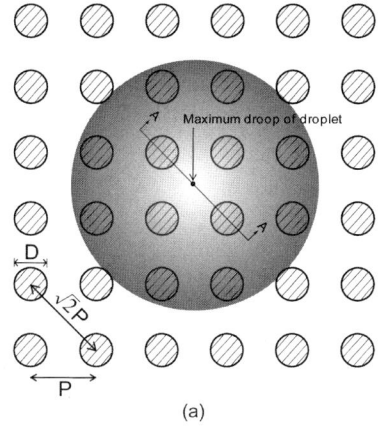

(a)

Section A-A

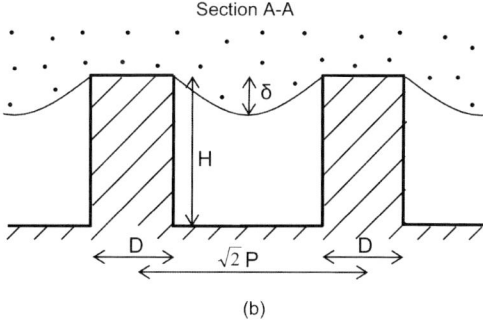

(b)

Fig. 11.7. A small water droplet suspended on a superhydrophobic surface consisting of a regular array of circular pillars. **a** Plan view. The maximum droop of droplet occurs in the center of square formed by four pillars. **b** Side view in section A-A. The maximum droop of droplet (δ) can be found in the middle of two pillars which are diagonally across [173, 174]

in Fig. 11.7. The local deformation for small droplets is governed by surface effects rather than gravity. The curvature of a droplet is governed by the Laplace equation, which relates the pressure inside the droplet to its curvature [6]. Therefore, the curvature is the same at the top and at the bottom of the droplet [197, 244]. For the patterned surface considered here, the maximum droop of the droplet occurs in the center of the square formed by the four pillars as shown in Fig. 11.7(a). Therefore, the maximum droop of the droplet, δ, in the recessed region can be found in the middle of two pillars which are diagonally across as shown in Fig. 11.7(b), which is $(\sqrt{2}P - D)^2/(8R)$. If the droop is much greater than the depth of the cavity

$$(\sqrt{2}P - D)^2/R \geq H, \tag{11.1}$$

then the droplet will just contact the bottom of the cavities between pillars, resulting in the transition from the Cassie–Baxter to Wenzel regime. Furthermore, in the case of large distances between the pillars, the liquid–air interface can easily be destabilized due to dynamic effects, such as surface waves that are formed at the liquid–air interface due to the gravitational or capillary forces. This leads to the formation of the homogeneous solid–liquid interface. However, whether the droplet droop or other mechanisms dominate the transition remains to be investigated, as we discussed in Sect. 8.2.

11.3.2.3 Contact Angle Measurements

The experiment performed with 1 mm in radius (5 μL volume) droplets on the patterned Si coated with PF$_3$ was designed to determine the static contact angle [40, 41, 173, 174]. The contact angles on the prepared surfaces are plotted as a function of pitch between the pillars in Fig. 11.8(a). A dotted line represents the transition criteria range obtained using (11.1). The flat Si coated with PF$_3$ showed a static contact angle of 109°. As the pitch increases up to 45 μm of series 1 and 126 μm of series 2, the static contact angle first increases gradually from 152° to 170°. Then, the contact angle starts decreasing sharply. Initial increase with an increase of pitch has to do with more open air space present which increases the propensity of air pocket formation. As predicted from the Jung and Bhushan [173] transition criteria (11.1), the decrease in contact angle at higher pitch values results due to the transition from composite interface to solid–liquid interface. In series 1, the value predicted from the transition criteria is a little higher than the experimental observations. However, in series 2, there is a good agreement between the experimental data and the theoretically predicted values by Jung and Bhushan [173] for the transition from Cassie and Baxter regime to Wenzel regime.

Figure 11.8(b) shows hysteresis and tilt angle as a function of pitch between the pillars [40, 41]. The flat Si coated with PF$_3$ showed a hysteresis angle of 34° and tilt angle of 37°. The patterned surfaces with low pitch increase the hysteresis and tilt angles compared to the flat surface due to the effect of sharp edges on the pillars, resulting in pinning [240]. Hysteresis for a flat surface can arise from roughness and surface heterogeneity. For a droplet moving down the inclined patterned surfaces, the line of contact of the solid, liquid, and air will be pinned at the edge point until it is able to move, resulting in increasing hysteresis and tilt angles. Figure 11.9 shows droplets on patterned Si with 5-μm diameter and 10-μm height pillars with different pitch values. The asymmetrical shape of the droplet signifies pinning. The pinning on the patterned surfaces can be observed as compared to the flat surface. The patterned surface with low pitch (7 μm) has more pinning than the patterned surface with high pitch (37.5 μm), because the patterned surface with low pitch has more sharp edges contacting with a droplet.

For various pitch values, hysteresis and tilt angles show the same trends with varying pitch between the pillars. After an initial increase as discussed earlier, they gradually decrease with increasing pitch (due to reduced number of sharp edges) and show an abrupt minimum in the value which has the highest contact angle. The lowest hysteresis and tilt angles are 5° and 3°, respectively, which were observed on the patterned Si with 45 μm of series 1 and 126 μm of series 2. As discussed earlier, an increase in the pitch value allows the formation of composite interface. At higher pitch values, it is difficult to form the composite interface. The decrease in

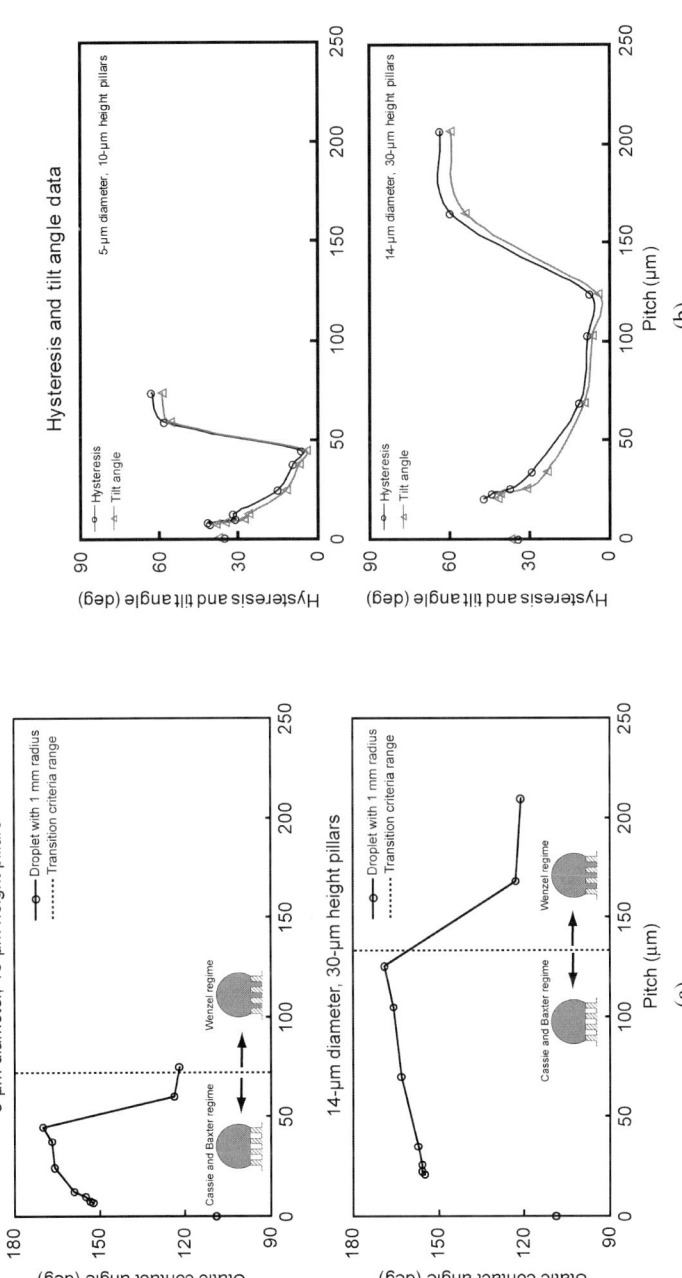

Fig. 11.8. a Static contact angle (a *dotted line* represents the transition criteria range obtained using (15)), and **b** hysteresis and tilt angles as a function of geometric parameters for two series of the patterned surfaces with different pitch values for a droplet with 1 mm in radius (5 μL volume). Data at zero pitch correspond to a flat sample [40, 173, 174]

Patterned surfaces with 5-µm diameter and
10-µm height pillars with different pitch values

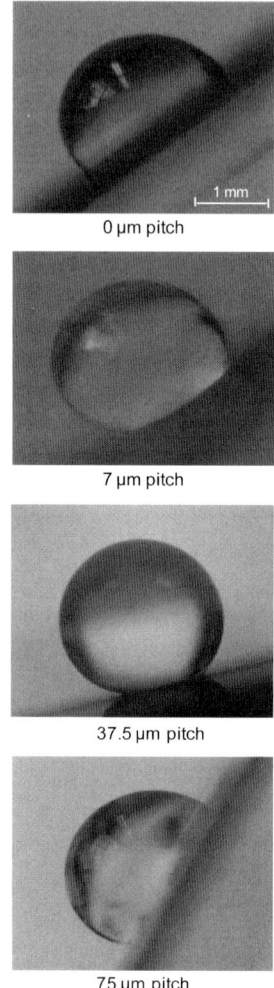

0 µm pitch

7 µm pitch

37.5 µm pitch

75 µm pitch

Fig. 11.9. Optical micrographs of droplets on the inclined patterned surfaces with different pitch values. The images were taken when the droplet started to move down. Data at zero pitch correspond to a flat sample [40]

hysteresis and tilt angles occurs due to the formation of composite interface at pitch values ranging from 7 µm to 45 µm in series 1 and from 21 µm to 126 µm in series 2. The hysteresis and tilt angles start to increase again due to the lack of formation of air pockets at pitch values raging from 60 µm to 75 µm in series 1 and from 168 µm to 210 µm in series 2. These results suggest that the air pocket formation and the reduction of pinning in the patterned surface play an important role for a surface with both low hysteresis and tilt angle [40]. Hence, to create superhydrophobic surfaces, it is important that they are able to form a stable composite interface with air pockets between the solid and the liquid. Capillary waves, nanodroplet condensation, hydrophilic spots due to chemical surface inhomogeneity, and liquid pressure can destroy the composite interface. Nosonovsky and Bhushan [244] suggested that these factors that make the composite interface unstable have different characteristic length scales, so nanostructures, or the combination of microstructures and nanostructures, is required to resist them.

11.3.2.4 Observation of the Transition during the Droplet Evaporation

Jung and Bhushan [173, 174] performed droplet evaporation experiments to observe the Cassie–Wenzel regime's transition on two different patterned Si surfaces coated with PF_3. The series of six images in Fig. 11.10 show the successive photos of a droplet evaporating on two patterned surfaces. The initial radius of the droplet is about 700 μm, and the time interval between successive photos is 30 s. In the first five photos, the drop is first in a hydrophobic state, and its size gradually decreases with time. However, as the radius of the droplet reaches 360 μm on the surface with 5-μm diameter, 10-μm height, and 37.5-μm pitch pillars, and 423 μm on the surface with 14-μm diameter, 30-μm height, and 105-μm pitch pillars, the Cassie–Wenzel regime's transition occurs, as indicated by the arrow. Figure 11.10 also shows a close-up of water droplets on two different patterned Si surfaces coated with PF_3 before and after the transition. The light passes below the left droplet, indicating that air pockets exist, so that the droplet is in the Cassie–Baxter state. However, an air pocket is not visible below the bottom right droplet, so it is in Wenzel state. This could result from an impalement of the droplet on the patterned surface, characterized by a smaller contact angle.

To find the contact angle before and after transition, the values of the contact angle are plotted against the theoretically predicted value, based on the Wenzel and Cassie–Baxter models. Figure 11.11 shows the static contact angle as a function of geometric parameters for the experimental contact angles before (circle) and after (triangle) the transition compared with the Wenzel and Cassie–Baxter equations (solid lines) with a given value of θ_0 for two series of the patterned Si with different pitch values coated with PF_3 [174]. The fit is good between the experimental data and the theoretically predicted values for the contact angles before and after transition.

To verify the validity of the transition criteria in terms of droplet size, the critical radius of the droplet deposited on the patterned Si with different pitch values coated with PF_3 was measured during the evaporation experiment. Figure 11.12 shows the radius of a droplet as a function of geometric parameters for the experimental results (circle) compared with the Cassie–Wenzel regime's transition (solid lines) for two series of the patterned Si with different pitch values coated with PF_3. It was found that the critical radius of impalement is indeed in good quantitative agreement with our predictions. The critical radius of the droplet increases linearly with the geometric parameter (pitch). For the surface with small pitch, the critical radius of a droplet can become quite small. Based on this trend, one can design superhydrophobic surfaces, even for small droplets.

Evaporation of a droplet on patterned surface

5-μm diameter, 10-μm height, and 37.5-μm pitch pillars

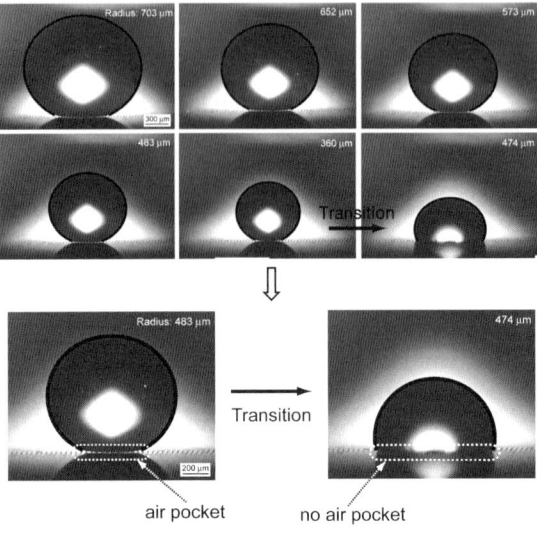

14-μm diameter, 30-μm height,and 105-μm pitch pillars

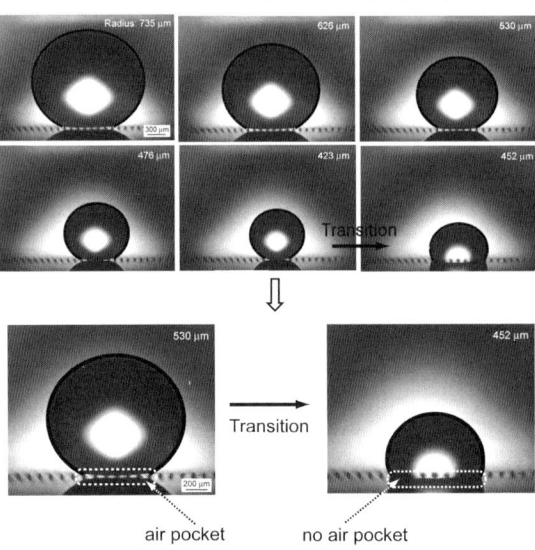

Fig. 11.10. Evaporation of a droplet on two different patterned surfaces. The initial radius of the droplet is about 700 μm, and the time interval between successive photos is 30 s. As the radius of droplet reaches 360 μm on the surface with 5-μm diameter, 10-μm height, and 37.5-μm pitch pillars, and 420 μm on the surface with 14-μm diameter, 30-μm height, and 105-μm pitch pillars, the transition from Cassie and Baxter regime to Wenzel regime occurs, as indicated by the arrow. Before the transition, air pocket is clearly visible at the bottom area of the droplet, but after the transition, air pocket is not found at the bottom area of the droplet [174]

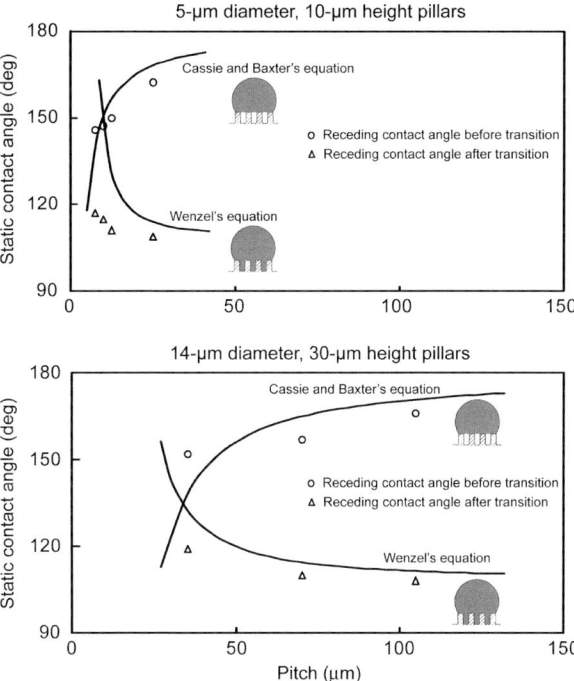

Contact angles before and after transition

Fig. 11.11. Receding contact angle as a function of geometric parameters before (*circle*) and after (*triangle*) transition compared with predicted static contact angle values obtained using Wenzel and Cassie and Baxter equations (*solid lines*) with a given value of θ_0 for two series of the patterned surfaces with different pitch values [174]

To verify their transition criterion, Jung and Bhushan [173, 174] used another approach using the dust mixed in water. Figure 11.13 presents the dust trace remaining after droplet with 1 mm radius (5 µL volume) evaporation on the patterned Si surface with pillars of 5-µm diameter and 10-µm height with 37.5-µm pitch in which the transition occurred at 360 µm radius of the droplet, and with 7-µm pitch in which the transition occurred at about 20 µm radius of the droplet during the process of evaporation. As shown in the top image, after the Cassie–Wenzel regime's transition, the dust particles remained not only at the top of the pillars but also at the bottom with a footprint size of about 450 µm. However, as shown in the bottom image, the dust particles remained on only a few pillars until the end of the evaporation process. The transition occurred at about 20 µm radius of droplet and the dust particles left a footprint of about 25 µm. From Fig. 11.12, we observe that the transition occurs at about 300 µm radius of droplet on the 5-µm diameter and 10-µm height pillars with 37.5-µm pitch, but the transition does not occur on the patterned Si surface with pitch of less than about 5 µm. These experimental observations are consistent with model predictions. In the literature, it has been shown that on super-

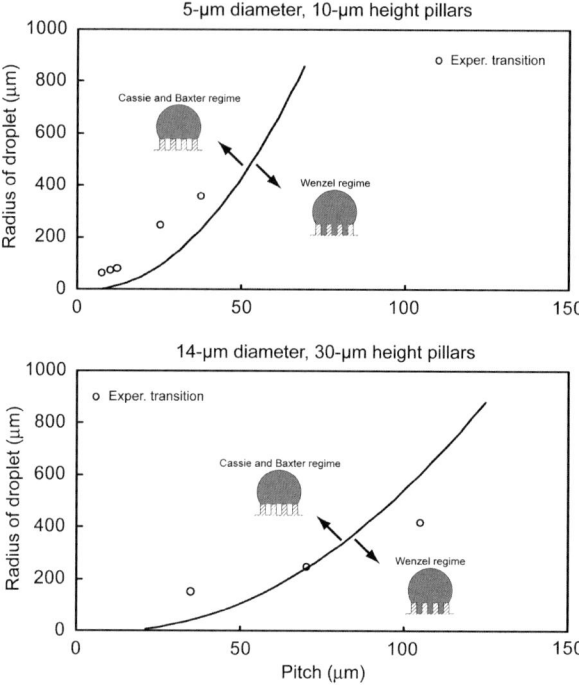

Fig. 11.12. Radius of droplet as a function of geometric parameters for the experimental results (*circle*) compared with the transition criteria from Cassie and Baxter regime to Wenzel regime (*solid lines*) for two series of the patterned surfaces with different pitch values [174]

hydrophobic natural lotus, the droplet remains almost in the Cassie–Baxter regime during the evaporation process [357]. This indicates that the distance between the pillars should be minimized enough to improve the ability of the droplet to resist sinking.

11.3.2.5 Observation and Measurement of Contact Angle Using ESEM

Figure 11.14 shows how water droplets grow and merge under ESEM [174]. ESEM is used as a contact angle analysis tool to investigate superhydrophobicity on the patterned surfaces. Microdroplets (in dimension of less than 1 mm diameter) are distributed on the patterned surface coated with PF_3 during increasing condensation by decreasing temperature. Even if the microdroplets are not the same size, they show the hydrophobic characteristics of the patterned surface. At the beginning, some small water droplets appear, i.e., water droplets at locations 1, 2, and 3 in the left image. During increasing condensation by decreasing temperature, water droplets at locations 1 and 3 gradually increase in size and water droplets at location 2 merge together to form one big droplet in the middle image. With further condensation, water

Dust trace after droplet evaporation

5-µm diameter, 10-µm height,and 37.5-µm pitch pillars

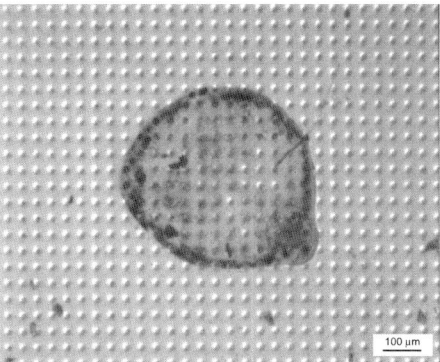

5-µm diameter, 10-µm height,and 7-µm pitch pillars

Fig. 11.13. Dust trace remained after droplet evaporation for the patterned surface. In the top image, the transition occurred at 360 µm radius of droplet, and in the bottom image, the transition occurred at about 20 µm radius of droplet during the process of droplet evaporation. The footprint size is about 450 and 25 µm for the top and bottom images, respectively [174]

droplets at locations 1 and 2 increase in size and water droplets at location 3 merge together to form one big droplet in the right image. In all cases condensation was initiated at the bottom, therefore, as can be observed, the droplets are in the Wenzel regime. This could also be evidence that the droplet on the macroscale used in the conventional contact angle measurement comes from the merging of smaller droplets [174].

Compared with the conventional contact angle measurement, ESEM is able to provide detailed information about the contact angle of microdroplets on patterned surfaces. The diameter of the water droplets used for the contact angle measurement is more than 10 µm such that the size limit pointed out by Stelmashenko et al. [303] was avoided. For droplet size less than 1 µm, substrate backscattering can distort the intensity profile such that the images are inaccurate [174].

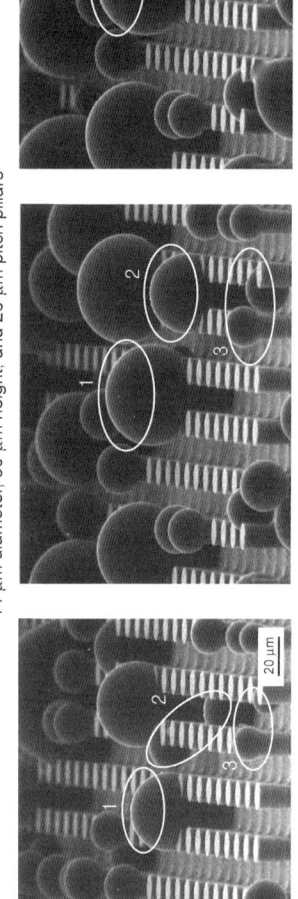

Process of growing droplets on patterned surface in an ESEM

During increasing condensation

14-μm diameter, 30-μm height, and 26-μm pitch pillars

Water droplets in 1, 2, 3 appear   Water droplets in 2 merge   Water droplets in 3 merge

Fig. 11.14. Microdroplet (in dimension of less than 1 mm diameter) growing and merging process under ESEM during increasing condensation by decreasing temperature. *Left image:* Some small water droplets appear at the beginning, i.e. water droplets 1, 2, 3. *Middle image:* Water droplets at locations 1 and 3 increase in size and water droplets at location 2 merge together to form one big droplet. *Right image:* Water droplets at locations 1 and 2 increase in size and water droplets at location 3 merge together to form one big droplet [174]

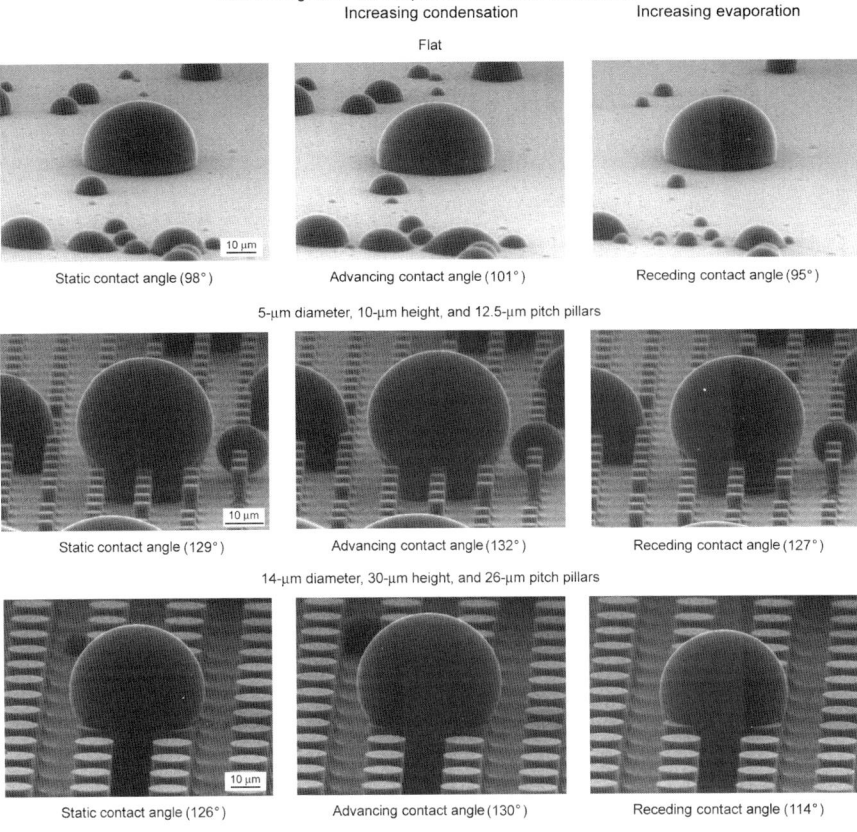

Fig. 11.15. Microdroplets on flat and two patterned surfaces using ESEM. Second set of images were taken during increasing condensation, and the third set of images were taken during increasing evaporation. Static contact angle was measured when the droplet was stable. Advancing contact angle was measured after increasing condensation by decreasing the temperature of the cooling stage. Receding contact angle was measured after decreasing evaporation by increasing the temperature of the cooling stage [174]

As shown in Fig. 11.15, the static contact angle and hysteresis angle of the microdroplets condensed on a flat surface and on two different patterned surfaces were obtained from the images and corrected using methodology mentioned earlier. The difference between the data estimated from the images and corrected θ is about 3%. Once the microdroplet's condensation and evaporation has reached a dynamic equilibrium, static contact angles are determined. The flat Si coated with PF_3 showed a static contact angle of 98°. The patterned surfaces coated with PF_3 increase the static contact angle compared to the flat surface coated with PF_3 due to the effect of roughness. The advancing contact angle was taken after increasing condensation by decreasing the temperature of the cooling stage. The receding contact angle was

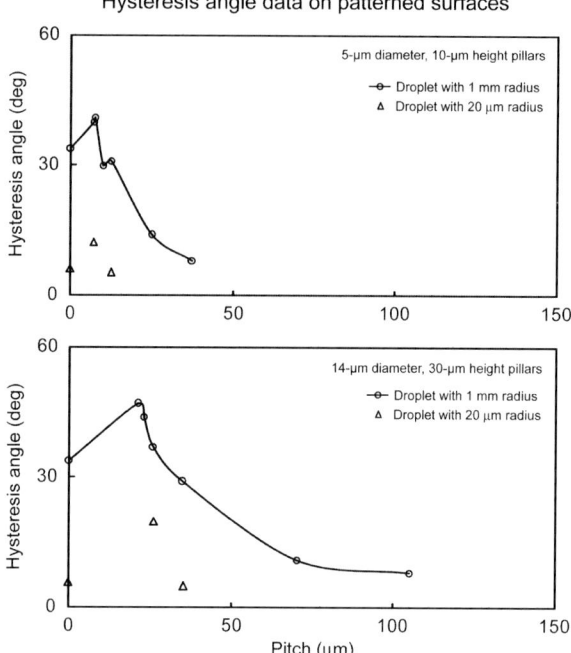

Fig. 11.16. Hysteresis angle as a function of geometric parameters for the microdroplet with about 20 μm radius from ESEM (triangle) compared with the droplet with 1 mm radius (5 μL volume) (circle and solid lines) for two series of the patterned surfaces with different pitch values. Data at zero pitch correspond to a flat sample [174]

taken after increasing evaporation by increasing the temperature of the cooling stage. The hysteresis angle was then calculated [174].

Figure 11.16 shows hysteresis angle as a function of geometric parameters for the microdroplets formed in the ESEM (triangle) for two series of the patterned Si with different pitch values coated with PF$_3$. Data at zero pitch correspond to a flat Si sample. The droplets with about 20 μm radii that are larger than the pitch were selected in order to look at the effect of pillars in contact with the droplet. These data were compared with conventional contact angle measurements obtained with the droplet with 1 mm radius (5 μL volume; circle and solid lines) [40]. When the distance between pillars increases above a certain value, the contact area between the patterned surface and the droplet decreases, resulting in the decrease of the hysteresis angle. Both the droplets with 1 mm and 20 μm radii show the same trend. The hysteresis angles for the patterned surfaces with low pitch are higher compared to the flat surface due to the effect of sharp edges on the pillars, resulting in pinning [240]. Hysteresis for a flat surface can arise from roughness and surface heterogeneity. For a droplet advancing on the patterned surfaces, the line of contact of the solid, liquid, and air will be pinned at the edge point until it is able to move, resulting in increasing hysteresis angle. The hysteresis angle for the microdroplet from ESEM is lower compared to that for the droplet with 1 mm radius. The difference of hysteresis angle between a microdroplet and a droplet with 1 mm radius could come from the different pinning effects, because the latter has more sharp edges contacting with a droplet compared with the former. The results show how droplet size can affect the wetting properties of patterned Si surfaces [174].

11.4 Self-cleaning

Thus far there has been proposed no quantitative theory of self-cleaning that would relate, for example, the size and contact angle of a droplet with the size of a contaminating particle being washed away. There is a qualitative understanding that water-repellent surfaces do also repel other contaminants and that dust can easily be washed from them by flowing water. A number of experimental studies have been conducted. The self-cleaning abilities of patterned surfaces were investigated by Fürstner et al. [118]. They studied Si wafer specimens with regular patterns of spikes that were manufactured by X-ray lithography. The specimens were hydrophobized with Au thiol. For comparison, they also studied replicates of plant surfaces, made by a two-component silicon molding mass applied to the leaf surface. The negative replica is flexible and rubber-like. A melted hydrophobic polyether was applied onto this mold. They also studied several metal foil specimens, hydrophobized by means of a fluorinated agent. In order to investigate the self-cleaning effect, a luminescent and hydrophobic powder was used as a contaminant. Following contamination, the specimens were subjected to an artificial fog and rain [118].

Droplets of water rolled off easily from Si samples with a microstructure consisting of rather slender and sufficiently high spikes; this is attributed to the fact that the Cassie wetting state occurred. These samples could be cleaned almost completely after artificial contamination by means of the fog treatment. The behavior of water drops was different upon surfaces with low spikes and a rather high pitch. The researchers found a considerable decrease of the contact angles and a distinct rise in the sliding angles apparently corresponding to the Wenzel state. Some metal foils and some replicates had two levels of roughness. These specimens did not show a total removal of all contaminating particles when they were subjected to artificial fog, but water drops impinging with sufficient kinetic energy could clean them perfectly. A substrate without structures smaller than 5 μm could not be cleaned by means of fog consisting of water droplets with diameter 8–20 μm because this treatment resulted in a continuous water film on the samples. However, artificial rain removed all the contamination. On the other hand, smooth specimens made of the same material could not be cleaned completely by impinging droplets. This is a clear indication of the different contact phenomena on smooth hydrophobic surfaces in contrast to self-cleaning microstructured surfaces. Another interesting observation of this group was that despite the missing structure of the wax crystals, the water contact angle of the lotus replica was the highest of all the replicates, indicating that the microstructure formed by the papillae alone is already optimized with regard to water repellency [118].

11.5 Commercially Available Lotus-Effect Products

A number of products that use the lotus-effect are already commercially available or being developed. In addition, many patents have been granted for various possible applications of self-cleaning surfaces [45]. Most of these applications use the self-cleaning effect, especially in the case of glasses (for architecture, automotive, optical sensor, and other applications), roof tiles, and other architectural applications. Additionally, sprays and paints that create clean surfaces (e.g., graffiti-resistant) have been suggested, as well as water-repellent textiles. Some agricultural applications are also discussed (e.g., pesticide additives that can decrease bouncing off plant surfaces or increase penetration into the soil).

From a commercial point of view, cleaning of windows is expensive and cumbersome, especially if the windows are on a skyscraper. Self-cleaning windows using the Lotus-effect have been released to the market by several companies. How far these windows will be a commercial success remains to be seen [57]. The Germany-based Web site Lotus-Effekt.de, dedicated to the commercial application of patented self-cleaning superhydrophobic micro-to-nano structured surfaces, states the following:

"*Lotusan*®, an exterior paint from the firm *Sto* is marketed already with the greatest success since 1999. It is used by professional firms of house painters and is, not yet, available for the general building trade. Up to the present *Lotusan*® has been used to paint c. 300,000 buildings. In 2004 *Degussa* (daughter company of *Goldschmidt*) has introduced the first spray: *Tegotop*® 105 which can be used to impregnate surfaces. Self-cleaning textiles are being tested at present and will be available commercially from summer 2005. Marquees will probably the first to receive such treatment. Optical sensors in public high impact areas (for instance, toll bridge sensors on highways) are furnished already throughout Germany with Lotus-*Effect*® glass manufactured by *Ferro AG*. A series of further products is being tested, among these *Aeroxide*® LE of the *Degussa* for plastics. For years self-cleaning glasses have figured in advertisement. Frequently this is about so-called photo catalytic stratification. The firm *Ferro* keeps prototypes of architectural glass with Lotus-*Effect*® in permanent test conditions. In the region of optical sensors (toll bridges) Lotus-*Effect*® glasses are already used successfully. For architectural glass and rear windows of cars applications will probably follow soon. With *Erlus-Lotus*® the first self-cleaning roof in the world came on the market. For demonstrations we employ a spoon with a perfect Lotus-*Effect*® surface. Honey and many other substances roll off without a hint of residue. The spoon is a prototype that is, unfortunately, not for sale. Firms can already order the first spray Tegotop® 105 for testing from the *Degussa-Goldschmidt AG*. The properties of the new-fangled intelligent textiles are astonishing. Not only does water roll off, but ketchup and red wine do likewise. The area of use will hardly lie with suits, ties or shirts, but rather with outdoor clothing, marquees, tents and with tarpaulins for lorries" [10].

In addition to the household and "conventional" products, possible use of roughness- or heterogeneity-induced superhydrophobicity in nano- and biotechnology applications is often discussed. This includes, for example, nonsticky surfaces for the components of micro/nanoelectromechanical systems (NEMS/MEMS). Since adhesion plays an important role for small devices, the so-called "stiction" of two component surfaces is a significant problem in that industry, which may lead to device failure. Making a surface hydrophobic can reduce meniscus force and stiction [249].

Controlling droplets containing biologically relevant molecules (DNA and proteins) is important in biotechnology. Superhydrophobicity is useful for these applications: the almost fully spherical droplets on a superhydrophobic surface can shrink exactly like a drop in free air. Furthermore, the positioning and shape of water droplets can be controlled by a pattern that combines hydrophilic and hydrophobic elements. Interestingly, some desert beetles capture their drinking water by a hydrophobic/philic structured back [259]. At a patterned heterogeneous substrate, hydrophilic regions can help to contain small liquid volumes of DNA, which may improve spotting and analyzing DNA and proteins by avoiding wall contact [57, 130].

In micro/nanofluidics, a guided motion of droplets on heterogeneous hydrophobic/philic surfaces gives the opportunity to develop droplet-based microfluidics systems, as opposed to the classical concept based on microfluidic channels. Droplets moving freely on open surfaces and bulk liquids flowing in channels constitute two extremes, with the patterned heterogeneous hydrophobic/philic surfaces being the intermediate between these two [102]. Driving the liquids along the channels and making them merge at predefined locations offers a novel way to mix reactants or steer biochemical reactions, defining the concept of a "liquid microchip" [128] or "surface-tension confined microfluidics" [199]. These open structures have advantages over capillaries, because blocking of the capillary by unforeseen chemical reactions cannot occur. Droplets have very low contact areas with the substrate, and they are easy to move by external fields, for example, electrostatic forces or surface capillary waves. Systems that make use of a droplet-based actuation mechanism are also being developed, and their aim is to control droplet positioning and motion on the substrates with as little surface contact as possible, and to turn the droplet-based system into a programmable reactor, by which the liquid positions are prescribed and tuned [57].

11.6 Summary

In this chapter, we discussed artificial superhydrophobic surfaces. There are several ways to manufacture these surfaces, and new methods continue to emerge. Some methods (such as lithography) allow scientists to create patterned surfaces with clearly defined and controlled geometrical features. These features have a typical size ranging from $1\,\mu m$ to $100\,\mu m$. Other (and often cheaper) methods lead to self-assembled or random rough surfaces. This includes extending, etching, polymer solution evaporation, sol-gel, and other methods. There are technologies available to produce transparent superhydrophobic materials; hierarchical surfaces and switchable surfaces that can change from hydrophobic to hydrophilic under an external control. The difference in the superhydrophobic properties of surfaces with pattern and random structure still has to be investigated.

A proper control of roughness constitutes the main challenge to producing a reliable superhydrophobic surface, while if the initial material is hydrophilic, a surface treatment or coating is required that will decrease the surface energy. While the two factors—roughness and low surface energy—are required for superhydrophobicity, the role of roughness clearly dominates. For example, it is not really important whether the low energy surface is built of typical for paraffin $-CH_2-$ groups or having much lower energy $-CH_3-$ groups of fluorocarbons. Furthermore, many rough surfaces without any lower energy characteristics still exhibit superhydrophobicity. The role of hierarchical roughness still remains to be investigated. While many suggestions have been made regarding why superhydrophobic surfaces in nature are hierarchical, experiments with nonhierarchical patterned surfaces demonstrate superhydrophobicity as well.

In addition, we discussed a number of emerging applications of the lotus-effect, superhydrophobicity and controlled hydrophobicity, ranging from the household applications (glasses, paints) to nanotechnology and microfluidics.

12

Gecko-Effect and Smart Adhesion

Abstract The "smart adhesion" of gecko is discussed. It is shown that gecko achieves very high adhesion as well as the ability to detach easily at will due to a hierarchical organization of attachment pads. Experimental data and theoretical models are presented. Properties of a biomimetic adhesive tape using the gecko-effect are reviewed.

In the preceding chapters we have studied in detail how roughness affects wetting. Here we will discuss the so-called adaptive or "smart" adhesion between a gecko foot and a solid surface and engineered surfaces that mimic the smart adhesion of the gecko foot. The topic was investigated in detail by Autumn et al. [16], Autumn [15], Gorb [132], and Bhushan and coworkers [37, 47, 48, 52, 181–184].

12.1 Gecko

Several creatures, including insects, spiders, and lizards, have a unique ability to cling to and detach from walls using their attachment systems. Although these creatures have different foot morphology, in most cases they have small hairs that cover the surfaces of their feet, called setae. Using setae, animals develop close contact with a substrate that provides enough attachment force to cling to and crawl on a wide range of natural and artificial surfaces. It also provides reversible adhesion, since they retain the ability to remove their feet from the attachment surface at will by peeling. This universal ability for attachment and detachment is called "smart adhesion" [52]. The most advanced attachment ability in lizards is found in the Tokay (T.) gecko or *Gekko gecko*. This ability was known even in ancient times; almost 2500 years ago the ability of geckos "to run up and down a tree in any way, even with the head downwards" was observed by Aristotle [37]. However, little was understood about the mechanism of this phenomenon until the microscopic hairs covering the gecko's toe were discovered in the late nineteenth century and a hierarchical morphology of the gecko toe was revealed after the advent of the SEM in the 1950s (Fig. 12.1).

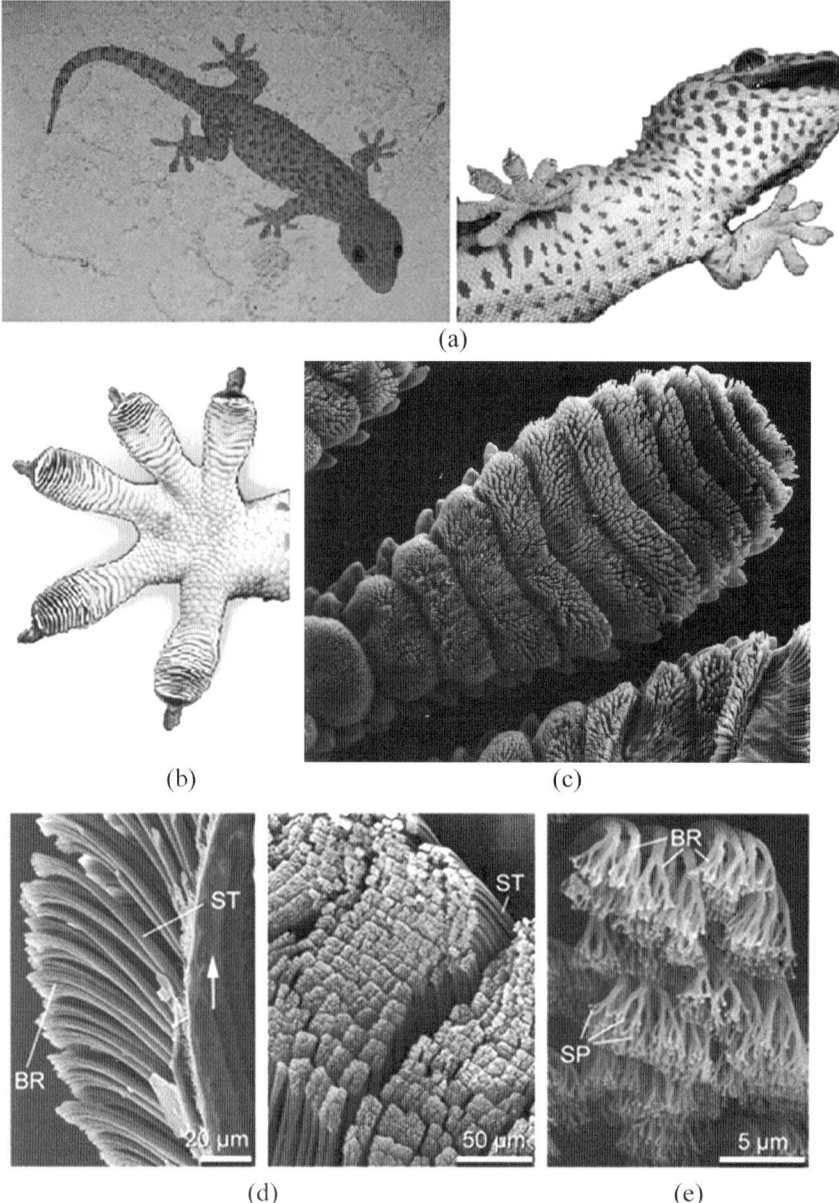

(a)

(b)   (c)

(d)   (e)

Fig. 12.1. a Tokay gecko looking top-down (*left*) and bottom-up (*right*) [16]. The hierarchical structures of a gecko foot; **b** a gecko foot [16] and **c** a gecko toe [15]. Each toe contains hundreds of thousands of setae and each seta contains hundreds of spatulae. Scanning electron microscopy (SEM) micrographs two (at different magnifications) of **d** the setae [126] and **e** the spatulae [126]. ST seta, SP spatula, BR branch

Table 12.1. Surface characteristics of Tokay gecko feet [37]

Component	Size	Density	Adhesive force
Seta	30–130/5–10 length/diameter (μm)	~14000 setae/mm^2	194 μN (in shear) ~20 μN (normal)
Branch	20–30/1–2 length/diameter (μm)	–	–
Spatula	2–5/0.1–0.2 length/diameter (μm)	100–1000 spatulae per seta	–
Tip of spatula	~0.5/0.2–0.3/~0.01 length/width/thickness (μm)	–	11 nN (normal)

The attachment pads of a T. gecko's feet consist of a complicated hierarchy of structures beginning with lamellae, soft ridges approximately 1–2 mm in length located on the attachment pads (toes). Tiny curved hairs (setae) extend from the lamellae. These setae are typically 30–130 μm in length and 5–10 μm in diameter (Table 12.1). The setae of a gecko have several branches. Each seta branches into several hundred substructures, called spatulae. A branched seta looks like a broom and has a length of about 20–30 μm and a diameter of about 1–2 μm. The tips of the spatulae have a typical size on the order of 500 nm in length, 200–300 nm in width and about 10 nm in thickness. Spatulae are oriented at an angle with respect to the contacting surface to facilitate peeling [52]. Attachment systems in other creatures, including insects and spiders, have a structure similar to that of gecko feet. As the mass of the creature increases, the radius of the terminal attachment element decreases (Fig. 12.2). This allows a greater number of setae to be packed in an area [37].

12.2 Hierarchical Structure of the Attachment Pads

Setae are composed of β-keratin with an elastic modulus in the range 1–20 GPa. Kim and Bhushan [181–184] approximated a gecko's seta with a hierarchical spring model (Fig. 12.3). Each level of springs in the model corresponds to a level of seta hierarchy. The upper springs correspond to the thicker part of the gecko's seta, the middle part of spring corresponds to the branches, and the lower part of spring corresponds to the spatulae. The upper level is the thickest branch of seta. It is 75 μm in length and 5 μm in diameter. The middle level (branch) has a length of 25 μm and diameter of 1 mm. The lower level (spatula) is the thinnest branch with a length of 2.5 μm and a diameter of about 0.1 μm (Table 12.2). Autumn et al. [16] showed that the optimal attachment angle between the substrate and a gecko's seta is 30° in the single seta pull-off experiment. This finding is supported by the adhesion models of setae as cantilever beams. Therefore, in this study, the value of the angle was fixed at 30° [37].

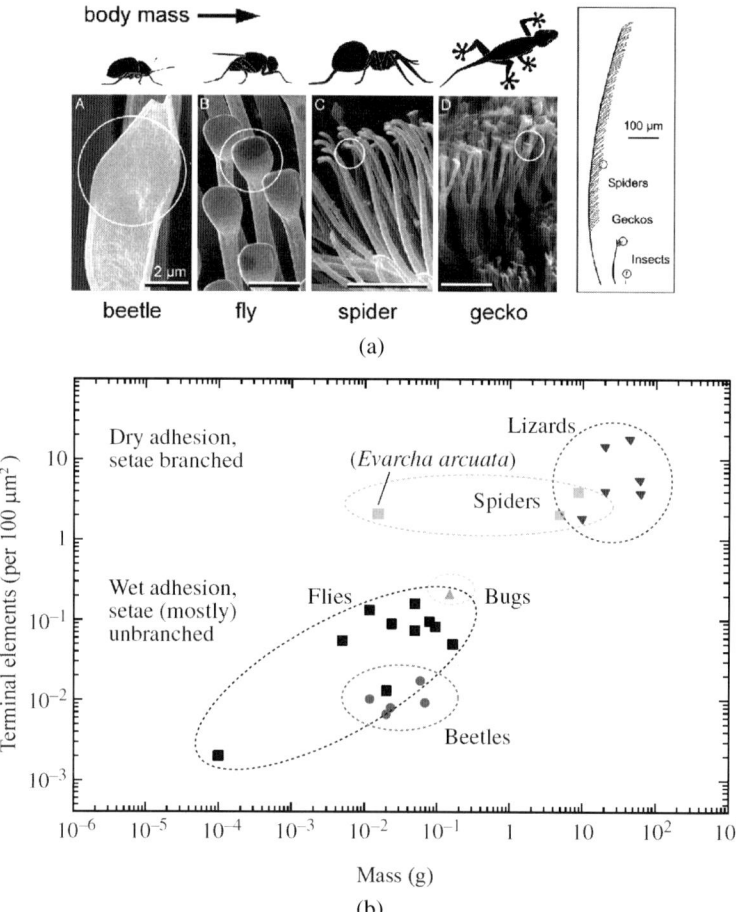

Fig. 12.2. a Terminal elements of the hairy attachment pads of a beetle, fly, spider and gecko [13] shown at different scales (*left* and *right*), and **b** the dependence of terminal element density on body mass [111]. Data are from Artz et al. [13] and Kesel et al. [177]

The adhesion force of a single seta was measured by Autumn et al. [16]. The attachment pads of a T. gecko have the total area of two feet of the order of 200 mm^2, which can produce a clinging ability of approximately 20 N (vertical force required to pull a lizard down from a nearly vertical surface). In isolated gecko setae, a 2.5 mN preload yielded adhesion of 20–40 mN and thus an adhesion coefficient, which represents the strength of adhesion with respect to preload, was approximately 8–16. With regard to the natural living conditions of the animals, we can separate the mechanics of gecko attachment into two parts: the mechanics of adhesion of a single contact with a flat surface and an adaptation of a large number of spatulae to a natural rough surface [37].

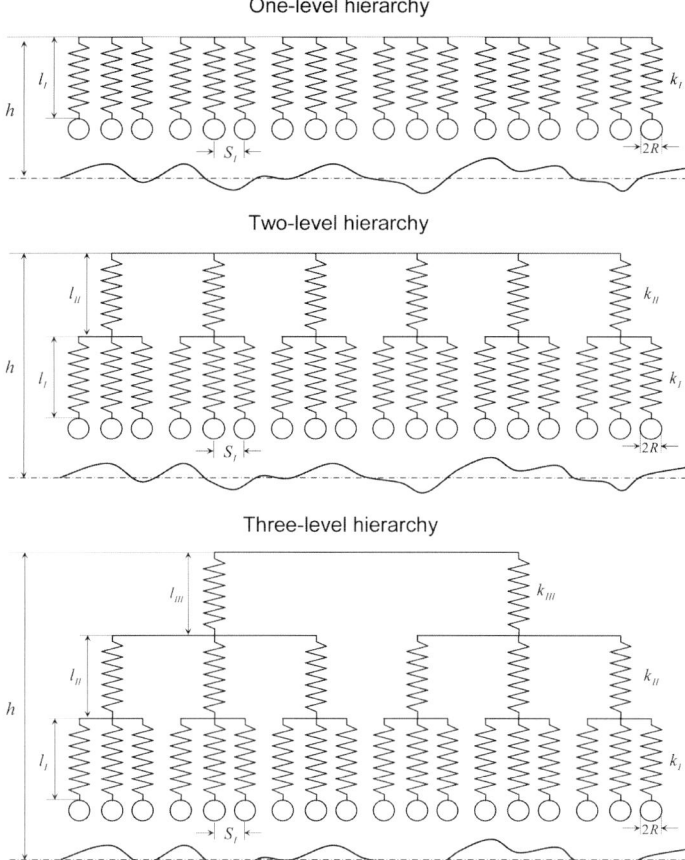

Fig. 12.3. One-, two- and three-level hierarchical spring models for simulating the effect of hierarchical morphology on interaction of a seta with a rough surface. In this figure, l_1, l_2, l_3 are lengths of structures, s_I is space between spatulae, k_I, k_{II}, k_{III} are stiffnesses of structures, I, II and III are level indexes, R is radius of tip, and h is distance between upper spring base of each model and mean line of the rough profile [181]

Table 12.2. Geometrical size, calculated stiffness, and typical densities of branches of seta for Tokay gecko [181]

Level of seta	Length (μm)	Diameter (μm)	Bending stiffness[a] (N/m)	Typical density (per mm^2)
III upper	75	5	2.908	14000
II middle	25	1	0.126	–
I lower	2.5	0.1	0.0126	$1.4–14 \times 10^6$

[a] For elastic modulus of 10 GPa with load applied at 60° to spatula long axis

12.3 Model of Hierarchical Attachment Pads

Kim and Bhushan [181–184] suggested a three-level hierarchical model of the gecko foot. They simulated numerically adhesion of the three-level spring hierarchy for gecko seta in contact with a randomly rough surface. The springs on every level of hierarchy had the same stiffness as the bending stiffness of the corresponding branches of seta. If the beam is oriented at an angle ϕ to the substrate and the contact load F is aligned normal to the substrate, the stiffness of seta branches k_m is calculated as

$$k_m = \frac{\pi R_m^2 E}{l_m \sin^2 \theta \left(1 + \frac{4l_m^2 \cot^2 \theta}{3R_m}\right)}, \tag{12.1}$$

where l_m and R_m are the length and the radius of seta branches, respectively, and m is the level number [37]. Their three-level model had springs with length $l_1 = 2.5\,\mu\text{m}$, $l_2 = 25\,\mu\text{m}$, and $l_3 = 75\,\mu\text{m}$ for levels I, II, and III, respectively. For an assumed elastic modulus $E = 10\,\text{GPa}$ of seta material with a load applied at an angle of $60°$ to spatulae long axis, the stiffness of every level of seta was calculated as $k_1 = 0.0126\,\text{N/m}$, $k_2 = 0.126\,\text{N/m}$, and $k_3 = 0.126\,\text{N/m}$, respectively.

The base of the springs and the connecting plate between the levels are assumed to be rigid. The distance s_1 between neighboring structures of level 1 was 0.35 mm, obtained from the average value of measured spatula density, $8 \times 10^6\,\text{mm}^{-2}$, and was obtained by multiplying 14 000 setae per mm^2 by an average of 550 spatulae per seta [184]. Assuming a 1 : 10 proportion of the number of springs in the upper level to that in the lower level, one spring at level III is connected to 10 springs on level II and each spring on level II also has 10 springs on level I. The number of springs, N_I, was calculated by dividing the scan length (2000 μm) by the distance s_1 (0.35 mm) and corresponds to 5700.

When springs approach the rough surface, the spring force was calculated for the springs approaching the rough surface and for pulling off. However, when the applied load is equal to zero, the springs do not detach due to adhesion attraction. Springs are pulled apart until the net force (pull-off force minus attractive adhesion force) at the interface is equal to zero. The adhesion force is the lowest value of elastic force F_{el} when the seta has detached from the contacting surface.

The random rough surfaces used for simulation were generated by a computer program [30, 32]. The roughness parameters are scale dependent, and therefore, adhesion values also are expected to be scale dependent. Increase in the scan length led to an increase in both RMS amplitude and correlation length, β^* [52]. For modeling the contact of the attachment system with random rough surfaces, the range of values of s from 0.01 μm to 30 μm and a fixed value of $\beta^* = 200\,\mu\text{m}$ were taken. The chosen range covers values of roughness for relatively smooth artificial surfaces to natural rough surfaces. A typical scan length of 2000 μm was chosen, which is comparable to a lamella length of a gecko.

The authors considered the capillary force consisting of the Laplace force and the surface tension force as well as the solid-to-solid interaction by the JKR and DMT theory. First, the adhesion forces exerted by a single gecko spatula in contact with planes with different contact angles for various relative humidity are calculated and compared with experimental data. Next, the adhesion analysis for three-level hierarchical model for gecko seta in contact with rough surfaces was performed with different s values.

They found that the Laplace force as well as the DMT adhesion force gives the larger effect on total adhesion force. Total adhesion force decreases with an increase in the contact angle on the substrate, and the difference of the total adhesion force among contact angles is larger in the intermediate humidity regime. As the relative humidity increases, total adhesion force for the surfaces with contact angle less than 60° has a higher value than the DMT adhesion force not considering wet contact; whereas above a value of 60°, total adhesion force has lower values at most levels of relative humidity.

DMT adhesion force constitutes a large portion in total adhesion force, and the capillary force is comparable with DMT force. Total adhesion force decreases with an increase in the contact angle on the substrate, and the difference of the total adhesion force among contact angles is larger in the intermediate humidity regime. In addition, we showed that the simulation results are in good agreement with the experimental results for a single spatula in contact with hydrophilic and hydrophobic surfaces.

12.4 Biomimetic Fibrillar Structures

Based on their model of adhesion of hierarchical structures in gecko feet, Kim and Bhushan [181–184], in a series of articles, investigated constraints and optimum design considerations for artificial fibrillar structures. They created an adhesion database by modeling the fibers as oriented cylindrical cantilever beams with spherical tips [182]. Following that, they carried out a numerical simulation of the attachment system in contact with random rough surfaces considering three constraint conditions—buckling, fracture, and sticking of fiber structure. For a given applied load and roughness of contacting surface and fiber material, a procedure to find the optimal fiber radius and aspect ratio for the desired adhesion coefficient was developed.

The design variables for an attachment system are as follows: fiber geometry (radius and aspect ratio of fibers, tip radius), fiber material, fiber density, and fiber orientation. The optimal values for the design variables to achieve the desired properties should be selected for fabrication of a biomimetic attachment system.

In the design of fibrillar structures a trade-off exists between the aspect ratio of the fibers and their adaptability to a rough surface. If the aspect ratio of the fibers is too large, they can adhere to each other or even collapse under their own weight as shown in Fig. 12.4(a). If the aspect ratio is too small (Fig. 12.4(b)), the structures will lack the necessary compliance to conform to a rough surface. Spacing between the individual fibers is also important. If the spacing is too small, adjacent fibers can attract each other through intermolecular forces which will lead to bunching [37].

Kim and Bhushan [182] identified the following three constraints on the parameters of the fibrillar structure that they investigated (Fig. 12.5):

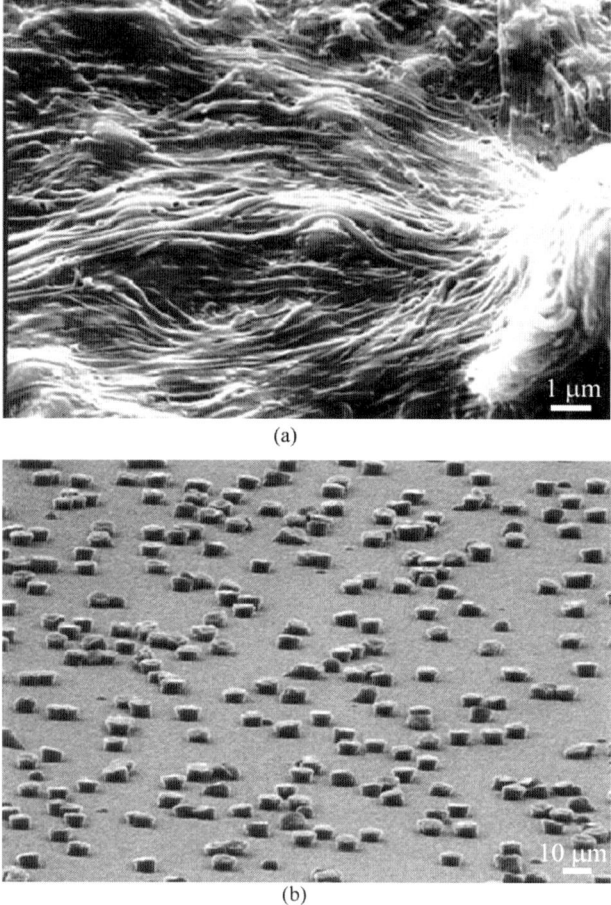

(a)

(b)

Fig. 12.4. SEM micrographs of **a** high aspect ratio polymer fibrils that have collapsed under their own weight and **b** low aspect ratio polymer fibrils that are incapable of adapting to rough surfaces [299]

1. *Nonbuckling condition.* A fibrillar interface can deliver a compliant response while still employing stiff materials because of bending and microbuckling of fibers (Fig. 12.6).
2. *Nonfiber fracture condition.* For small contacts, the strength of the system will eventually be determined by fracture of the fibers.
3. *Nonsticking condition.* A high density of fibers is also important for high adhesion. However, if the space S between neighboring fibers is too small, the adhesion forces between them become stronger than the forces required to bend the fibers. Then, fibers might stick to each other and get entangled. Therefore, to prevent fibers from sticking to each other, they must be spaced apart and be stiff enough to prevent sticking or bunching [37].

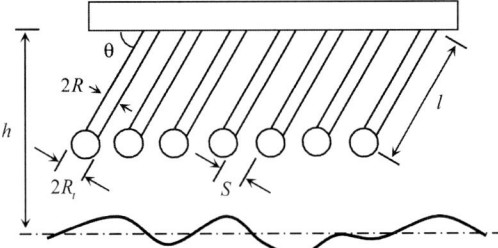

Fig. 12.5. Single-level attachment system with oriented cylindrical cantilever beams with spherical tip. In this figure, l is the length of fibers; θ is the fiber orientation, R is the fiber radius; R_t is the tip radius; S is the spacing between fibers; and h is distance between base of model and mean line of the rough profile [182]

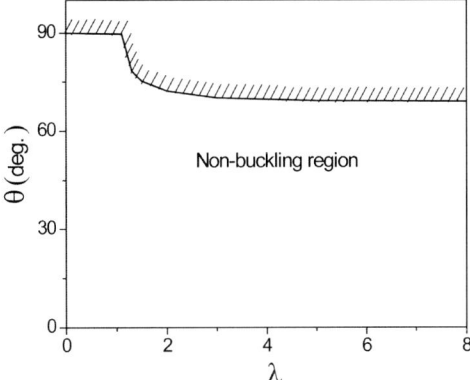

Fig. 12.6. Critical fiber orientation as a function of aspect ratio λ for non-buckling condition for pinned-clamped micro-beams ($bc = 2$) [182]

12.5 Self-cleaning

The hierarchical structure of the attachment pads allows geckos to combine smart-adhesion with self-cleaning using the lotus-effect. Natural contaminants (dirt and dust) as well as man-made pollutants are unavoidable and have the potential to interfere with geckos' clinging ability. Particles found in the air consist of particulates that are typically less than $10 \, \mu m$ in diameter while those found on the ground can often be larger. Intuitively, it seems that the great adhesion strength of gecko feet would cause dust and other particles to become trapped in the spatulae and that they would have no way of being removed without some sort of cleaning action on behalf of the gecko. However, geckos are not known to groom their feet like beetles, nor do they secrete sticky fluids to remove adhering particles like ants and tree frogs, yet they retain adhesive properties [37]. One potential source of cleaning is during the time when the lizards undergo molting, or the shedding of the superficial layer of epidermal cells. However, this process only occurs approximately once per month. If molting were the sole source of cleaning, the gecko would rapidly lose its adhesive properties as it is exposed to contaminants in nature [143].

Hansen and Autumn [143] suggested the hypothesis that gecko setae become cleaner with repeated use—a phenomenon known as self-cleaning. The cleaning ability of gecko feet was first tested experimentally by applying 2.5 μm radius silica–alumina ceramic microspheres to clean setal arrays. It was found that a significant fraction of the particles was removed from the setal arrays with each step taken by the gecko.

12.6 Biomimetic Tape Made of Artificial Gecko Skin

It has been suggested that the gecko-effect is applied to superadhesive tape. According to the literature studies, the dominant adhesion mechanism used by geckos and spider attachment systems is the van der Waals adhesion forces [37]. The complex divisions of the gecko skin (lamellae-setae-branches-spatulae) enable a large real area of contact between the gecko skin and mating surface. Hence, a hierarchical fibrillar micro/nanostructure is desirable for dry superadhesive tapes. The development of nanofabricated surfaces capable of replicating this adhesion force developed in nature is limited by current fabrication methods. Several techniques have been used in an attempt to create and characterize bioinspired adhesive tapes.

Two poly(vinylsiloxane) (PVS) samples were produced at the Max Planck Institute for Metals Research (Stuttgart, Germany), one consisting of mushroom-shaped pillars (Fig. 12.7(a)) and the other sample was an unstructured control surface (Fig. 12.7(b)). The samples were studied and characterized by Bhushan and Sayer [48]. The structured sample is composed of pillars that are arranged in a hexagonal order to allow maximum packing density. They are approximately 100 μm in height, 60 μm in base diameter, 35 μm in middle diameter, and 25 μm in diameter at the narrowed region just below the terminal contact plates. These plates were of about 40 μm in diameter and 2 μm in thickness at the lip edges. The adhesion force of the two samples in contact with a smooth flat glass substrate was measured by Gorb et al. [133] using a homemade microtribometer. Results revealed that the structured specimens featured an adhesion force more than twice that of the unstructured specimens. The adhesion force was also found to be independent of the preload. Moreover, it was found that the adhesion force of the structured sample was more tolerant to contamination compared to the control, and it could be easily cleaned with a soap solution [37].

Bhushan and Sayer [48] characterized the surface roughness, friction force, and contact angle of the structured sample and compared the results to an unstructured control. The macroscale coefficient of kinetic friction of the structured sample was found to be almost four times greater than the unstructured sample. This increase was determined to be a result of the structured roughness of the sample and not the random nanoroughness. It is also noteworthy that the static and kinetic coefficients of friction are approximately equal for the structured sample. It is believed that the

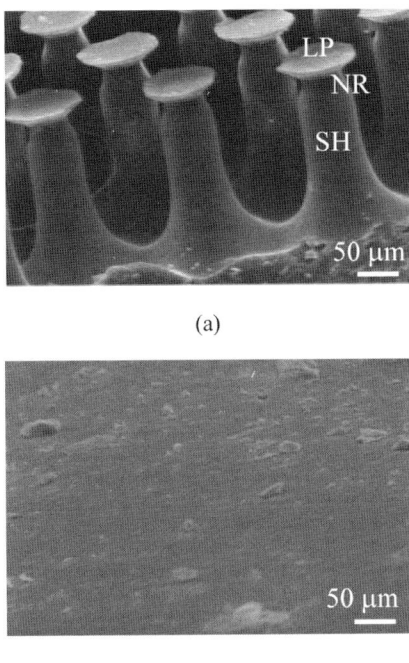

(a)

(b)

Fig. 12.7. SEM micrographs of the **a** structured and **b** unstructured PVS samples. SH: shaft, NR: neck region, LP: lip [48]

divided contacts allow the broken contacts of the structured sample to constantly re-create contact. The pillars also increased the hydrophobicity of the structured sample in comparison to the unstructured sample as expected due to increased surface roughness. A large contact angle is important for self-cleaning, which indicates that the structured sample is more tolerant of contamination than the unstructured sample [37].

12.7 Summary

The adhesive properties of geckos and other creatures such as flies, beetles, and spiders, are due to the hierarchical structures present on each creature's attachment pads. Geckos have developed the most intricate adhesive structures of any of the aforementioned creatures. The attachment system consists of ridges called lamellae that are covered in microscale setae that branch off into nanoscale spatulae. Each structure plays an important role in adapting to surface roughness bringing the spatulae in close proximity with the mating surface. These structures as well as material properties allow the gecko to obtain a much larger real area of contact between its feet and a mating surface than is possible with a nonfibrillar material. Two feet of a Tokay gecko have about $220\,mm^2$ of attachment pad area on which the gecko is able to generate approximately $20\,N$ of adhesion force. Although capable of generating high adhesion forces, a gecko is able to detach from a surface at will—an ability known as smart adhesion. Detachment is achieved by a peeling motion of the gecko's feet from a surface. The adhesion strength of gecko setae is dependent on the orientation; maximum adhesion occurs at $30°$. During walking a gecko is able to peel its foot from surfaces by changing the angle at which its setae contact a surface.

Recent creation of a three-level hierarchical model for a gecko lamella consisting of setae, branches, and spatulae has brought more insight into adhesion of biological attachment systems. One-, two-, and three-level hierarchically structured spring models for simulation of a seta contacting with random rough surfaces were considered. The simulation results show that the multilevel hierarchical structure has a higher adhesion force as well as adhesion energy than the one-level structure for a given applied load, due to better adaptation and attachment ability. It is concluded that the multilevel hierarchical structure produces adhesion enhancement, and this enhancement increases with an increase in the applied load and a decrease in the stiffness of springs.

There is great interest among the scientific community to create surfaces that replicate the adhesion strength of gecko feet. These surfaces would be capable of reusable dry adhesion and would have uses in a wide range of applications from everyday objects such as tapes, fasteners, and toys to microelectronic and space applications and even wall-climbing robots. In the design of fibrillar structures, it is necessary to ensure that the fibrils are compliant enough to easily deform to the mating surface's roughness profile, yet rigid enough to not collapse under their own weight. Spacing between the individual fibrils is also important. If the spacing is too small, adjacent fibrils can attract each other through intermolecular forces which will lead to bunching.

13

Other Biomimetic Surfaces

Abstract The issues of hierarchical organization in biomaterials and surfaces are discussed. Various biomimetic surfaces and effects are reviewed, including the shark skin, darkling beetle, water strider, spider web, and several others.

The concept of bionics or biomimetics emerged in the 1960s, however, it has been developing very dynamically in the past decade due to advancements in nano- and biotechnologies. Many major challenges of modern engineering science are related to miniaturization [286]. Studying natural organisms and biological systems provides insights on how these problems can be solved, while emerging technologies give an opportunity to mimic the biological systems. A successful transition of these ideas into the technical world requires more than just observation, but also detailed analysis and possible modification in view of materials and technologies available to an engineer. In this chapter, we will review biomimetic surfaces with the emphasis on the surfaces with hierarchical structure.

13.1 Hierarchical Organization in Biomaterials

Biomimetics is the application of methods and systems found in living nature to the study and design of engineering systems. In the biomimetic design of materials, a number of ideas have been suggested. This includes the study of biological self-assembly, receptors, protein machines, muscle filaments, and microstructured surfaces. The attention of engineers was driven to such diverse areas as artificial cartilage for shock absorption, the mucus for the solid-fluid transformation, collagen, use of insect cuticle microstructure for advanced composites, biomimetic surfaces to control cell adhesion, and drug delivery [69]. Most technical materials, such as steel, metals, silicon, and plastics, require high temperatures and/or pressures to be manufactured, whereas biological organisms do not have access to these high temperatures or pressures. Nevertheless, nature has developed many materials with remarkable functional properties that are often superior to engineered materials. Although sometimes fragile, biological organisms can often deal with extreme mechanical loads. The key is a complex hierarchical structure of the natural materials.

There are several important differences in the ways nature and an engineer use materials. An engineer has a much greater range of available elements, including iron and metals, while nature has to deal mostly with polymers and composites of polymers and ceramic structures built of light elements. Nature builds trees and skeletons by means of growth or biologically controlled self-assembly adapting to the environmental condition and not by the secure design and selection of the materials with required final properties, as engineers do. Biological materials are grown not according to the final "design specification," but using the recipes contained in the genetic code. As a result, biological materials and tissues are created by hierarchical structuring at all levels, adaptation of form and structure to the function, capability of adaptation to changing conditions, and self-healing [114]. The genetic algorithm interacts with the environmental condition, which provides flexibility. For example, a tree branch can grow differently in the direction of the wind and in the opposite direction. The only way to provide this adaptive self-assembly is a hierarchical self-organization of the material. Hierarchical structuring allows the adaptation and optimization of the material at each level.

A remarkable property of biological tissues is their ability for self-healing, which is also related to the hierarchical organization. There are several biological mechanisms of self-repair. At the molecular level, there are dynamically breaking and repairing "sacrificial" bonds, which allow for material to deform in a quasi-plastic manner without fracture. In bones, there is a cyclic replacement of material by specialized cells, which allows for a bone to adapt to changing conditions and to repair damage. Many fractured or critically damaged living tissues can heal themselves by formation of an intermediate tissue (based on the response to inflammation) followed by the scar tissue [114]. While there are almost no self-healing artificial materials available at this point, some interesting biomimetic solutions have been proposed. For example, one system under development contains a reservoir with a hydrophobic polymer that is intended to mimic the wax of the lotus leaf and thus combines superhydrophobic and self-healing properties [57].

We can conclude from this discussion that hierarchical structure is a consequence of the fact that biological materials are not designed in their final form, but self-assembled. This argument applies also to the surfaces, so the biomimetic surfaces are often hierarchical. In the preceding sections we have studied two well-known examples—superhydrophobic surfaces based on the lotus-effect and attachment mechanisms based on the gecko-effect. Here we will discuss several additional examples of biomimetic surfaces that have been suggested as possible engineering solutions.

13.2 Moth-Eye-Effect

The moth-eye-effect is the ability of nanostructured optical surfaces not to reflect light, that is, to remain invisible. The effect was discovered in the 1960s as a re-

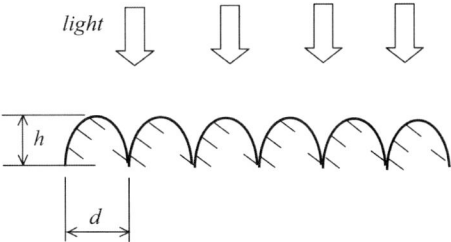

Fig. 13.1. Schematic of the moth-eye-effect

sult of the study of insect eyes. For nocturnal insects, it is important to not reflect the moonlight, since the reflection makes the insect vulnerable to predators. The light reflection is avoided by a continuously increasing refractive index of the optical medium. The little protuberances upon the cornea surface increase the refractive index. These protuberances are very small microtrichia (about 200 nm in diameter). For an increase in transmission and reduced reflection, a continuous matching of the refraction index at the boundary of the adjacent materials (cornea and air) is required. If the periodicity of the surface pattern is smaller than the light wavelength, the light is not reflected [132]. If this condition is satisfied, it may be assumed that at any depth the effective refraction index is the mean of that of air and the bulk material, weighted in proportion to the amount of material present at that depth (Fig. 13.1). For a moth-eye surface with the height of the protuberances of h and the spacing of d, it is expected that the reflectance is very low for wavelengths less than about $2.5h$ and greater than d at normal incidence, and for wavelengths greater than $2d$ for oblique incidence. For protuberances with 220 nm depth and the same spacing (typical values for the moth eye), a very low reflectance is expected for the wavelengths between 440 and 550 nm [340].

This moth-eye-effect should not be confused with reducing of the specular reflectance by roughening of a surface. Roughness merely redistributes the reflected light as diffuse scattering. In the moth eye's case, there is no increase in diffuse scattering, the transmitted wavefront is not degraded, and the reduction in reflection gives rise to a corresponding increase in transmission [340].

In addition to nanostructures that lead to nonreflective surfaces, many insects, such as butterflies, use structural coloration due to the presence of scales and bristles. Scales of scarab beetles bear additional microtrichia responsible for the coloration of their surfaces. This coloration serves for camouflage, mimicry, and species and sex recognition. Some insects use a mechanism called iridescence, using a complex multilayer structure for optical interference. Such structures can produce complicated optical effects including strong polarization, color mixing, and reflection angle broadening [132].

This effect has been suggested for use in the displays of various devices, such as cell phones. Unlike conventional displays that require an internal source of light and become bleak in bright light (e.g., in sunlight), the displays using the biomimetic technology would work from reflected light, they would be seen well in the sunlight and consume little energy. A company in San Jose, CA, is developing such displays, calling the principle the "interference modulation (IMod)." The product is expected to appear on the market in 2008. To display individual pixels, the IMod displays use simple MEMS gap-closing actuators that consist of a plate that can deflect the half-wavelength distance when an electric signal is applied.

A number of attempts have been made to design nanopatterned nonreflective surfaces based on the moth-eye-effect. Hadobás et al. [140] prepared patterned silicon surfaces with 300 nm periodicity and depth up to 190 nm using interference lithography. They found a significant reduction in reflectivity, partially due to the moth-eye-effect. Gao et al. [127] used epoxy and resin to replicate the antireflective surface of cicada's eye. It is also possible to create transparent surfaces using the moth-eye-effect [83]. The moth-eye-effect can be combined with the lotus-effect so that self-cleaning, nonreflective glass can be created.

Due to recent developments in the nanophotonics and plasmonics physics, it became possible to manipulate light at the nanoscale using arrays of nanoparticles [307] and other plasmonic nanodevices [306]. The plasmonic nanodevices are designed for phase, polarization, and feedback control. This allows them to guide light in the nanoscale via nanoparticle arrays. The moth-eye-effect may be viewed as a special case of these devices with the particles forming the moth eye structure to prevent reflection.

13.3 Shark Skin

Shark skin is covered by a special type of scales, called placoid, that form small V-shaped bumps, made from the same material as sharks' teeth. The rough surface reduces friction when the shark glides through water, which makes sharks very quick and efficient swimmers. Shark skin is so rough that it can be used as sanding paper. The "shark-skin-effect" is based on the fact that a body in a stream is provided with small ridges aligned in the local flow direction; a significant drag reduction can be reached in turbulent flow conditions due to the control of the streamwise vortices in the turbulent flow. Wainwright et al. [331] also found that the internal pressure of the shark increases more than ten-fold from slow to fast swimming. This pressure increase causes the shark's skin to deform faster [20, 163, 265]. Several commercial products use the shark-skin-effect. This includes boat and aircraft surfaces, swimming suits, and other applications.

13.4 Darkling Beetle

Some beetles in the Namib Desert in South Africa, such as the darkling beetle, collect drinking water from fog-laden wind on their backs. Droplets form on the top (front) fused "wings" (elytra) and roll down the beetle's surface to its mouthparts. These large droplets form by virtue of the insect's bumpy surface, which consists of alternating hydrophobic, wax-coated regions and hydrophilic, nonwaxy regions [259].

Experiments with artificial fog showed the feasibility of this mechanism, and artificial surfaces consisting of altering hydrophobic and hydrophilic regions have been suggested [130]. While microfluidics with patterned surfaces, as opposed to micro/nanochannels, is a big and promising field, as discussed in the preceding chapters the origin of it may be traced to the "darkling-beetle-effect."

13.5 Water Strider

A water strider (*Gerris remigis*) has the ability to walk upon a water surface without getting wet. Even the impact of rain droplets with a size greater than the strider's size does not make it immerse into water. Gao and Jiang [123] showed that the special hierarchical structure of strider legs, which are covered by large numbers of oriented tiny hairs (microsetae) with fine nanogrooves, may be responsible for the water resistance. According to their measurements, a leg does not pierce the water surface until a dimple of 4.38 mm depth is formed. They found that the maximal supporting force of a single leg is 1.52 mN, or about 15 times the total body weight of the insect. The corresponding volume of water ejected is roughly 300 times that of the leg itself. Gao and Jiang [123] suggested that superhydrophobicity of the water strider leg is responsible for these abilities. They measured the contact angle of the insect's legs with water and found it equal to 167.6°. Scanning electronic micrographs revealed numerous oriented setae on the legs. The setae are needle-shaped hairs, with diameters ranging from three micrometers down to several hundred nanometers. Most setae are roughly 50 µm in length and arranged at an inclined angle of about 20° from the surface of leg. Many elaborate nanoscale grooves were found on each microseta, and these form a unique hierarchical structure. This hierarchical micro- and nanostructuring on the leg's surface seems to be responsible for its water resistance and the strong supporting force.

13.6 Spider Web

Spider web gives an interesting example of a structure built of a one-dimensional fiber. Spiders fabricate a very strong, continuous, insoluble fiber. The web can hold a significant amount of water droplets, and it is resistant to rain, wind, and sunlight. Spider silk is three times stronger than steel, having the tensile strength of 1.2 GPa. The spider generates the silk fiber and at the same time it is hanging on it. It has a sufficient supply of raw material for its silk to span great distances [25]. Some spider silks have high stiffness with a tensile modulus of about 10 GPa, while others are elastomeric with a stiffness of about 1 GPa and extension to rupture of 200%. The spider web fibers are produced by spinning concentrated aqueous protein solutions. Research in genetic engineering suggests that the synthesis of structural proteins in microbial culture to produce polymers for fibers may become commercially useful in the future [134]. Dzenis [103] suggested an electrospinning technique to produce

2 μm diameter continuous fibers from polymer solutions that are somewhat similar to spider silk fibers.

While mechanical properties of the web fiber are remarkable, it is also quite interesting how a spider creates a two-dimensional web out of its silk fiber. Krink and Vollrath [193] analyzed spider web-building behavior using a computer model that constructed artificial webs with a rule-based simulation. They found that web characteristics like spiral distances, eccentricities, and vertical hub location could to a large degree be accurately simulated with the model. They later proposed a "virtual spider robot" that builds virtual webs which mimic perfectly the visual architecture of real webs of the garden cross spider *Araneus diadematus*. They suggested that the garden spider uses web-building decision rules which are strictly local and based on the interactions with previously placed threads to generate global architecture [194]. This may be interesting for modeling biological self-assembly of complex material using local rules for the overall structure that is still adaptive to external conditions.

13.7 Other Biomimetic Examples

Biomimetic surfaces are not limited to the above-mentioned examples. Many other ideas in this area have been suggested. For example, the impact sensitive paint mimicking bruised skin [25]. Our skin is sensitive to impact, leading to the purple color in areas that are hit. This idea inspired researchers in mid-1980s to develop a coating that indicates impact damage. Such a coating, on the basis of a paint mixed with micro-capsules (sized 1 to 10 μm) with a certain chemical reagent, was used in the air industry to indicate possible damage to components made of impact-sensitive composite materials. An impact may lead to a significant strength loss in such a material without visible structural damage, so the change of color indicates the damage that may be potentially dangerous for the aircraft [25].

The desert sand fish skink (*Scincus scincus*) is a lizard that has adapted to an underground existence. The skink can virtually dive and "swim" beneath the surface of loose sand due to special properties of its scales—they have very low friction and abrasion. This skink specie's scales are covered by "nanothresholds," long ridges with submicron height and distance of 10 μm or less. Rechenberg and El Khyeri [279], who brought attention to this lizard and to what they called the "sandfish-effect," suggested that electrostatic charge created by submicron sized thresholds on the scale plays a role in friction reduction by creating a repulsive force between the scale and sand grains.

Other examples of functional biomimetic surfaces include surfaces with periodic roughness for sound generation mimicking certain spiders and insects, thermoregulation and prevention of drying, grooming, sampling, filtrating, and grinding [132]. Recent studies of the artificial skin design principles should also be mentioned, due to the importance of these studies for medicine [347].

Another interesting object of biomimetic research is diatoms and marine sponges. Diatoms are a common type of sea phytoplankton. Most diatoms are unicellular, and their characteristic feature is that they are encased within a unique cell wall made of silica (hydrated silicon dioxide) called a frustule. These frustules show a wide diversity in form, some quite beautiful and ornate, but usually consist of two asymmetrical sides with a split between them. The biogenic silica of the cell walls is synthesized intracellularly and then extruded to the cell exterior and added to the wall.

Hildebrand et al. [153] examined silica cell wall synthesis in the diatom *Thalassiosira pseudonana*. The innate capabilities of diatoms to form complex three-dimensional silica structures on the nano- to microscale exceed current synthetic approaches because they use a fundamentally different formation process. Understanding the molecular details of the process requires identifying structural intermediates and correlating their formation with the genes and proteins involved. They observed distinct silica morphologies during the formation of different cell wall substructures, and identified three different scales of structural organization. At all levels, structure formation correlated with optimal design properties for the final product.

Since silicon-based materials are widely used in nanotechnology, researchers are looking for ways to mimic how these materials are synthesized in diatoms. This requires understanding the corresponding parts of the diatom's DNA code and the involved biochemical mechanisms. Thamatrakoln et al. [315] studied silicon transport mechanisms in diatoms and involved proteins and amino acids. They also identified genes involved in the silicon transporters. Poulsen et al. [270] shows that the biomineral-forming machinery of diatoms can be genetically tailored to incorporate functional proteins into diatom silica in vivo.

Another interesting and promising field involves the application of an array of sensors that analyze chemical structure and can lead to an "artificial nose" or an "artificial tongue." Various techniques, such as the AFM cantilever arrays, have been suggested for this purpose. Single-cantilever sensors can determine quantities below the detection limits of equivalent "classical" methods, thus catalytic processes can be observed with picojoule sensitivity in nanocalorimetry [131]. Baller et al. [21] used a microfabricated array of silicon cantilevers for the detection of vapors. Each of the cantilevers was coated with a specific sensor layer to transduce a physical process or a chemical reaction into a nanomechanical response. The response pattern of eight cantilevers was analyzed with principal component analysis and artificial neural network techniques, which facilitates the application of the device as an artificial chemical nose. Interestingly, they reported that natural flavors, such as bitter almond, cherry, orange, artificial rum, vanilla, and lemon, could be recognized with high reliability [21]. Attempts have also been made to design an "artificial tongue" in a similar manner [328].

13.8 Summary

We have reviewed several characteristics of micro- and nanopatterns found on biological surfaces. Hierarchical organization of biological tissues and surfaces is a consequence of the way these objects are built: not according to a final "blueprint," but using a genetic algorithm. Therefore, biological surfaces can use the advantages of the hierarchical structure, such as the ability to control processes at several scale levels. The most interesting examples of this are the lotus-effect and the gecko-effect. Other effects associated with micron and submicron scale patterns, such as the moth-eye-, shark-skin-, darkling-beetle-, and sandfish-effects, are also, at least to a certain extent, based on this hierarchical design.

Interestingly, the evolutionary study of these functional surfaces shows that similar solutions have multiple origins, rather than emerge from the same lineage of living organisms. For example, attachment systems consisting of a pair of surfaces with microscopic hairs appeared independently in different organisms. Similar optical systems are found in insect, birds, and plants and have an independent origin [132]. Instead of evolutionary divergence, we deal here with convergence. In a similar way, convergence is found in the mechanical design of lamnid sharks and tunas, which irrespective to the evolutionary lineage provide similar solutions to similar mechanical problems [100]. This is especially interesting from the biomimetic point of view because they correspond to optimum solutions for a particular problem.

Historically, the technology on the scale size comparable with the human body, that is, the macroscale, was first developed. To obtain technologically important materials such as iron and other metals, people used heat and pressure, and details and devices built of these materials were comparable with the size of human body. In the twentieth century, with the development of microelectronics, it became clear that miniaturization is possible and advantageous. Richard Feynman stated in his famous lecture some 50 years ago that there are no physical laws that prohibit the creation of very small machines. However, it was not clear then how to build such machines in practice and whether it was possible to build them using existing tools. Consequent discoveries in molecular biology and biochemistry demonstrated that organic molecules provide "building blocks" that allow us to create very sophisticated systems and materials with properties that are not attainable at the macroscale. This attracted attention to the biological systems as a source of inspiration for engineers and to the hierarchical organization of these systems. As nanotechnology matures, more sophisticated and complex ways to organize multiscale hierarchical structures will be found.

14

Outlook

Abstract In this chapter, the conclusions of the preceding chapters and the book as a whole are formulated and presented.

In this book we studied many systems and phenomena which may not seem to have much in common: random and fractal engineering surfaces, mechanisms of dry friction and adhesion, stick-slip, nonlinear dynamic friction effects, wetting and phase transitions, motion of liquid along rough surfaces, superhydrophobicity, and various biological and biomimetic surfaces. However, besides the great diversity of these systems and phenomena, many of them have a lot in common. This is hierarchical organization that may lead to many unusual and unexpected properties. The hierarchy requires a complex, holistic approach to the study of these systems in order to understand and use them. Such an approach implies that properties of a given system cannot be determined or explained by the sum of its component parts alone. Instead, the system as a whole determines in an important way how the parts behave.

Since Newton and Leibniz, science has concentrated on analysis and identification of the parts of various systems and the mechanisms of their work. This allowed scientists to find fundamental laws of nature, such as Newton's laws, and to identify many processes that are governed by simple deterministic rules (such as rotation of the planets around the Sun) and can be approximated by smooth continuous functions and differential equations that prescribe local interaction rules. In the last third of the twentieth century it became evident that such an approach is limited, and that many systems cannot be reduced to local interactions, equilibrium, deterministic behavior, and smooth functions. This is especially true with respect to complex biological systems. At that time various new concepts of the complex dynamic systems emerged, including nonequilibrum thermodynamics, dissipative systems, synergetics, fractals, deterministic chaos, the theory of catastrophes, self-organized criticality, and so on. Many of these approaches have been accepted with enthusiasm by physicists, since they provide unconventional and promising insights to existing phenomena. It was recognized that many important or interesting systems operate far from equilibrium, demonstrate critical behavior, instability, self-affinity, self-organization, and hierarchical structure.

The authors of this book are engineers and not physicists. We do appreciate the beauty of fine theoretical mathematical and physical concepts; however, we are interested in the application of these concepts for new technology. From the viewpoint of an engineer, two great advances in technology happened in the past few decades: the emergence of nanotechnology and of biotechnology. Those were the advances that caused engineers working at the cutting edge of emerging fields to reconsider many traditional concepts. In particular, today we understand biological systems much better than before and we are able to use these properties for a range of technological problems. Biomimetics offers a journey from engineering to biology and back when an engineer approaches a biological object from the engineering perspective, adjusts his tools and methods for these systems, and then applies new tools and methods back to the engineering problems. In this book we applied these methods to the study of surfaces.

The most striking feature of a biological system is its hierarchical self-organization that allows it to acquire the required functionality in an adjustable manner without specifying the final design of the system. Furthermore, it often provides extraordinary properties (such as very high strength, small size, low friction, etc.) and the ability for self-repair. Clearly, hierarchical systems require us to find new ways to describe and analyze them, keeping in mind that the hierarchy and complexity tend to emerge from critical behavior and the loss of symmetry. Our assumption was that the hierarchical structure of successful biological and other surfaces is a consequence of the hierarchical organization of friction and adhesion mechanisms.

Advances in nanoscience and nanotechnology emphasize the importance of the mesoscale as an intermediate scale between the conventional continuum description of the matter and the atomic or molecular scale. At the meso- or nanoscale many properties of materials are different from the macroscale. Small thermal fluctuations can play a significant role; in this sense the nanoscale behavior is similar to the near-critical behavior, characterized by large correlation lengths and power laws. Various stability issues, such as metastability, can play a different role at the nanoscale than they play at the macroscale. In addition, nanosystems have high surface-to-volume ratios, which make surface phenomena, such as friction and adhesion, very important for these systems.

Analyzing friction mechanisms, we paid attention to the fact that despite the diversity of these mechanisms, they have many common features. In particular, all these mechanisms lead to an almost linear dependence of the friction force on the normal load, referred to as "Amontons–Coulomb's law." The common feature of different mechanisms of dry friction is the presence of a small parameter, which is equal to the ratio of the magnitudes of interface forces to bulk forces. The two quantities can also be seen as a ratio of characteristic lengths for these two types of forces or as the "vertical" extent of contact and "horizontal" size of contact. Mathematically, the small parameter leads to linearization of the force dependence on the normal load. In a similar manner, a small parameter also plays an important role in critical phenomena leading to the power law critical exponents. We suggested mapping friction regimes on the basis of these two parameters.

Despite the apparent linearity, friction exhibits nonlinear behavior. Such nonlinearity leads to various dynamic effects, such as the stick–slip phase transition, dynamic instabilities, and the possibility of self-organized criticality and hierarchical organization of friction mechanisms. We also investigated wetting or dissipation during solid–liquid friction and found many similarities with solid–solid friction. In particular, the energy dissipation during the wetting–dewetting cycle is dependent upon both the contact area and the triple line length. In a similar manner, for solid–solid friction, both adhesion at the contact area and deformation at the edge of contact lead to dissipation.

We also found that superhydrophobicity requires a synergy of several mechanisms with various characteristic length scales that lead to a stable composite solid–liquid–air interface. Our assumption was that the hierarchy of wetting and friction mechanisms leads to the hierarchical structures of surfaces for low/high friction, adhesion, and nonwetting. Such hierarchical structures are found in natural rough surfaces with the corresponding functionalities.

Understanding the functionality of these surfaces requires new approaches to surface roughness and roughness descriptions. These new approaches will be different from that of conventionally engineered rough surfaces, but at the same time they will need to be broad and universal enough to include various forms of functional rough surfaces found in nature and produced in laboratories. Our approach involved hierarchical surfaces with multiscale roughness. The main difference between the hierarchical and fractal surfaces is that the latter do not involve any characteristic scale length parameters, whereas the first involve a set of such parameters. We also used extensively modern concepts of the theory of dynamic and hierarchical systems.

The major motivation behind our study is the development of new design principles and new technologies for manufacturing functional surfaces. These new developments are especially significant in the fields of nano- and biotechnology. Design of new functional micro- and nanostructured surfaces, especially those based on the biomimetic principles requires, on one hand, a new look on the hierarchical organization of the dissipation mechanisms, and on the other hand, an extensive study of functional surfaces in living biological objects. Our book provides recent results in these fields.

References

[1] M.E. Abdelsalam, P.N. Bartlett, T. Kelf, J. Baumberg, Wetting of regularly structured gold surfaces. Langmuir **21**, 1753–1757 (2005)

[2] G.G. Adams, Self-excited oscillations of the two elastic half-spaces sliding with a constant coefficient of friction. J. Appl. Mech. **62**, 867–872 (1995)

[3] G.G. Adams, M. Nosonovsky, Contact modeling—forces. Tribol. Int. **33**, 441–442 (2000)

[4] G.G. Adams, S. Muftu, N. Mohd Azhar, A scale-dependent model for multiasperity contact and friction. J. Tribol. **125**, 700–708 (2003)

[5] G.G. Adams, J.R. Barber, M. Ciavarella, J.R. Rice, A paradox in sliding contact with friction. J. Appl. Mech. **72**, 450–452 (2005)

[6] A.V. Adamson, *Physical Chemistry of Surfaces* (Wiley, New York, 1990)

[7] C.A. Angell, Approaching the limits. Nature **331**, 206 (1988)

[8] M.A. Anisimov, Thermodynamics at the meso- and nanoscale, in *Dekker Encyclopedia of Nanoscience and Technology* (Dekker, New York, 2004), pp. 3893–3904

[9] M.A. Anisimov, Divergence of Tolman's length for a droplet near the critical point. Phys. Rev. Lett. **98**, 035702 (2007)

[10] Anonymous, Lotus effect FAQ, http://www.lotus-effekt.de. Accessed 10 Aug 2007

[11] R. Apfel, M. Smith, The tensile strength of di-ethyl ether using Briggs's method. J. Appl. Phys. **48**, 2077 (1977)

[12] J.F. Archard, Elastic deformation and the laws of friction. Proc. R. Soc. Lond. A **243**, 190–205 (1957)

[13] E. Arzt, S. Gorb, R. Spolenak, From micro to nano contacts in biological attachment devices. Proc. Natl. Acad. Sci. U.S.A. **100**, 10603–10606 (2003)

[14] D.B. Asay, S.H. Kim, Effects of adsorbed water layer structure on adhesion force of silicon oxide nanoasperity contact in humid ambient. J. Chem. Phys. **124**, 174712 (2006)

[15] K. Autumn, How gecko toes stick. Am. Sci. **94**, 124–132 (2006)

[16] K. Autumn, Y.A. Liang, S.T. Hsieh, W. Zesch, W.P. Chan, T.W. Kenny, R. Fearing, R.J. Full, Adhesive force of a single gecko foot-hair. Nature **405**, 681–685 (2000)

[17] V. Bahadur, S.V. Garimella, Electrowetting-based control of static droplet states on rough surfaces. Langmuir **23**, 4918–4924 (2007)

[18] P. Bak, *How Nature Works: The Science of Self-Organized Criticality* (Springer, New York, 1996)

[19] E.A. Baker, Chemistry and morphology of plant epicuticular waxes, in *The Plant Cuticle*, ed. by D.F. Cutler, K.L. Alvin, C.E. Price (Academic Press, New York, 1982), pp. 139–165

[20] P. Ball, Engineering shark skin and other solutions. Nature **400**, 507–509 (1999)

[21] M.K. Baller, H.P. Lang, J. Fritz, C. Gerber, J.K. Gimzewski, U. Drechsler, H. Rothuizen, M. Despont, P. Vettiger, F.M. Battiston, J.P. Ramseyer, P. Fornaro, E. Meyer, H.-J. Güntherodt, A cantilever array-based artificial nose. Ultramicroscopy **82**, 1–9 (2000)

[22] L. Barbieri, E. Wagner, P. Hoffmann, Water wetting transition parameters of perfluorinated substrates with periodically distributed flat-top microscale obstacles. Langmuir **23**, 1723–1734 (2007)

[23] F.E. Bartell, J.W. Shepard, Surface roughness as related to hysteresis of contact angles. J. Phys. Chem. **57**, 455–458 (1953)

[24] W. Barthlott, C. Neinhuis, Purity of the sacred lotus, or escape from contamination in biological surfaces. Planta **202**, 1–8 (1997)

[25] Y. Bar-Cohen (ed.), *Biomimetics: Biologically Inspired Technologies* (Taylor and Francis, Boca Raton, 2005)

[26] V. Bergeron, D. Bonn, J.Y. Martin, L. Vovelle, Controlling droplet deposition with polymer additives. Nature **405**, 772–775 (2000)

[27] B. Bhushan, Magnetic slider/rigid disk substrate materials and disk texturing techniques—status and future outlook. Adv. Inf. Storage Syst. **5**, 175–210 (1993)

[28] B. Bhushan, *Tribology and Mechanics of Magnetic Storage Systems*, 2nd edn. (Wiley, New York, 1996)

[29] B. Bhushan, *Tribology Issues and Opportunities in MEMS* (Kluwer Academic, Dordrecht, 1998)

[30] B. Bhushan, *Principles and Applications of Tribology* (Wiley, New York, 1999)

[31] B. Bhushan (ed.), *Handbook of Micro/Nanotribology*, 2nd edn. (CRC Press, Boca Raton, 1999)

[32] B. Bhushan, *Introduction to Tribology* (Wiley, New York, 2002)

[33] B. Bhushan, Adhesion and stiction: mechanisms, measurement techniques and methods for reduction. J. Vac. Sci. Technol. B **21**, 2262–2296 (2003)

[34] B. Bhushan, *Nanotribology and Nanomechanics—An Introduction* (Springer, Heidelberg, 2005)

[35] B. Bhushan, Nanotribology and nanomechanics. Wear **259**, 1507–1531 (2005)

[36] B. Bhushan (ed.), *Springer Handbook of Nanotechnology*, 2nd edn. (Springer, Heidelberg, 2007)

[37] B. Bhushan, Adhesion of multi-level hierarchical attachment systems in gecko feet. J. Adhes. Sci. Technol. **21**, 1213–1258 (2007)

[38] B. Bhushan, Nanotribology and nanomechanics of MEMS/NEMS and bioMEMS/bioNEMS materials and devices. Microelectron. Eng. **84**, 387–412 (2007)

[39] B. Bhushan, Y.C. Jung, Micro and nanoscale characterization of hydrophobic and hydrophilic leaf surface. Nanotechnology **17**, 2758–2772 (2006)

[40] B. Bhushan, Y.C. Jung, Wetting study of patterned surfaces for superhydrophobicity. Ultramicroscopy **107**, 1033–1041 (2007)

[41] B. Bhushan, Y.C. Jung, Wetting, adhesion and friction of superhydrophobic and hydrophilic leaves and fabricated micro/nanopatterned surfaces. J. Phys. Condens. Matter **20**, 225010 (2008)

[42] B. Bhushan, M. Nosonovsky, Scale effects in friction using strain gradient plasticity and dislocation-assisted sliding (microslip). Acta Mater. **51**, 4331–4345 (2003)

[43] B. Bhushan, M. Nosonovsky, Comprehensive model for scale effect in friction. Acta Mater. **52**, 2461–2474 (2004)

[44] B. Bhushan, M. Nosonovsky, Scale effects in dry and wet friction, wear, and interface temperature. Nanotechnology **15**, 749–761 (2004)

[45] B. Bhushan, M. Nosonovsky, Hydrophobic surface with geometric roughness pattern, U.S. Patent Pending 2004

[46] B. Bhushan, M. Nosonovsky, Scale effect in tribology and mechanical properties, in *Springer Handbook of Nanotechnology*, 2nd edn., ed. by B. Bhushan (Springer, Berlin, 2007), pp. 1167–1197

[47] B. Bhushan, R.A. Sayer, Gecko feet: natural attachment systems for smart adhesion, in *Applied Scanning Probe Methods*, vol. 7, ed. by B. Bhushan (Springer, Berlin, 2007), pp. 41–76

[48] B. Bhushan, R.A. Sayer, Surface characterization and friction of bio-inspired reversible adhesive tape. Microsyst. Technol. **13**, 71–78 (2007)

[49] B. Bhushan, S. Sundararajan, Micro/nanoscale friction and wear mechanisms of thin films using atomic force and friction force microscopy. Acta Mater. **46**, 3793–3804 (1998)

[50] B. Bhushan, J. Israelachvili, U. Landman, Nanotribology: friction, wear and lubrication at the atomic scale. Nature **374**, 607–616 (1995)

[51] B. Bhushan, A.V. Kulkarni, W. Bonin, J.T. Wyrobek, Nanoindentation and picoindentation measurements using a capacitance transducer system in atomic force microscope. Philos. Mag. A **74**, 1117–1128 (1996)

[52] B. Bhushan, A.G. Peressadko, T.W. Kim, Adhesion analysis of two-level hierarchical morphology in natural attachment systems for smart adhesion. J. Adhes. Sci. Technol. **20**, 1475–1491 (2006)

[53] B. Bhushan, M. Nosonovsky, Y.C. Jung, Towards optimization of patterned superhydrophobic surfaces. J. R. Soc. Interface **4**, 643–648 (2007)

[54] B. Bhushan, M. Nosonovsky, Y.C. Jung, Lotus effect: roughness-induced superhydrophobic surfaces, in *Nanotribology and Nanomechanics: An Introduction*, 2nd edn., ed. by B. Bhushan (Springer, Berlin, 2008)

[55] J. Bico, U. Thiele, D. Quéré, Wetting of textured surfaces. Colloids Surf. A **206**, 41–46 (2002)

[56] P.J. Blau, Scale effects in steady-state friction. Tribol. Trans. **34**, 335–342 (1991)

[57] R. Blossey, Self-cleaning surfaces—virtual realities. Nature Mater. **2**, 301–306 (2003)

[58] E. Bormashenko, T. Stein, G. Whyman, Y. Bormashenko, E. Pogreb, Wetting properties of the multiscaled nanostructured polymer and metallic superhydrophobic surfaces. Langmuir **22**, 9982–9985 (2006)

[59] E. Bormashenko, Y. Bormashenko, T. Stein, G. Whyman, R. Pogreb, Z. Barkay, Environmental scanning electron microscope study of the fine structure of the triple line and Cassie–Wenzel wetting transition for sessile drops deposited on rough polymer substrates. Langmuir **23**, 4378–4382 (2007)

[60] E. Bormashenko, R. Pogreb, G. Whyman, M. Erlich, Cassie–Wenzel wetting transition in vibrated drops deposited on the rough surfaces: is dynamic Cassie–Wenzel transition 2D or 1D affair? Langmuir **23**, 6501–6503 (2007)

[61] E. Bormashenko, Y. Bormashenko, T. Stein, G. Whyman, E. Bormashenko, Why do pigeon feathers repel water? Hydrophobicity of pennae, Cassie–Baxter wetting hypothesis and Cassie–Wenzel capillarity-induced wetting transition. J. Colloid Interface Sci. **311**, 212–216 (2007)

[62] L. Boruvka, A.W. Neumann, Generalization of the classical theory of capillarity. J. Chem. Phys. **66**, 5464–5476 (1977)

[63] F.P. Bowden, D. Tabor, *The Friction and Lubrication of Solids* (Clarendon, Oxford, 1950)

[64] G.A.D. Briggs, B.J. Briscoe, The effect of surface topography on the adhesion of elastic solids. J. Phys. D **10**, 2453–2466 (1977)

[65] M. Brugnara, C. Della Volpe, S. Siboni, D. Zeni, Contact angle analysis on polymethyl-methacrylate and commercial wax by using an environmental scanning electron microscope. Scanning **28**, 267–273 (2006)

[66] S.V. Buldyrev, J. Ferrante, F.R. Zypman, Dry friction avalanches: experiments and theory. Phys. Rev. E **74**, 066110 (2006)

[67] Z. Burton, B. Bhushan, Hydrophobicity, adhesion, and friction properties of nanopatterned polymers and scale dependence for micro- and nanoelectromechanical systems. Nano Lett. **5**, 1607–1613 (2005)

[68] Z. Burton, B. Bhushan, Surface characterization and adhesion and friction properties of hydrophobic leaf surfaces. Ultramicroscopy **106**, 709–719 (2006)

[69] M.E. Byrne, D.B. Henthorn, Y. Huang, N.A. Peppas, Micropatterning biomimetic materials for bioadhesion and drug delivery, in *Biomimetic Materials and Design*, ed. by A.K. Dillow, A.M. Lowman (Dekker, New York, 2002), pp. 443–470

[70] R.W. Carpick, N. Agrait, D.F. Ogletree, M. Salmeron, Measurement of interfacial shear (friction) with an ultrahigh vacuum atomic force microscope. J. Vac. Sci. Technol. B **14**, 1289–1295 (1996)

[71] A. Carpinteri, Paggi, M. Size-scale effect on the friction coefficient. Int. J. Solids Struct. **42**, 2901–2910 (2005)

[72] A. Carpinteri, N. Pugno, Are scaling laws on strength of solids related to mechanics or to geometry? Nature Mater. **4**, 421–423 (2005)

[73] A. Cassie, S. Baxter, Wettability of porous surfaces. Trans. Faraday Soc. **40**, 546–551 (1944)

[74] W.R. Chang, I. Etzion, D.B. Bogy, An elastic-plastic model for the contact of rough surfaces. J. Tribol. **109**, 257–263 (1987)

[75] A. Checco, P. Guenoun, J. Daillant, Nonlinear dependence of the contact angle of nanodroplets on contact line curvatures. Phys. Rev. Lett. **91**, 186101 (2003)

[76] Y.L. Chen, C.A. Helm, J. Israelachvili, Molecular mechanisms associated with adhesion and contact-angle hysteresis of monolayer surfaces. J. Phys. Chem. **95**, 10736–10747 (1991)

[77] Y.T. Cheng, D.E. Rodak, A. Angelopoulos, T. Gacek, Microscopic observation of condensation of water on lotus leaves. Appl. Phys. Lett. **87**, 194112 (2005)

[78] Y.T. Cheng, D.E. Rodak, C.A. Wong, C.A. Hayden, Effect of micro- and nanostructures on the self-cleaning behavior of lotus leaves. Nanotechnology **17**, 1359–1362 (2006)

[79] C.-H. Choi, C.-J. Kim, Large slip of aqueous liquid flow over a nanoengineered superhydrophobic surface. Phys. Rev. Lett. **96**, 066001 (2006)

[80] S.E. Choi, P.J. Yoo, S.J. Baek, T.W. Kim, H.H. Lee, An ultraviolet-curable mold for sub-100-nm lithography. J. Am. Chem. Soc. **126**, 7744–7745 (2004)

[81] M.A.S. Chong, Y.B. Zheng, H. Gao, L.K. Tan, Combinational template-assisted fabrication of hierarchically ordered nanowire arrays on substrates for device applications. Appl. Phys. Lett. **89**, 233104 (2006)

[82] T.S. Chow, Nanoadhesion between rough surfaces. Phys. Rev. Lett. **86**, 4592–4595 (2001)

[83] P.B. Clapham, M.C. Hutley, Reduction of length reflection by moth eye principle. Nature **244**, 281–282 (1973)

[84] C. Cottin-Bizonne, J.-L. Barrat, L. Bocquet, E. Charlaix, Low-friction flows of liquid at nanopatterned interfaces. Nature Mater. **2**, 237–240 (2003)

[85] S.R. Coulson, I. Woodward, J.P.S. Badyal, S.A. Brewer, C. Willis, Super-repellent composite fluoropolymer surfaces. J. Phys. Chem. B **104**, 8836–8840 (2000)

[86] R.G. Craig, G.C. Berry, F.A. Peyton, Wetting of poly-(methyl) methacrylate and polystyrene by water and saliva. J. Phys. Chem. **64**, 541–543 (1960)

[87] H. Czichos, Tribology and its many facets: from macroscopic to microscopic and nanoscale phenomena. Meccanica **36**, 605–615 (2001)

[88] C. Daly, J. Zhang, J.B. Sokoloff, Friction in the zero sliding velocity limit. Phys. Rev. E **68**, 066118 (2003)

[89] B.V. Derjaguin, N.V. Churaev, Structural component of disjoining pressure. J. Colloid Interface Sci. **49**, 249–255 (1974)

[90] B.V. Derjaguin, V.M. Muller, Y.P. Toporov, Effect of contact deformation on the adhesion of particles. J. Colloid Interface Sci. **53**, 314–326 (1975)

[91] V.S. Deshpande, A. Needleman, E. Van der Giessen, Discrete dislocation plasticity analysis of static friction. Acta Mater. **52**, 3135–3149 (2004)

[92] P.G. de Gennes, *Soft Interfaces* (Cambridge University Press, Cambridge, 1997)

[93] P.G. de Gennes, On fluid/wall slippage. Langmuir **18**, 3413–3414 (2002)

[94] P.G. de Gennes, F. Brochard-Wyart, D. Quéré, *Capillarity and Wetting Phenomena* (Springer, Berlin, 2003)

[95] A. del Campo, C. Greiner, SU-8: a photoresist for high-aspect-ratio and 3D submicron lithography. J. Micromechanics Microengineering **17**, R81–R95 (2007)

[96] J.M. di Meglio, Contact angle hysteresis and interacting surface defects. Europhys. Lett. **17**, 607–612 (1992)

[97] M. Dienwiebel, G.S. Verhoeven, P. Namboodiri, J.W.M. Frenken, J.A. Hemberg, H.W. Zandbergen, Superlubricity of graphite. Phys. Rev. Lett. **92**, 126101 (2004)

[98] J.H. Dieterich, Modeling of rock friction. J. Geophys. Res. **84**, 2161–2168 (1979)

[99] J.H. Dieterich, B.D. Kilgore, Direct observation of frictional contacts: new insights for state-dependent properties. Pure Appl. Geophys. **143**, 283–302 (1994)

[100] J.M. Donley, C.A. Sepulveda, P. Konstantinidis, S. Gemballa, R.E. Shadwick, Convergent evolution in mechanical design of lamnid sharks and tunas. Nature **429**, 61–65 (2004)

[101] D. Dowson, *History of Tribology*, 2nd edn. (Wiley, New York, 1998)

[102] J. Drelich, J.L. Wilbur, J.D. Miller, G.M. Whitesides, Contact angles for liquid drops at a model heterogeneous surface consisting of alternating and parallel hydrophobic/hydrophilic strips. Langmuir **12**, 1913–1922 (1996)

[103] Y. Dzenis, Spinning continuous fibers for nanotechnology. Science **25**, 1917–1919 (2004)

[104] M. Einax, M. Schulz, S. Trimper, Friction and second-order phase transition. Phys. Rev. B **70**, 046113 (2004)

[105] H.Y. Erbil, A.L. Demirel, Y. Avci, Transformation of a simple plastic into a superhydrophobic surface. Science **299**, 1377–1380 (2003)

[106] I. Etsion, State of the art in laser surface texturing. J. Tribol. **127**, 248–253 (2005)

[107] N. Eustathopoulos, M.G. Nicholas, B. Drevet, *Wettability at High Temperatures* (Pergamon, Amsterdam, 1999)

[108] C.W. Extrand, Model for contact angles and hysteresis on rough and ultraphobic surfaces. Langmuir **18**, 7991–7999 (2002)

[109] C.W. Extrand, Contact angle hysteresis on surfaces with chemically heterogeneous islands. Langmuir **19**, 3793–3796 (2003)

[110] H. Eyring, *Statistical Mechanics and Dynamics* (Wiley, New York, 1964)

[111] W. Federle, Why are so many adhesive pads hairy? J. Exp. Biol. **209**, 2611–2621 (2006)

[112] X.J. Feng, L. Feng, M.H. Jin, J. Zhai, L. Jiang, D.B. Zhu, Reversible superhydrophobicity to super-hydrophilicity transition of aligned ZnO nanorod films. J. Am. Chem. Soc. **126**, 62–63 (2004)

[113] G.S. Fox-Rabinovich, G.E. Totten (eds.), *Self-Organization During Friction* (CRC Press, Boca Raton, 2006)

[114] P. Fratzl, Biomimetic materials research: what can we really learn from nature's structural materials? J.R. Soc. Interface **4**, 637–642 (2007)

[115] Y.I. Frenkel, On the behavior of liquid drops on a solid surface. 1. The sliding of drops on an inclined surface. J. Exp. Theor. Phys. (USSR) **18**, 659 (1948)

[116] Y.I. Frenkel, T. Kontorova, On the theory of plastic deformation and twinning. Sov. Phys. USSR **1**, 137–152 (1939)

[117] C.G.L. Furmidge, Studies at phase interfaces. I. The sliding of liquid drops on solid surfaces and a theory for spray retention. J. Colloid Sci. **17**, 309–324 (1962)

[118] R. Fürstner, W. Barthlott, C. Neinhuis, P. Walzel, Wetting and self-cleaning properties of artificial superhydrophobic surfaces. Langmuir **21**, 956–961 (2005)

[119] J.J. Gagnepain, C. Roques-Carnes, Fractal approach to two-dimensional and three-dimensional surface roughness. Wear **109**, 119–126 (1986)

[120] S. Ganti, B. Bhushan, Generalized fractal analysis and its applications to engineering surfaces. Wear **180**, 17–34 (1995)

[121] C. Gao, X.F. Tian, B. Bhushan, A meniscus model for optimization of texturing and liquid lubrication of magnetic thin-film rigid disks. Tribol. Trans. **38**, 201–212 (1995)

[122] H. Gao, Y. Huang, W.D. Nix, J.W. Hutchinson, Mechanism-based strain-gradient plasticity. J. Mech. Phys. Solids **47**, 1239–1263 (1999)

[123] X.F. Gao, L. Jiang, Biophysics: water-repellent legs of water striders. Nature **432**, 36 (2004)

[124] L. Gao, T.J. McCarthy, The lotus effect explained: two reasons why two length scales of topography are important. Langmuir **22**, 2966–2967 (2006)

[125] L. Gao, T.J. McCarthy, How Wenzel and Cassie were wrong. Langmuir **23**, 3762–3765 (2007)

[126] H. Gao, X. Wang, H. Yao, S. Gorb, E. Artz, Mechanics of hierarchical adhesion structures in geckos. Mech. Mater. **37**, 275–285 (2005)

[127] H. Gao, Z. Liu, J. Zhang, G. Zhang, G. Xie, Precise replication of antireflective nanostructures from biotemplates. Appl. Phys. Lett. **90**, 123115 (2007)

[128] H. Gau, S. Herminghaus, P. Lenz, R. Lipowsky, Liquid morphologies on structured surfaces: from microchannels to microchips. Science **283**, 46–49 (1999)

[129] E. Gerde, M. Marder, Friction and fracture. Nature **413**, 285–288 (2001)

[130] S.D. Gillmor, A.J. Thiel, T.C. Strother, L.M. Smith, M.G. Lagally, Hydrophilic/hydrophobic patterned surfaces as templates for DNA arrays. Langmuir **16**, 7223–7228 (2000)

[131] J.K. Gimzewski, C. Gerber, E. Meyer, R.R. Schlittler, Observation of a chemical reaction using a micromechanical sensor. Chem. Phys. Lett. **217**, 589–594 (1994)

[132] S. Gorb, Functional surfaces in biology: mechanisms and applications, in *Biomimetics: Biologically Inspired Technologies*, ed. by Y. Bar-Cohen (Taylor and Francis, Boca Raton, 2005), pp. 381–397

[133] S. Gorb, M. Varenberg, A. Peressadko, J. Tuma, Biomimetic mushroom-shaped fibrillar adhesive microstructure. J. R. Soc. Interface **4**, 271–275 (2007)

[134] J. Gosline, C. Nichols, A. Guerette, P. Chang, S. Katz, The macromolecular design of spiders silks, in *Biomimetics: Design and Processing of Materials*, ed. by M. Sarikaya, I.A. Aksay (American Institute of Physics, Woodbury, 1995), pp. 237–261

[135] M.D. Greenberg, *Foundations of Applied Mathematics* (Prentice-Hall, Englewood Cliffs, 1978)

[136] M. Greenspan, C. Tschiegg, Radiation-induced acoustic cavitation: apparatus and some results. J. Res. Natl. Bur. Stand. C **71**, 299–312 (1967)

[137] J.A. Greenwood, J.B.P. Williamson, Contact of nominally flat surfaces. Proc. R. Soc. Lond. A **295**, 300–319 (1966)

[138] J.A. Greenwood, J.J. Wu, Surface roughness and contact: an apology. Meccanica **36**, 617–630 (2001)

[139] P. Gupta, A. Ulman, F. Fanfan, A. Korniakov, K. Loos, Mixed self-assembled monolayer of alkanethiolates on ultrasmooth gold do not exhibit contact angle hysteresis. J. Am. Chem. Soc. **127**, 4–5 (2005)

[140] K. Hadobás, S. Kirsch, A. Carl, M. Acet, E.F. Wasserman, Reflection properties of nanostructure-arrayed silicon surfaces. Nanotechnology **11**, 161–164 (2000)

[141] H. Haken, *Advanced Synergetics: Instability Hierarchies of Self-Organizing Systems and Devices* (Springer, New York, 1993)

[142] J.T. Han, Y. Jang, D.Y. Lee, J.H. Park, S.H. Song, D.Y. Ban, K. Cho, Fabrication of a bionic superhydrophobic metal surface by sulfur-induced morphological development. J. Mater. Chem. **15**, 3089–3092 (2005)

[143] W.R. Hansen, K. Autumn, Evidence for self-cleaning in gecko setae. Proc. Natl. Acad. Sci. U.S.A. **102**, 385–389 (2005)

[144] G.S. Hartley, R.T. Brunskill, Reflection of water Drops from surfaces, in *Surface Phenomena in Chemistry and Biology*, ed. by J.F. Danielli (Pergamon, London, 1958), pp. 214–223

[145] G. He, M.O. Robbins, Scale effects and the molecular origins of tribological behavior, in *Nanotribology: Critical Assessment and Research Needs*, ed. by S.M. Hsu, Z.C. Ying (Kluwer Academic, Dordrecht, 2003), pp. 29–44

[146] G. He, M.H. Müser, M.O. Robbins, Adsorbed layers and the origin of static friction. Science **284**, 1650–1652 (1999)

[147] M. He, A.S. Blum, D.E. Aston, C. Buenviaje, R.M. Overney, R. Luginbühl, Critical phenomena of water bridges in nanoasperity contacts. J. Chem. Phys. **114**, 1355–1360 (2001)

[148] B. He, N.A. Patankar, J. Lee, Multiple equilibrium droplet shapes and design criterion for rough hydrophobic surfaces. Langmuir **19**, 4999–5003 (2003)

[149] S.J. Henderson, R.J. Speedy, A Berthelot–Bourdon tube method for studying water under tension. J. Phys. E **13**, 778–782 (1980)

[150] E. Herbert, F. Caupin, The limit of metastability of water under tension: theories and experiments. J. Phys. Condens. Matter **17**, S3597–S3602 (2005)

[151] S. Herminghaus, Roughness-induced non-wetting. Europhys. Lett. **52**, 165–170 (2000)

[152] M. Hikita, K. Tanaka, T. Nakamura, T. Kajiyama, A. Takahara, Superliquid-repellent surfaces prepared by colloidal silica nanoparticles covered with fluoroalkyl groups. Langmuir **21**, 7299–7302 (2005)

[153] M. Hildebrand, E. York, J.I. Kelz, A.K. Davis, L.G. Frigeri, D.P. Allison, M.J. Doktycz, Nanoscale control of silica morphology and three-dimensional structure during diatom cell wall formation. J. Mater. Res. **21**, 2689–2698 (2006)

[154] A.W. Homola, J.N. Israelachvili, P.M. McGuiggan, M.L. Gee, Fundamental experimental studies in tribology: the transition from "interfacial" friction of undamaged molecularly smooth surfaces to normal friction with wear. Wear **136**, 65–83 (1990)

[155] E. Hosono, S. Fujihara, I. Honma, H.S. Zhou, Superhydrophobic perpendicular nanopin film by the bottom-up process. J. Am. Chem. Soc. **127**, 13458–13459 (2005)

[156] Y. Huang, H. Gao, W.D. Nix, J.W. Hutchinson, Mechanism-based strain-gradiant plasticity: analysis. J. Mech. Phys. Solids **48**, 99–128 (2000)

[157] L. Huang, S.P. Lau, H.Y. Yang, E.S.P. Leong, S.F. Yu, S. Prawer, Stable superhydrophobic surface via carbon nanotubes coated with a ZnO thin film. J. Phys. Chem. B **109**, 7746–7748 (2005)

[158] J. Hurtado, Kim K.-S., Scale effect in friction of single-asperity contacts. I. From concurrent slip to single-dislocation-assisted slip. II. Multiple-dislocations-cooperated slip. Proc. R. Soc. Lond. A **455**, 3363–3400 (1999)

[159] J.W. Hutchinson, Plasticity at the micron scale. Int. J. Solids Struct. **37**, 225–238 (2000)

[160] C. Ishino, K. Okumura, Nucleation scenarios for wetting transition on textured surfaces: the effect of contact angle hysteresis. Europhys. Lett. **76**, 464–470 (2006)

[161] J.N. Israelachvili, *Intermolecular and Surface Forces*, 2nd edn. (Academic Press, London, 1992)

[162] J.N. Israelachvili, M.L. Gee, Contact angles on chemically heterogeneous surfaces. Langmuir **5**, 288–289 (1989)

[163] M. Itoh, S. Tamano, R. Iguchi, K. Yokota, N. Akino, R. Hino, S. Kubo, Turbulent drag reduction by the seal fur surface. Phys. Fluids **18**, 065102 (2006)

[164] H. Jansen, M. de Boer, R. Legtenberg, M. Elwenspoek, The black silicon method: a universal method for determining the parameter setting of a fluorine-based reactive ion etcher in deep silicon trench etching with profile control. J. Micromechanics Microengineering **5**, 115–120 (1995)

[165] R. Jetter, L. Kunst, A.L. Samuels, Composition of plant cuticular waxes, in *Biology of the Plant Cuticle*, ed. by M. Riederer, C. Müller (Blackwell, Oxford, 2006), pp. 145–181

[166] K.B. Jinesh, J.W.M. Frenken, Capillary condensation in atomic scale friction: how water acts like a glue. Phys. Rev. Lett. **96**, 166103 (2006)

[167] J.F. Joanny, P.G. de Gennes, A model for contact angle hysteresis. J. Chem. Phys. **81**, 552–562 (1984)

[168] K.L. Johnson, Adhesion and friction between a smooth elastic spherical asperity and a plane surface. Proc. R. Soc. Lond. A **453**, 163–179 (1997)

[169] K.L. Johnson, Mechanics of adhesion. Tribol. Int. **31**, 413–418 (1998)

[170] K.L. Johnson, K. Kendall, A.D. Roberts, Surface energy and the contact of elastic solids. Proc. R. Soc. Lond. A **324**, 301–313 (1971)

[171] R.E. Johnson, R.H. Dettre, Contact angle hysteresis, in *Contact Angle, Wettability, and Adhesion*. Adv. Chem. Ser., vol. 43, ed. by F.M. Fowkes (American Chemical Society, Washington, 1964), pp. 112–135

[172] Y.C. Jung, B. Bhushan, Contact angle adhesion and friction properties of micro and nanopatterned polymers for superhydrophobicity. Nanotechnology **17**, 4970–4980 (2006)

[173] Y.C. Jung, B. Bhushan, Wetting transition of water droplets on superhydrophobic patterned surfaces. Scripta Mater. **57**, 1057–1060 (2007)

[174] Y.C. Jung, B. Bhushan, Wetting behavior during evaporation and condensation of water microdroplets on superhydrophobic patterned surfaces. J. Microsc. **229**, 127–140 (2008)

[175] H. Kamusewitz, W. Possart, D. Paul, The relation between Young's equilibrium contact angle and the hysteresis on rough paraffin wax surfaces. Colloids Surf. A Physicochem. Eng. Asp. **156**, 271–279 (1999)

[176] T. Kasai, B. Bhushan, G. Kulik, L. Barbieri, P. Hoffmann, Micro/nanotribological study of perfluorosilane SAMs for antistiction and low wear. J. Vac. Sci. Technol. B **23**, 995–1003 (2005)

[177] A.B. Kesel, A. Martin, T. Seidl, Adhesion measurements on the attachment devices of the jumping spider evarcha arcuata. J. Expl. Biol. **206**, 2733–2738 (2003)

[178] D.A. Kessler, Surface physics: a new crack on friction. Nature **413**, 260–261 (2001)

[179] M.T. Khorasani, H. Mirzadeh, Z. Kermani, Wettability of porous polydimethylsiloxane surface: morphology study. Appl. Surf. Sci. **242**, 339–345 (2005)

[180] J. Kijlstra, K. Reihs, A. Klami, Roughness and topology of ultra-hydrophobic surfaces. Colloids Surf. A Physicochem. Eng. Asp. **206**, 521–529 (2002)

[181] T.W. Kim, B. Bhushan, Adhesion analysis of multi-level hierarchical attachment system contacting with a rough surface. J. Adhes. Sci. Technol. **21**, 1–20 (2007)

[182] T.W. Kim, B. Bhushan, Optimisation of biomimetic attachment system contacting with a rough surface. J. Vac. Sci. Technol. A **25**, 1003–1012 (2007)

[183] T.W. Kim, B. Bhushan, Effect of stiffness of multi-level hierarchical attachment system on adhesion enhancement. Ultramicroscopy **107**, 902–912 (2007)

[184] T.W. Kim, B. Bhushan, The adhesion model considering capillarity for gecko attachment system. J.R. Soc. Interface **6**, 319–327 (2008)

[185] D. Kim, W. Hwang, H.C. Park, K.H. Lee, Superhydrophobic micro- and nanostructures based on polymer sticking. Key Eng. Mat. **334–335**, 897–900 (2007)

[186] S.B. Kiselev, J.M.H. Levelt Sengers, Q. Zheng, Physical limit to the stability of superheated and stretched water, in *Proceedings of the 12th International Conference on the Properties of Water and Steam*, ed. by H.J. White et al. (Begell, New York, 1995), pp. 378–385

[187] R.J. Klein, P.M. Biesheuvel, B.C. Yu, C.D. Meinhart, F.F. Lange, Producing superhydrophobic surfaces with nano-silica spheres. Z. Met.kd. **94**, 377–380 (2003)

[188] G.W. Koch, S.C. Sillet, G.W. Jennings, S.D. Davis, The limits of tree height. Nature **428**, 851 (2004)

[189] L. Kogut, I. Etsion, A static friction model for elastic-plastic contacting rough surfaces. J. Tribol. **126**, 34–40 (2004)

[190] V.N. Koinkar, B. Bhushan, Scanning and transmission electron microscopies of single-crystal silicon microworn/machined using atomic force microscopy. J. Mater. Res. **12**, 3219–3224 (1997)

[191] A. Kovalchenko, O. Ajayi, A. Erdemir, G. Fenske, I. Etsion, The effect of laser texturing of steel surfaces and speed-load parameters on the transition of lubrication regime from boundary to hydrodynamic. Tribol. Trans. **47**, 299–307 (2004)

[192] B. Krasovitski, A. Marmur, Drops down the hill: theoretical study of limiting contact angles and the hysteresis range on a tilted plane. Langmuir **21**, 3881–3885 (2004)

[193] T. Krink, F. Vollrath, Spider web-building behavior with rule based simulation and genetic algorithms. J. Theor. Biol. **185**, 321–331 (1997)

[194] T. Krink, F. Vollrath, Emergent properties in the behavior of a virtual spider robot. Proc. R. Soc. Lond. B **265**, 2051–2055 (1998)

[195] T.N. Krupenkin, J.A. Taylor, T.M. Schneider, S. Yang, From rolling ball to complete wetting: the dynamic tuning of liquids on nanostructured surfaces. Langmuir **20**, 3824–3827 (2004)

[196] T.N. Krupenkin, J.A. Taylor, E.N. Wang, P. Kolodner, M. Hodes, T.R. Salamon, Reversible wetting-dewetting transitions on electrically tunable superhydrophobic nanostructured surfaces. Langmuir **23**, 9128–9133 (2007)

[197] A. Lafuma, D. Quéré, Superhydrophobic states. Nature Mater. **2**, 457–460 (2003)

[198] J. Lahann, S. Mitragotri, T. Tran, H. Kaido, J. Sundaram, I.S. Choi, S. Hoffer, G.A. Somorjai, R. Langer, A reversibly switching surface. Science **299**, 371–374 (2003)

[199] P. Lam, K.J. Wynne, G.E. Wnek, Surface-tension-confined microfluidics. Langmuir **18**, 948–951 (2002)

[200] L.D. Landau, E.M. Lifshitz, *Fluid Mechanics* (Addison-Wesley, Reading, 1959)

[201] K.K.S. Lau, J. Bico, K.B.K. Teo, M. Chhowalla, G.A.J. Amaratunga, W.I. Milne, G.H. McKinley, K.K. Gleason, Superhydrophobic carbon nanotube forests. Nano Lett. **3**, 1701–1705 (2003)

[202] E. Lauga, M.P. Brenner, Dynamic mechanisms for apparent slip on hydrophobic surfaces. Phys. Rev. E **70**, 026311 (2004)

[203] W. Lee, M. Jin, W. Yoo, J. Lee, Nanostructuring of a polymeric substrate with well-defined nanometer-scale topography and tailored surface wettability. Langmuir **20**, 7665–7669 (2004)

[204] W. Li, A. Amirfazli, A thermodynamic approach for determining the contact angle hysteresis for superhydrophobic surfaces. J. Colloid Interface Sci. **292**, 195–201 (2006)

[205] F.F. Ling, Scaling law for contour length of engineering surfaces. J. Appl. Phys., **62**, 2570–2572 (1987)

[206] S.T. Lou, Z.Q. Ouyang, Y. Zhang, X.J. Li, J. Hu, M.Q. Li, F.J. Yang, Nanobubbles on solid surface imaged by atomic force microscopy. J. Vac. Sci. Technol. B **18**, 2573–2575 (2000)

[207] M. Ma, R.M. Hill, Superhydrophobic surfaces. Curr. Opin. Colloid Interface Sci. **11**, 193–202 (2006)

[208] M. Ma, R.M. Hill, J.L. Lowery, S.V. Fridrikh, G.C. Rutledge, Electrospun poly(styrene-block-dimethylsiloxane) block copolymer fibers exhibiting superhydrophobicity. Langmuir **21**, 5549–5554 (2005)

[209] M. Ma, Y. Mao, M. Gupta, K.K. Gleason, G.C. Rutledge, Superhydrophobic fabrics produced by electrospinning and chemical vapor deposition. Macromolecules **38**, 9742–9748 (2005)

[210] G. Macdougall, C. Ockrent, Surface energy relations in liquid solid systems. Proc. R. Soc. Lond. **180**, 0151–0173 (1942)

[211] N. Maeda, N. Chen, N. Tirrell, J.N. Israelachvili, Adhesion and friction mechanisms of polymer-on-polymer surfaces. Science **297**, 379–382 (2002)

[212] A. Majumdar, B. Bhushan, Role of fractal geometry in roughness characterization and contact mechanics of surfaces. J. Tribol. **112**, 205–216 (1990)

[213] A. Majumdar, B. Bhushan, Fractal model of elastic-plastic contact between rough surfaces. J. Tribol. **113**, 1–11 (1991)

[214] B. Mandelbrot, *The Fractal Geometry of Nature* (Freeman, New York, 1983)

[215] B. Mandelbrot, D.E. Passoja, A.J. Paullay, Fractal character of fracture surfaces of metals. Nature **308**, 721–722 (1984)

[216] A. Marmur, Wetting on hydrophobic rough surfaces: to be heterogeneous or not to be? Langmuir **19**, 8343–8348 (2003)

[217] A. Marmur, The lotus effect: superhydrophobicity and metastability. Langmuir **20**, 3517–3519 (2004)

[218] A. Marmur, Underwater superhydrophobicity: theoretical feasibility. Langmuir **22**, 1400–1402 (2006)

[219] E. Martines, K. Seunarine, H. Morgan, N. Gadegaard, C.D.W. Wilkinson, M.O. Riehle, Superhydrophobicity and superhydrophilicity of regular nanopatterns. Nano Lett. **5**, 2097–3003 (2005)

[220] J.A.C. Martins, J.T. Oden, F.M.F. Simoes, A study of static and kinetic friction. Int. J. Eng. Sci. **28**, 29–92 (1990)

[221] D. Maugis, Adhesion of spheres: the JKR-DMT transition using a Dugdale model. J. Colloid Interface Sci. **150**, 243–269 (1992)

[222] D. Maugis, *Contact, Adhesion and Rupture of Elastic Solids* (Springer, Berlin, 1999)

[223] W. Ming, D. Wu, R. van Benthem, G. de With, Superhydrophobic films from raspberry-like particles. Nano Lett. **5**, 2298–2301 (2005)

[224] F.C. Moon, *The Machines of Leonardo da Vinci and Franz Reuleaux* (Springer, Berlin, 2007)

[225] M. Müser, M. Urbakh, M.O. Robbins, Statistical mechanics of static and low-velocity kinetic friction. Adv. Chem. Phys. **126**, 187–272 (2003)

[226] A. Nakajima, A. Fujishima, K. Hashimoto, T. Watanabe, Preparation of transparent superhydrophobic boehmite and silica films by sublimation of aluminum acetylacetonate. Adv. Mater. **11**, 1365–1368 (1999)

[227] C. Neinhuis, W. Barthlott, Characterization and distribution of water-repellent, self-cleaning plant surfaces. Ann. Bot. **79**, 667–677 (1997)

[228] J. Nicolis, *Dynamics of Hierarchical Systems: An Evolutionary Approach* (Springer, Berlin, 1986)

[229] S. Niederberger, D.H. Gracias, K. Komvopoulos, G.A. Somorjai, Transition from nanoscale to microscale dynamic friction mechanisms on polyethylene and silicon surfaces. J. Appl. Phys. **87**, 3143–3150 (2000)

[230] W.D. Nix, H. Gao, Indentation size effects in crystalline materials: a law for strain-gradient plasticity. J. Mech. Phys. Solids **46**, 411–425 (1998)

[231] M.T. Northen, K.L. Turner, A batch fabricated biomimetic dry adhesive. Nanotechnology **16**, 1159–1166 (2005)

[232] M. Nosonovsky, Multiscale roughness and stability of superhydrophobic biomimetic interfaces. Langmuir **23**, 3157–3161 (2007)

[233] M. Nosonovsky, Model for solid–liquid and solid–solid friction for rough surfaces with adhesion hysteresis. J. Chem. Phys. **126**, 224701 (2007)

[234] M. Nosonovsky, On the range of applicability of the Wenzel and Cassie equations. Langmuir **23**, 9919–9920 (2007)

[235] M. Nosonovsky, Modeling size, load, and velocity effect on friction at micro/nanoscale. Int. J. Surf. Sci. Eng. **1**, 22–37 (2007)

[236] M. Nosonovsky, Oil as a lubricant in Ancient Middle East. Tribol. Online **2**, 44–49 (2007)

[237] M. Nosonovsky, G.G. Adams, Dilatational and shear waves induced by the frictional sliding of two elastic half-spaces. Int. J. Eng. Sci. **39**, 1257–1269 (2001)

[238] M. Nosonovsky, G.G. Adams, Vibration and stability of frictional sliding of two elastic bodies with a wavy contact interface. J. Appl. Mech. **71**, 154–161 (2004)

[239] M. Nosonovsky, B. Bhushan, Scale effect in dry friction during multiple-asperity contact. J. Tribol. **127**, 37–46 (2005)

[240] M. Nosonovsky, B. Bhushan, Roughness optimization for biomimetic superhydrophobic surfaces. Microsys. Technol. **11**, 535–549 (2005)

[241] M. Nosonovsky, B. Bhushan, Stochastic model for metastable wetting of roughness-induced superhydrophobic surfaces. Microsyst. Technol. **12**, 231–237 (2006)

[242] M. Nosonovsky, B. Bhushan, Wetting of rough three-dimensional superhydrophobic surfaces. Microsyst. Technol. **12**, 273–281 (2006)

[243] M. Nosonovsky, B. Bhushan, Hierarchical roughness makes superhydrophobic surfaces stable. Microelectron. Eng. **84**, 382–386 (2007)

[244] M. Nosonovsky, B. Bhushan, Hierarchical roughness optimization for biomimetic superhydrophobic surfaces. Ultramicroscopy **107**, 969–979 (2007)

[245] M. Nosonovsky, B. Bhushan, Biomimetic superhydrophobic surfaces: multiscale approach. Nano Lett. **7**, 2633–2637 (2007)

[246] M. Nosonovsky, B. Bhushan, Multiscale friction mechanisms and hierarchical surfaces in nano- and bio-tribology. Mater. Sci. Eng. R **58**, 162–193 (2007)

[247] M. Nosonovsky, B. Bhushan, Lotus effect: roughness-induced superhydrophobicity, in *Applied Scanning Probe Methods*. Biomimetics and Industrial Applications, vol. VII, ed. by B. Bhushan, H. Fuchs (Springer, Heidelberg, 2007), pp. 1–40

[248] M. Nosonovsky, B. Bhushan, Roughness-induced superhydrophobicity: a way to design non-adhesive surfaces. J. Phys. Condens. Matter **20**, 225009 (2008)

[249] M. Nosonovsky, B. Bhushan, Capillary effects and instabilities in nanocontacts. Ultramicroscopy, (2008, in press)

[250] M. Nosonovsky, B. Bhushan, Patterned non-adhesive surfaces: superhydrophobicity and wetting regime transitions. Langmuir **24**, 1525–1533 (2008)

[251] M. Nosonovsky, B. Bhushan, Do hierarchical mechanisms of superhydrophobicity lead to self-organized criticality? Scripta Mater. (2008, submitted)

[252] M. Nosonovsky, B. Bhushan, Biologically-inspired surfaces: broadening the scope of roughness. Adv. Func. Mater. **18**, 843–855 (2008)

[253] M. Nosonovsky, S.M. Hsu, Model for adhesion-based energy dissipation during friction. in *Proceedings of STLE/ASME International Joint Tribology Conference, IJTC 2006* (2006)

[254] M. Nosonovsky, S.H. Yang, H. Zhang, Sensitivity of adhesion force between rough surfaces to roughness distribution. J. Colloids Interface Sci. (2008, submitted)

[255] T. Onda, S. Shibuichi, N. Satoh, K. Tsujii, Super-water-repellent fractal surfaces. Langmuir **12**, 2125–2127 (1996)

[256] D. Oner, T.J. McCarthy, Ultrahydrophobic surfaces. Effects of topography length scales on wettability. Langmuir **16**, 7777–7782 (2000)

[257] R.A. Onions, J.F. Archard, The contact of surfaces having a random structure. J. Phys. D **6**, 289–304 (1973)

[258] P. Painlevé, Sur le lois du frottement de glissemment. C. R. Acad. Sci. Paris **121**, 112–115 (1895)

[259] A.R. Parker, C.R. Lawrence, Water capture by a desert beetle. Nature **414**, 33–34 (2001)

[260] N.A. Patankar, On the modeling of hydrophobic contact angles on rough surfaces. Langmuir **19**, 1249–1253 (2003)

[261] N.A. Patankar, Transition between superhydrophobic states on rough surfaces. Langmuir **20**, 7097–7102 (2004)

[262] N.A. Patankar, Mimicking the lotus effect: influence of double roughness structures and slender pillars. Langmuir **20**, 8209–8213 (2004)

[263] B.N.J. Persson, *Sliding Friction. Physical Principles and Applications* (Springer, Berlin, 2000))

[264] B.N.J. Persson, Contact mechanics for randomly rough surfaces: new results and some comments. Surf. Sci. Rep. **61**, 201–227 (2006)

[265] G. Polidori, R. Taïar, S. Fohanno, T. Mai, A. Lodini, Skin-friction drag analysis from the forced convection modeling in simplified underwater swimming. J. Biomech. **39**, 2535–2541 (2006)

[266] T. Pompe, A. Fery, S. Herminghaus, Measurement of contact line tension by analysis of the three-phase boundary with nanometer resolution, in *Apparent and Microscopic Contact Angles*, ed. by J. Drelich, J.S. Laskowski, K.L. Mittal (VSP, Utrecht, 2000), pp. 3–12

[267] C.P. Poole, F.J. Owens, *Introduction to Nanotechnology* (Wiley, New York, 2003)

[268] C.Y. Poon, B. Bhushan, Comparison of surface roughness measurements by stylus profiler, AFM and non-contact optical profiler. Wear **190**, 76–88 (1995)

[269] D.A. Porter, K.E. Easterling, *Phase Transformations in Metals and Alloys*, 2nd edn. (Chapman & Hall, New York, 1993)

[270] N. Poulsen, C. Berne, J. Spain, N. Kroger, Silica immobilization of an enzyme through genetic engineering of the diatom thalassiosira pseudonana. Angew. Chem. (Int. Ed.) **461**, 1843–1846 (2007)

[271] I. Prigogine, *Thermodynamics of Irreversible Processes* (Wiley, New York, 1961)

[272] N.M. Pugno, A general shape/size effect law for nanoindentation. Acta Mater. **55**, 1947–1953 (2007)

[273] B.T. Qian, Z.Q. Shen, Fabrication of superhydrophobic surfaces by dislocation-selective chemical etching on aluminum, copper, and zinc substrates. Langmuir **21**, 9007–9009 (2005)

[274] D. Quéré, Surface wetting: model droplets. Nature Mater. **3**, 79–80 (2004)

[275] D. Quéré, Non-sticking drops. Rep. Prog. Phys. **68**, 2495–2532 (2005)

[276] D. Quéré, A. Ajdari, Liquid drops: surfing the hot spot. Nature Mater. **5**, 429–430 (2006)

[277] Y.I. Rabinovich, J.J. Adler, A. Ata, R.K. Singh, B.M. Moudgil, Adhesion between nanoscale rough surfaces. J. Colloid Interface Sci. **232**, 10–16 (2000)

[278] K. Ranjith, J. Rice, Slip dynamics at an interface between dissimilar materials. J. Mech. Phys. Solids **49**, 341–361 (2001)

[279] I. Rechenberg, A.R. El Khyeri, The sandfish of the Sahara. A model for friction and wear reduction, Department of Bionics and Evolution Techniques, Technical University of Berlin website http://www.bionik.tu-berlin.de/institut/safiengl.htm. Accessed 14 Aug 2007

[280] J. Rice, Dislocation nucleation from a crack tip: an analysis based on the Pierls concept. J. Mech. Phys. Solids **40**, 239–271 (1991)

[281] D. Richard, D. Quéré, Bouncing water drops. Europhys. Lett. **50**, 769–775 (2000)

[282] S.M. Rowan, M.I. Newton, G. McHale, Evaporation of microdroplets and the wetting of solid surfaces. J. Phys. Chem. **99**, 13268–13271 (1995)

[283] J.S. Rowlinson, B. Widom, *Molecular Theory of Capillarity* (Clarendon, Oxford, 1982)

[284] M. Ruths, J.N. Israelachvili, Surface forces and nanorheology of molecularly thin films, in *Springer Handbook of Nanotechnology*, 2nd edn., ed. by B. Bhushan (Springer, Berlin, 2007), pp. 859–924

[285] D. Satas (ed.), *Coating Technology Handbook* (Dekker, New York, 1991)

[286] M. Scherge, S. Gorb, *Biological Micro- and Nanotribology: Nature's Solutions*. Nano-Science and Technology Series (Springer, Berlin, 2001)

[287] U.D. Schwarz, O. Zwörner, P. Köster, R. Wiesendanger, Quantitative analysis of the frictional properties of solid materials at low loads. 1. Carbon compounds. Phys. Rev. B **56**, 6987–6996 (1997)

[288] S. Semal, T.D. Blake, V. Geskin, M.L. de Ruijter, G. Castelein, J. De Coninck, Influence of surface roughness on wetting dynamics. Langmuir **15**, 8765–8770 (1999)

[289] M.E.R. Shanahan, A. Carré, Viscoelastic dissipation in wetting and adhesion phenomena. Langmuir **11**, 1396–1402 (1995)

[290] H.M. Shang, Y. Wang, S.J. Limmer, T.P. Chou, K. Takahashi, G.Z. Cao, Optically transparent superhydrophobic silica-based films. Thin Solid Films **472**, 37–43 (2005)

[291] R. Sharma, D.S. Ross, Kinetics of liquid penetration into periodically constrained capillaries. J. Chem. Soc. Faraday Trans. **87**, 619–624 (1991)

[292] F. Shi, Y. Song, J. Niu, X. Xia, Z. Wang, X. Zhang, Facile method to fabricate a large-scale superhydrophobic surface by galvanic cell reaction. Chem. Mater. **18**, 1365–1368 (2006)

[293] S. Shibuichi, T. Onda, N. Satoh, K. Tsujii, Super-water-repellent surfaces resulting from fractal structure. J. Phys. Chem. **100**, 19512–19517 (1996)

[294] J. Shiu, C. Kuo, P. Chen, C. Mou, Fabrication of tunable superhydrophobic surfaces by nanosphere lithography. Chem. Mater. **16**, 561–564 (2004)

[295] N.J. Shirtcliffe, G. McHale, M.I. Newton, G. Chabrol, C.C. Perry, Dual-scale roughness produces unusually water-repellent surfaces. Adv. Mater. **16**, 1929–1932 (2004)

[296] N.J. Shirtcliffe, G. McHale, M.I. Newton, C.C. Perry, P. Roach, Porous materials show superhydrophobic to superhydrophilic switching. Chem. Commun. **31**, 3135–3137 (2005)

[297] R. Shuttleworth, G.L.J. Bailey, The spreading of liquid over a rigid solid. Discuss. Faraday Soc. **3**, 16–22 (1948)

[298] S. Singh, J. Houston, F. van Swol, C.J. Brinker, Superhydrophobicity: drying transition of confined water. Nature **442**, 526 (2006)

[299] M. Sitti, R.S. Fearing, Synthetic gecko foot-hair micro/nanostructures as dry adhesives. J. Adhes. Sci. Technol. **17**, 1055–1073 (2003)

[300] J.B. Sokoloff, Possible microscopic explanation of the virtually universal occurrence of static friction. Phys. Rev. B **65**, 115415 (2002)

[301] J.B. Sokoloff, Theory of the effects of multiscale surface roughness and stiffness on static friction. Phys. Rev. B **73**, 016104 (2006)

[302] R. Stark, G. Schitter, A. Stemmer, Velocity dependent friction laws in contact mode atomic force microscopy. Ultramicroscopy **100**, 309–317 (2004)

[303] N.A. Stelmashenko, J.P. Craven, A.M. Donald, E.M. Terentjev, B.L. Thiel, Topographic contrast of partially wetting water droplets in environmental scanning electron microscopy. J. Microsc. **204**, 172–183 (2001)

[304] D.E. Stewart, Finite dimensional contact mechanics. Phil. Trans. R. Soc. Lond. A **359**, 2467–2482 (2001)

[305] T. Stifter, O. Marti, B. Bhushan, Theoretical investigation of the distance dependence of capillary and van der Waals forces in scanning force microscopy. Phys. Rev. B **62**, 13667–13673 (2000)

[306] M. Sukharev, T. Siedman, Phase and polarization control as a route to plasmonic nanodevices. Nano Lett. **6**, 715–719 (2006)

[307] M. Sukharev, T. Siedman, Coherent control of light propagation via nanoparticle arrays. J. Phys. B, **40**, S283–S298 (2007)

[308] M. Sun, C. Luo, L. Xu, H. Ji, Q. Ouyang, D. Yu, Y. Chen, Artificial lotus leaf by nanocasting. Langmuir **21**, 8978–8981 (2005)

[309] P.S. Swain, R. Lipowsky, Contact angles on heterogeneous surfaces: a new look at Cassie's and Wenzel's laws. Langmuir **14**, 6772–6780 (1998)

[310] R. Szoszkiewicz, B. Bhushan, B.D. Huey, A.J. Kulik, G. Gremaud, Correlations between adhesion hysteresis and friction at molecular scales. J. Chem. Phys. **122**, 144708 (2005)

[311] D. Tabor, Surface forces and surface interactions. J. Colloid Interface Sci. **58**, 2–13 (1976)

[312] N.S. Tambe, B. Bhushan, Scale dependence of micro/nano-friction and adhesion of MEMS/NEMS materials, coatings and lubricants. Nanotechnology **15**, 1561–1570 (2004)

[313] N.R. Tas, P. Mela, T. Kramer, J.W. Berenschot, A. van der Berg, Capillarity induced negative pressure of water plugs in nanochannels. Nano Lett. **3**, 1537 (2003)

[314] K. Teshima, H. Sugimura, Y. Inoue, O. Takai, A. Takano, Transparent ultra water-repellent poly(ethylene terephthalate) substrates fabricated by oxygen plasma treatment and subsequent hydrophobic coating. Appl. Surf. Sci. **244**, 619–622 (2005)

[315] K. Thamatrakoln, A.J. Alverson, M. Hildebrand, Comparative sequence analysis of diatom silicon transport. J. Phycol. **42**, 822–834 (2006)

[316] T.R. Thomas, *Rough Surfaces* (Longman, New York, 1982)

[317] X. Tian, B. Bhushan, A numerical three-dimensional model for the contact of rough surfaces by variational principle. J. Tribol. **118**, 33–42 (1996)

[318] D.M. Tolstoi, Significance of the normal degree of freedom and natural normal vibrations in contact friction. Wear **10**, 199–213 (1967)

[319] G.A. Tomlinson, The molecular theory of friction. Philos. Mag. **7**, 905–916 (1929)

[320] O.N. Tretinnikov, Wettability and microstructure of polymer surfaces: stereochemical and conformational aspects, in *Apparent and Microscopic Contact Angles*, ed. by J. Drelich, J.S. Laskowski, K.L. Mittal (VSP, Utrecht, 2000), pp. 111–128

[321] D.H. Trevena, *Cavitation and Tension in Liquids* (Hilger, Bristol, 1987)

[322] Y. Tsori, Discontinuous liquid rise in capillaries with varying cross-sections. Langmuir **22**, 8860–8863 (2006)

[323] D.L. Turcotte, Self-organized criticality. Rep. Prog. Phys. **62**, 1377–1429 (1999)

[324] M.T. Tyree, Accent of water. Nature **423**, 923 (2003)

[325] J.W.G. Tyrrell, P. Attard, Images of nanobubbles on hydrophobic surfaces and their interaction. Phys. Rev. Lett. **87**, 176104 (2001)

[326] M. Urbakh, J. Klafter, D. Gourdon, J. Israelachvili, The nonlinear nature of friction. Nature **430**, 525–528 (2004)

[327] S. Vedantam, M.V. Panchagnula, Phase field modeling of hysteresis in sessile drops. Phys. Rev. Lett. **99**, 176102 (2007)

[328] Y. Vlasov, A. Legin, Non-selective chemical sensors in analytical chemistry: from 'electronic nose' to 'electronic tongue'. Fresenius J. Anal. Chem. **361**, 255–260 (1998)

[329] R.S. Voronov, D.V. Papavassiliou, L.L. Lee, Boundary slip and wetting properties of interfaces: correlation of the contact angle with the slip length. J. Chem. Phys. **124**, 204701 (2006)

[330] P. Wagner, R. Furstner, W. Barthlott, C. Neinhuis, Quantitative assessment to the structural basis of water repellency in natural and technical surfaces. J. Exp. Bot. **54**, 1295–1303 (2003)

[331] S.A. Wainwright, F. Vosburgh, J.H. Hebrank, Shark skin: function in locomotion. Science **202**, 747–749 (1978)

[332] S. Wang, L. Feng, L. Jiang, One-step solution-immersion process for the fabrication of stable bionic superhydrophobic surfaces. Adv. Mater. **18**, 767–770 (2006)

[333] Y. Wang, Q. Zhu, H. Zhang, Fabrication and magnetic properties of hierarchical porous hollow nickel microspheres. J. Mater. Chem. **16**, 1212–1214 (2006)

[334] S. Wang, H. Liu, D. Liu, X. Ma, X. Fang, L. Jiang, Enthalpy driven three state switching of a superhydrophilic/superhydrophobic surface. Angew. Chem. Int. Ed. Engl. **46**, 3915–3917 (2007)

[335] X. Wang, H. Zhang, S.M. Hsu, The effect of dimple size and depth on friction reduction under boundary lubrication pressure, in *Proceedings of the STLE/ASME International Joint Tribology Conference IJTC2007*, San Diego, CA, 2007

[336] L. Wenning, M.H. Müser, Friction laws for elastic nanoscale contacts. Europhys. Lett. **54**, 693–699 (2001)

[337] R.N. Wenzel, Resistance of solid surfaces to wetting by water. Ind. Eng. Chem. **28**, 988–994 (1936)

[338] D.J. Whitehouse, J.F. Archard, The properties of random surfaces in significance to their contact. Proc. R. Soc. Lond. A **316**, 97–121 (1970)

[339] K.A. Wier, T.J. McCarthy, Condensation on ultrahydrophobic surfaces and its effect on droplet mobility: ultrahydrophobic surfaces are not always water repellent. Langmuir **22**, 2433–2436 (2006)

[340] S.J. Wilson, M.C. Hutley, The optical properties of the 'moth-eye' antireflective surfaces. J. Mod. Opt. **29**, 993–1009 (1982)

[341] X.D. Wu, L.J. Zheng, D. Wu, Fabrication of superhydrophobic surfaces from microstructured ZnO-based surfaces via a wet-chemical route. Langmuir **21**, 2665–2667 (2005)

[342] L. Xu, W. Chen, A. Mulchandani, Y. Yan, Reversible conversion of conducting polymer films from superhydrophobic to superhydrophilic. Angew. Chem. (Int. Ed.) **44**, 6009–6012 (2005)

[343] H. Yabu, M. Shimomura, Single-step fabrication of transparent superhydrophobic porous polymer films. Chem. Mater. **17**, 5231–5234 (2005)

[344] V. Yaminsky, Molecular mechanisms of hydrophobic transitions, in *Apparent and Microscopic Contact Angles*, ed. by J. Drelich, J.S. Laskowski, K.L. Mittal (VSP, Utrecht, 2000), pp. 47–95

[345] S.H. Yang, H. Zhang, S.M. Hsu, Correction for random surface roughness on colloidal probes in measuring adhesion. Langmuir **23**, 1195–1202 (2007)

[346] S.H. Yang, M. Nosonovsky, H. Zhang, K.-H. Chung, Nanoscale water capillary bridges under deeply negative pressure. Chem. Phys. Lett. **451**, 88–92 (2008)

[347] I.V. Yannas, J.F. Burke, Design of an artificial skin. Basic design principles. J. Biomed. Mater. Res. **14**, 65–81 (2004)

[348] Z.C. Ying, S.M. Hsu, First observation of elastic plowing in nanofriction, in *Proceedings of the World Tribology Congress III—2005*, pp. 339–340

[349] H. Yoshizawa, Y.L. Chen, J. Israelachvili, Fundamental mechanisms of interfacial friction—relation between adhesion and friction. J. Phys. Chem. **97**, 4128–4140 (1993)

[350] F.G. Yost, J.R. Michael, E.T. Eisenmann, Extensive wetting due to roughness. Acta Metall. Mater. **45**, 299–305 (1995)

[351] H.B. Zeng, M. Tirrell, J. Israelachvili, Limit cycles in dynamic adhesion and friction processes: a discussion. J. Adhes. **82**, 933–943 (2006)

[352] L. Zhai, F.C. Cebeci, R.E. Cohen, M.F. Rubner, Stable superhydrophobic coatings from polyelectrolyte multilayers. Nano Lett. **4**, 1349–1353 (2004)

[353] L.C. Zhang, K.L. Johnson, W.C.D. Cheong, A molecular dynamics study of scale effects on the friction of single-asperity contacts. Tribol. Lett. **10**, 23–28 (2001)

[354] X. Zhang, F. Shi, X. Yu, H. Liu, Y. Fu, Z.Q. Wang, L. Jiang, X. Li, Polyelectrolyte multilayer as matrix for electrochemical deposition of gold clusters: toward superhydrophobic surface. J. Am. Chem. Soc. **126**, 3064–3065 (2004)

[355] J.L. Zhang, J.A. Li, Y.C. Han, Superhydrophobic PTFE surfaces by extension. Macromol. Rapid Commun. **25**, 1105–1108 (2004)

[356] X.H. Zhang, N. Maeda, V.S.J. Craig, Physical properties of nanobubbles on hydrophobic surfaces in water and aqueous solutions. Langmuir **22**, 5025–5035 (2006)

[357] X. Zhang, S. Tan, N. Zhao, X. Guo, X. Zhang, Y. Zhang, J. Xu, Evaporation of sessile water droplets on superhydrophobic natural lotus and biomimetic polymer surfaces. Chem. Phys. Chem. **7**, 2067–2070 (2006)

[358] N. Zhao, Q.D. Xie, L.H. Weng, S.Q. Wang, X.Y. Zhang, J. Xu, Superhydrophobic surface from vapor-induced phase separation of copolymer micellar solution. Macromolecules **38**, 8996–8999 (2005)

[359] Y. Zhu, S. Granick, Limits of the hydrodynamic no-slip boundary condition. Phys. Rev. Lett. **88**, 106102 (2002)

[360] L. Zhu, Y. Xiu, J. Xu, P.A. Tamirisa, D.W. Hess, C. Wong, Superhydrophobicity on two-tier rough surfaces fabricated by controlled growth of aligned carbon nanotube arrays coated with fluorocarbon. Langmuir **21**, 11208–11212 (2005)

[361] F. Zypman, J. Ferrante, M. Jansen, K. Scanlon, P. Abel, Evidence of self-organized criticality in dry sliding friction. J. Phys. Condens. Matter **15**, L191–L196 (2003)

Index

acetone 186

actuator 246
  gap-closing 246

adhesion 4, 5, 9–12, 15, 21, 25, 27, 28, 31,
  32, 34, 35, 37, 38, 42, 56, 57, 62, 71,
  75, 77–79, 81, 91, 110, 111, 113, 115,
  134, 141, 150, 169, 184, 185, 192,
  194, 195, 197, 206, 211, 229, 231,
  233, 234, 236–240, 242, 243

adhesion hysteresis 32, 34, 37, 38, 56, 78,
  108, 110–112, 137, 154, 155, 158,
  163, 164, 167, 177

adsorbent 17

AFM modes
  contact 15, 191, 192
  noncontact 15
  tapping 184, 190, 192

air 4, 15, 25, 69, 73, 76, 78, 82–87, 89,
  90, 98, 101, 105, 108, 109, 113, 119,
  122, 127–131, 133, 140–143, 151,
  170–173, 175, 187, 196, 197, 204,
  210, 212, 216, 219, 226, 229, 239,
  245, 248

alkanoic acids 205

alkoxysilane 204

alkylketene dimer 203

annealing 205, 208

apparent (nominal) area of contact 50, 51

architecture 228, 248

area of contact 23, 31, 32, 39, 41–43, 45, 49,
  50, 98, 150, 167

Arrhenius method 59

artificial

fog 183, 227, 247
  nose 249
  rain 183, 227
  tongue 249

atomic force microscope (AFM) 11, 15, 52,
  53, 69, 78, 172, 184, 187, 190–192,
  195–197, 207, 249

atomic scale 5, 28, 29, 39, 40, 52

attachment mechanisms 244

autocorrelation function 19, 20, 104

automotive industry 25

barbs 197

barbules 197

bearings 10, 25, 40

biomaterials 243

bionics 11, 243

biotechnology 229

biotribology 10

boiling 3, 5, 9, 66, 176

bouncing droplets 169, 172, 173

brucit-type cobalt hydroxide (BCH) 205

bubble 3–5, 67, 68, 73, 84, 113, 133, 134,
  142, 151, 154, 169–172, 178

buckling 237

camouflage 245

cantilever 11, 15, 78, 184, 207, 233, 237,
  249

cantilever array 249

capillarity 4

capillary
  bridge 69, 71, 77, 172

force 4, 69, 71, 77–79, 116, 174, 216, 236, 237
length 4, 107, 140, 145
number 174
waves 113, 115, 116, 138, 140, 148, 157, 219, 229
carbon nanotubes (CNT) 11, 201, 204
cartilage 243
Cassie equation 86, 90, 92, 94–96, 153, 166
Cassie–Baxter equation 83, 87, 89, 92, 112, 138, 196, 214, 219, 221
Cassie–Wenzel transition 153, 214
cavitation 67–69, 71, 133, 154, 172, 178
chemical bath deposition (CBD) 205
chemical bonds 3, 5, 17, 25, 28, 58, 73
chemical vapor deposition (CVD) 201, 203, 204
coating 15, 25, 81, 107, 140, 150, 181, 183, 187, 201, 203–205, 208, 209, 211, 230, 248
cobblestone mechanism 28, 39, 44, 62
coercivity 206
collagen 243
colloidal systems 205
colocasia 84, 185, 187, 190, 192, 194, 196
complexity science 9
condensation 5, 66, 69, 76, 115, 134, 135, 138, 140, 142, 148, 172, 204, 219, 223, 225
contact angle 71, 73–76, 78, 82, 84–87, 89–92, 94–96, 98, 99, 101, 103–107, 109–113, 122, 124–126, 129–131, 133, 134, 138, 140, 141, 143, 144, 150, 153–155, 157, 158, 160, 161, 163, 166, 167, 169, 171, 173, 176, 182, 184, 186, 187, 196, 197, 201–203, 205–212, 215, 216, 219, 222, 223, 226, 227, 237, 240, 241, 247
    advancing 82, 90, 107, 110, 141, 161, 177, 206, 225
    apparent 75, 91, 95, 96, 166
    dynamic 110, 207
    hysteresis 76, 82, 84, 91, 92, 105, 107, 108, 110–113, 115, 125, 133, 134, 138, 140, 141, 150, 153–155, 157–159, 161, 163–165, 167, 169, 176, 177, 183, 197, 206
    most stable 91, 155

receding 82, 90, 107, 110, 111, 141, 155, 157, 161, 163, 167, 177, 204, 206, 225
    static 75, 76, 81, 82, 85, 86, 91, 103, 104, 110, 184, 206, 209, 216, 219, 225
contact line 4, 74, 76, 78, 82, 91, 100, 107, 108, 113, 153, 163, 165, 166
cornea 245
correlation length 5, 6, 19, 24, 41, 52, 53, 104, 236
Coulomb–Amonton's law 27, 28, 31, 32, 40, 47, 48, 50, 51, 252
covalent bonds 17
creep 49, 50
critical point 6, 52, 67, 69, 212
crystal 13, 31, 59, 67, 69, 183, 203, 227
cuticle 181, 182, 243

darkling beetle 12, 246
deionized water 104, 184, 206
deposition 25, 66, 199, 201–205, 207, 208, 212
dewetting 91, 154, 159
diatom 248, 249
diffraction 17
dilatational waves 48
dimethylformamide (DMF) 203
disjoining pressure 75, 91, 113
dislocations 28, 30, 31, 39, 53, 60
disorder 8, 29, 55, 59
dispersion 3
display 245, 246
dissipation 7, 8, 27–29, 34, 38, 52, 54–56, 58, 62, 78, 91, 108, 112, 113, 161, 169, 173–175
dissipative system 9
DMT model 32, 78, 79, 236
DNA 12, 166, 229, 249
drag reduction 246
droplet 4, 73, 75, 76, 82, 84–87, 89–92, 94–96, 103, 104, 107, 108, 110–113, 115, 123–125, 129, 133, 134, 138, 140–142, 145, 150, 151, 153–155, 158, 159, 161, 163, 165–167, 169–178, 183, 184, 187, 196, 197, 204, 206–208, 212, 214–216, 219, 221–223, 226–229, 246, 247
Dugdale approximation 57
dust 207, 221, 227, 239
dynamic instability 48–51

elastic deformation 32, 34, 38, 48, 56
elasticity 48
electrospinning 200, 203, 247
electrostatic force 5, 229
elytra 246
energy functional 59, 165
entropy 7, 9, 29, 55
epicuticular wax 183, 186
epidermis 182
epoxy 205, 246
equilibrium 6, 36, 37, 52, 54, 58, 60, 66,
    67, 69, 73, 75, 76, 78, 79, 82, 83, 85,
    95, 109, 115, 116, 120, 126, 127, 130,
    142, 144, 145, 148, 150, 157, 165,
    212, 225
etching 25, 199, 201–203, 230
  chemical 200, 203
  laser 203
  plasma 200, 202
evaporation 25, 66, 76, 91, 135, 142, 158,
    161, 172, 197, 200, 203, 207, 219,
    221, 222, 226, 229

fagus 185, 187, 190, 192, 194–196
fiber 203, 237, 238, 247, 248
fibrillar structure 237, 242
flow 9, 13, 29, 65, 82, 91, 108, 133, 140,
    141, 163, 165, 169–171, 177, 200, 246
fluctuation 3, 5–7, 13, 52, 55, 58, 60, 67, 68,
    72, 121, 130
fluoroalkylsilane 203, 204
fluorocarbons 230
foil 227
force calibration 185, 207
fractal 7, 20–24, 26, 42, 43, 53, 83, 164, 203
fracture 27, 39, 237, 238, 244
Frank–Read dislocation sources 31
free energy 3, 4, 60, 72, 75, 142, 145, 147,
    148, 165, 171
  Gibbs 4, 75
  Helmholtz 4, 75
freezing 5, 66
friction
  Coulombian 48, 50
  dry 9, 27, 31, 40, 43, 47, 49, 53, 54, 81,
    110
  dynamic 48, 49
  rolling 34, 38
  sliding 25, 38, 49, 50

solid–liquid 38, 65, 82
solid–solid 27, 28, 38, 65, 111
state-and-rate 49
frustule 249
fulleren (C_{60}) 11

Gaussian distribution 19, 104
gecko 12, 231, 233, 234, 236, 237, 239–242,
    244, 250
geometrically necessary dislocations 29
glass 78, 84, 203, 228, 240, 246
goniometer 184, 197, 206
grains 5, 52, 108, 163, 248
graphene 11
graphite 11, 13, 55
gravity 4, 82, 107, 116, 123–125, 128, 142,
    145, 177, 215

hardcore repulsion 66
hardness 5, 7, 24, 29, 38, 42, 51
heterogeneity (inhomogeneity) 27, 28, 34,
    40, 55, 56, 58, 59, 61, 62, 75, 91, 92,
    95, 96, 108, 112, 113, 138–140, 148,
    166, 216, 219, 226, 228
hierarchical structure 21, 40, 53, 55, 61, 197,
    205, 206, 233, 237, 239, 241–244,
    247, 250
hierarchy 7, 9, 27, 28, 53–56, 58, 61, 62,
    164, 183, 196, 233, 236
history 82, 90, 154
honey 228
honeycomb pattern 106, 204
hydrocarbons 17, 182
hydrogen 32, 65
hydrophobic interaction 171, 172, 178
hydrophobicity 82, 94, 115, 133, 150, 151,
    169, 172, 177, 181, 203, 211, 230, 241

ice 65, 67, 71
ideal gas 66, 121
ill-posedness 50
indium tin oxide (ITO) 205
insect 231, 233, 243, 245–248, 250
interface
  composite 82, 87, 89, 90, 96, 98, 99, 101,
    111, 115, 116, 120–122, 124–131,
    133, 138, 140–142, 145, 147, 150,
    151, 153, 154, 164, 167, 170, 171,
    183, 212, 214, 216, 218, 219

homogeneous 84, 85, 87, 90, 115,
    120–122, 125, 130, 131, 134, 138,
    142, 145, 150, 153, 212, 214, 216
liquid–vapor 68, 73–75, 77, 91, 92, 95,
    96, 110, 113, 116, 119–122, 126, 128,
    139–148, 150, 151, 155–157
solid–liquid 3, 4, 13, 74, 82, 94, 110, 111,
    119, 120, 134, 142, 147, 154, 171, 216
solid–vapor 3, 13
interference modulation 246
interferometer 15

JKR model 32, 78, 79, 236

Kelvin equation 69, 76, 77, 79
keratin 233
kinetics 59, 141, 163
Koch curve 22
kurtosis 18, 19

Landau–Ginzburg functional 6, 9, 59, 165
Langmuir–Blodgett method 25
Laplace equation 72, 73, 90, 151, 171, 176,
    215
Laplace pressure 5, 73, 76, 77, 136, 141, 144
lateral resolution 23, 184, 207
lauric acid (LA) 85–87, 89, 107, 116,
    117, 119–122, 128–131, 134–137,
    142–145, 147, 148, 155, 177, 205, 212
laws (rules) of friction 10, 51, 59
lay 13
layer-by-layer deposition 205, 206
Leidenfrost effect 175, 178
Lennard-Jones potential 35, 57
light wavelength 245
linearity/nonlinearity 28, 40, 47, 48, 53, 55,
    253
lipids 181
liquid microchip 229
lithography 25, 199, 201, 202, 230, 246
    E-beam 200
    soft 202
    X-ray 200, 202, 227
lizard 231, 234, 239, 248
long wavelength limit 21, 23, 26
lotus 12, 84, 85, 103, 104, 141, 150,
    181–183, 185, 187, 190–192, 194,
    196, 199, 202, 204–206, 208–211,
    222, 227, 228, 230, 239, 244, 246, 250

lotus-effect 84, 181–183, 228, 230, 239, 244,
    246, 250
lubrication 9–11, 25
    boundary 25

magnolia 185, 187, 190, 192, 194–196
mapping 27, 61
marquees 228
mean curvature 73, 76, 77, 90
melting 5, 52, 66
meniscus 5, 71, 73, 76–79, 85, 115, 134,
    136, 137, 192, 211
meniscus force 69, 85, 115, 134, 136, 137,
    151, 172, 195, 211, 229
mesoscale 5–7, 9, 28, 29, 59
metal 15, 41, 66, 203, 205, 227, 240, 243,
    244, 250
metastable state 67, 69, 84, 90, 111, 120,
    121, 129, 130, 151, 214
mica 13, 71
microelectromechanical systems (MEMS)
    25, 81, 208, 228, 246
microfluidics 229, 230, 247
microhairs 197
microscale 5, 6, 15, 53, 58, 84, 92, 113, 141,
    145, 150, 155, 158, 167, 195–197,
    206, 241, 249
microsetae 141, 247
microsphere 206, 240
microsyringe 184, 206, 207
microtrichia 245
molecular dynamics 31, 58, 133, 170, 171
molecular scale 23, 72, 91, 107, 155, 157,
    163
monolayer 11, 17, 25, 71, 75, 91, 157, 205,
    206
moth-eye-effect 244–246
mucus 243
multiscale 12, 13, 20, 26–29, 40, 42, 58, 65,
    92, 113, 141, 142, 144, 148, 150, 151,
    154, 163, 164, 166, 167, 250

nanobubble 133, 170–172, 178
nanocalorimetry 249
nanochannel 69, 247
nanodevices 166, 246
nanodroplets 5, 138, 140
nanoelectromechanical system 81, 228
nanofibers 204

nanofluidics 229
nanogrooves 141, 197, 247
nanoparticle 5, 200, 205, 206, 246
nanoparticle arrays 246
nanophotonics 246
nanorods 201, 204
nanotechnology 4, 10, 11, 164, 230, 249, 250
nanotribology 10, 11
nanoturf 171
Navier length 170
negative pressure 67–69, 71
negative viscosity 49
normal distribution 19

octadecyltrichlorosilane (OTS) 202
oleophobic 82
optical profiler 184, 187, 190, 192, 197, 206, 207, 212
optical sensors 228
order-parameter 6, 9, 59, 165
oxygen 17, 202, 209
oxygen plasma 202, 209

paints 199, 228, 230
papilla(e) 84, 103, 106, 140, 141, 150, 182, 183, 185, 186, 228
papillose epidermal cells 84, 182, 183, 185
paraffin 150, 182, 230
pathogens 183
patterned surface 84, 111, 160, 164, 202, 207, 212, 215, 216, 218, 219, 222, 223, 225–227, 230, 247
pendulum 6
pennae 197
percolation 6
perfluorodecyltriethoxysilane 202, 209
perfluorooctanesulfonate (PFOS) 200, 203
phase 3, 5, 6, 52, 53, 59, 65–69, 71, 73–77, 91, 165, 166, 202, 204, 208, 212, 246
    diagram 6, 66, 67, 69, 72, 79
    equilibrium 67
    field 165, 166
    transformation 59
    transition 5, 6, 52, 53, 59, 66, 67, 79, 165
physical vapor deposition (PVD) 203
pixel 246
placoid 246
plasmonics 246

plastic yield 31, 39
plasticity 6, 29, 31, 39, 42
plowing 27, 28, 38–40, 44, 45, 54, 62
poly(acrylic acid) (PAA) 205, 206
poly(allylamine hydrochloride) (PAH) 205
polydimethylsiloxane (PDMS) 85, 200, 201, 203, 204, 205, 208
polyester 181
poly(ethylene terephthalate) (PET) 200, 202
polymer 5, 59, 90, 170, 173, 200, 202–204, 206, 208–210, 229, 244, 247, 248
polymethylmethacrylate (PMMA) 200, 202, 203, 208, 209, 211
polyoxyethylene 173
polypyrrole 203
polystyrene 203, 204, 208
polytetrafluoroethylene 85, 206
poly(vinylsiloxane) (PVS) 240
porosity 206
porous anodic alumina (PAA) 206
potential 3, 4, 29, 34, 35, 37, 54–58, 61, 75, 78, 85, 166, 196, 199, 200, 203, 204, 239
pressure 4, 5, 15, 42, 65–69, 71, 73, 75–77, 85, 90, 91, 124, 125, 128, 133, 138, 140, 143, 151, 153, 154, 169, 171, 172, 176, 204, 207, 208, 215, 219, 243, 246, 250
probability density function 19
probability distribution function 18
protein 229, 247, 249
protein machines 243
PUA mold 208
pull-off force 71, 236

quantum dots 11

Raman 17
ratchet mechanism 28, 39, 43, 44, 62
Rayleigh waves 48
real area of contact 21, 23, 24, 32, 37, 39–42, 45, 49, 50, 193–195, 211, 240, 241
receptors 243
reflection 15, 245, 246
refraction 245
relaxation 49
renormalization-group theory 6
resolution 13, 15, 20, 21, 187, 192, 207, 212
reversible superhydrophobicity 166

Reynolds number 174
root-mean square (RMS) 17, 18, 20, 212, 236
roughness 12, 13, 15, 17, 19–21, 23–26, 28, 32, 38, 41–43, 56, 58, 75, 78, 79, 81–85, 91, 92, 94–96, 98–101, 103–105, 107, 108, 111–113, 115, 122, 124, 131, 133, 134, 136–138, 140–142, 144, 145, 148, 150, 151, 154, 155, 158, 161, 163, 164, 166, 169, 181, 183, 186, 187, 195, 199, 201–205, 212, 216, 225–228, 230, 231, 236, 237, 240, 245, 248
roughness factor 86, 89, 92, 96, 98, 100, 104, 105, 122, 124, 129, 133, 136, 137, 150, 164, 182, 187, 196, 197, 201, 209–212
rum 249

sand fish skink 248
scale 5–7, 12, 20–23, 26–31, 34, 51, 53–56, 58, 59, 61, 62, 83, 92, 95, 96, 104, 108, 113, 121, 129, 130, 136, 139–142, 148, 150, 151, 158, 163, 164, 166, 167, 183, 195–197, 204, 205, 207, 211, 219, 236, 245, 246, 248–250
scaling laws 6, 30, 31, 51, 59, 62, 164
scanning electron microscopy (SEM) 15
scanning probe microscopy (SPM) 15
scanning tunneling microscope (STM) 15
scattering 15, 173, 245
self-affinity 21, 22
self-assembled monolayer 25, 202, 205, 208, 211, 212
self-assembly 204, 205, 243, 244, 248
self-cleaning 12, 84, 181–183, 186, 199, 202, 227, 228, 239–241, 246
self-excited oscillations 48
self-healing 244
self-organization 9, 244
self-organized criticality 9, 52, 108
self-repair 244
self-similarity 7, 22, 23
seta(e) 231, 233–237, 240–242, 247
shark skin 12, 246
shark-skin-effect 246
shear strength 32, 37–39, 42
short wavelength limit 20
SI system 3

silanes 157, 200, 202–205, 209, 212
silica 200, 201, 203–207, 240, 249
silicon 78, 85, 205, 211, 227, 243, 246, 249
   black 171, 202
   wafer 202, 205, 208
   single-crystal 211
skewness 18, 19
slip length 140, 170, 171
slip waves 48, 52, 53
smart adhesion 231, 242
sol-gel 200, 201, 204, 229
spatula(e) 233–237, 239–242
spectroscopy 17
spider 231, 233, 240, 241, 247, 248
spider silk 247, 248
spider web 247, 248
spinodal 68, 69, 71
spray 173, 228
spring constant 57, 184, 207
statistically stored dislocations 29
stick–slip 47–49, 52–54, 60, 251, 253
stiction 25, 81, 229
strain-gradient plasticity 5–7, 29, 31
stylus 15
sublimation 66
supercooled 67
superheated 67
superhydrophilicity 203, 204
superhydrophobic 12, 81–83, 92, 99, 107, 110, 113, 124, 133, 138, 140–142, 147, 148, 150, 163, 164, 166, 169–172, 175, 178, 181, 197, 199, 201–205, 212, 214, 218, 219, 222, 228–230, 244
   reversible 203
   transparent 199, 202, 203, 204, 229
superhydrophobicity 81, 83, 84, 112, 115, 133, 138, 141, 150, 164, 169–172, 177, 181, 183, 184, 187, 203–205, 222, 228–230, 247
superphobic 82
surface
   area 4, 37, 73, 74, 85, 86, 96, 100, 106, 107, 128, 130, 147, 173
   energy 4, 28, 34, 56, 58, 62, 72, 73, 75, 76, 82, 85, 89–92, 96, 119, 142, 153, 155, 164, 165, 173, 174, 192, 201, 202, 204, 208, 230
   nanopatterned 199, 208–210, 246

roughness 15, 17, 20, 34, 38–40, 51, 55, 56, 58, 61, 62, 81, 82, 84–86, 96, 105, 111, 116, 134, 136, 154, 166, 169, 184, 187, 192, 206, 207, 240–242
science 3, 4, 9
tension 4, 68, 74, 76, 85, 86, 91, 92, 107, 123–125, 128, 129, 141, 142, 153, 171, 175, 177, 236
texturing 10, 25
surfaces
concave 133
convex 150
nonreflective 245, 246
symmetry 9, 129–131

tarpaulins 229
teflon 200, 203
temperature 4, 6–8, 52, 65–67, 69, 71, 76, 85, 166, 176, 200, 202–204, 207, 208, 222, 225, 226, 243
tensile strength (negative pressure) 5, 69, 247
tetrahydroperfluorodecyltrichlorosilane (PF$_3$) 157, 202, 212
tetramethylsilane 203
textile 199, 228
thermodynamic equilibrium 7–9, 66, 75, 77, 78, 171, 176
thiol 227
third body mechanism 31, 40
tilt angle 75, 79, 91, 176, 177, 197, 205, 206, 216, 218
tip 11, 15, 39, 52, 69, 172, 184, 185, 192–195, 207, 210, 211, 233, 237

Tokay gecko 231, 233, 235, 242
tribology 9, 10, 22, 26
tribophysics 9
trichomes 183
triple line 4, 74–78, 83, 90–92, 94, 95, 98, 105, 108–113, 124, 126, 128, 133, 143–147, 151, 153, 158, 163, 164, 166, 167, 173, 175, 177, 183
triple point 67
turbulent flow 22, 246

van der Waals force 17, 25, 32, 34
vapor 3, 5, 17, 65–69, 73, 74, 76, 86, 92, 110, 154–156, 169, 170, 172, 176, 202, 204, 207, 208, 212, 249

water anomaly 65, 67
water strider 12, 141, 197, 247
water-repellency 107, 183
waviness 13
wax 84, 85, 103, 104, 140, 141, 150, 181–183, 186, 187, 203, 227, 244, 246
wear 9–11, 25, 39, 40
Weber number 174, 175
Wenzel equation 87, 92, 94, 122, 137, 153, 166, 182, 187, 196, 210
wine 229

X-ray 17, 25, 202

yield strength 5, 6, 29, 38, 42
Young equation 73–75, 77, 86, 90–92, 96, 115, 119, 142, 143

Printing: Krips bv, Meppel, The Netherlands
Binding: Stürtz, Würzburg, Germany